Mechanical Engineering Systems

Mechanical Engineering Systems

Richard Gentle

Peter Edwards

Bill Bolton

BUTTERWORTH HEINEMANN

OXFORD AUCKLAND BOSTON JOHANNESBURG MELBOURNE NEW DELHI

Butterworth-Heinemann
Linacre House, Jordan Hill, Oxford OX2 8DP
225 Wildwood Avenue, Woburn, MA 01801-2041
A division of Reed Educational and Professional Publishing Ltd

A member of the Reed Elsevier plc group

First published 2001

British Library Cataloguing in Publication Data
A catalogue record for this book is available from the British Library

ISBN 0 7506 5213 6

Composition by Genesis Typesetting, Laser Quay, Rochester, Kent
Printed and bound in Great Britain

Contents

Series Preface

'There is a time for all things: for shouting, for gentle speaking, for silence; for the washing of pots and the writing of books. Let now the pots go black, and set to work. It is hard to make a beginning, but it must be done' – Oliver Heaviside, *Electromagnetic Theory*, Vol 3 (1912), Ch 9, 'Waves from moving sources – Adagio. Andante. Allegro Moderato.'

Oliver Heaviside was one of the greatest engineers of all time, ranking alongside Faraday and Maxwell in his field. As can be seen from the above excerpt from a seminal work, he appreciated the need to communicate to a wider audience. He also offered the advice So be rigorous; that will cover a multitude of sins. And do not frown.' The series of books that this prefaces takes up Heaviside's challenge but in a world which is quite different to that being experienced just a century ago.

With the vast range of books already available covering many of the topics developed in this series, what is this series offering which is unique? I hope that the next few paragraphs help to answer that; certainly no one involved in this project would give up their time to bring these books to fruition if they had not thought that the series is both unique and valuable.

This motivation for this series of books was born out of the desire of the UK's Engineering Council to increase the number of incorporated engineers graduating from Higher Education establishments, and the Institution of Incorporated Engineers' (IIE) aim to provide enhanced services to those delivering Incorporated Engineering Courses. However, what has emerged from the project should prove of great value to a very wide range of courses within the UK and internationally – from Foundation Degrees or Higher Nationals through to first year modules for traditional 'Chartered' degree courses. The reason why these books will appeal to such a wide audience is that they present the core subject areas for engineering studies in a lively, student-centred way, with key theory delivered in real world contexts, and a pedagogical structure that supports independent learning and classroom use.

Despite the apparent waxing of 'new' technologies and the waning of 'old' technologies, engineering is still fundamental to wealth creation. Sitting alongside these are the new business focused, information and communications dominated, technology organisations. Both facets have an equal importance in the health of a nation and the prospects of individuals. In preparing this series of books, we have tried to strike a balance between traditional engineering and developing technology.

The philosophy is to provide a series of complementary texts which can be tailored to the actual courses being run – allowing the flexibility for course designers to take into account 'local' issues, such as areas of particular staff expertise and interest, while being able to demonstrate the depth and breadth of course material referenced to a common framework. The series is designed to cover material in the core texts which approximately corresponds to the first year of study with module texts focussing on individual topics to second and final year level. While the general structure of each of the texts is common, the styles are quite different, reflecting best practice in their areas. For example *Mechanical Engineering Systems* adopts a 'tell – show – do' approach, allowing students to work independently as well as in class, whereas *Business Skills for Engineers and Technologists* adopts a 'framework' approach, setting the context and boundaries and providing opportunities for discussion.

Another set of factors which we have taken into account in designing this series is the reduction in contact hours between staff and students, the evolving responsibilities of both parties and the way in which advances in technology are changing the way study can be, and is, undertaken. As a result, the lecturers' support material which accompanies these texts, is paramount to delivering maximum benefit to the student.

It is with these thoughts of Voltaire that I leave the reader to embark on the rigours of study:

'Work banishes those three great evils: boredom, vice and poverty.'

Alistair Duffy
Series Editor
De Montfort University, Leicester, UK

Further information on the IIE Textbook Series is available from
bhmarketing@repp.co.uk
www.bh.com/iie

Please send book proposals to:
rachel.hudson@repp.co.uk

Other titles currently available in the IIE Textbooks Series

Business Skills for Engineers and Technologists	0 7506 5210 1
Design Engineering	0 7506 5211 X

1 Introduction: the basis of engineering

Summary

The aim of this chapter is to set the scene for the rest of the book by showing how the content of the remaining chapters will form the basis of the technical knowledge that a professional mechanical engineer needs during a career. By considering a typical engineering problem it is shown that the four main subjects that make up this text are really all parts of a continuous body of knowledge that will need to be used in an integrated manner.

The chapter concludes by looking at the units that are used in engineering and showing the importance of keeping to a strict system of units.

Objectives

By the end of this chapter the reader should be able to:

- understand the seamless nature of basic engineering subjects;
- appreciate the way in which real engineering problems are tackled;
- recognize the correct use of SI units.

1.1 Real engineering

Cast your mind forward a few years; you have graduated successfully from your course, worked for a spell as a design engineer and now you are responsibile for a team which is being given a new project. Your job is to lead that team in designing a new ride-on lawnmower to fill a gap in the market that has been identified by the sales team. The sales people think that there is scope to sell a good number of low-slung ride-on lawnmowers to places which use a lot of barriers or fences for crowd control, such as amusement parks. Their idea is that the new mower could be driven under the fences, cutting the grass as it goes, without the time wasting activity of having to drive to a gateway in order to move from one area to another. They have produced something they call a 'concept specification' which is really a wish list of features that they would like the new lawnmower to have.

(1) It must be very low, like a go-kart, to go under the barriers.
(2) It must be fast when not mowing so that it can be driven quickly around the park.
(3) It should dry and collect the grass cuttings as it goes so that the park customers do not get their shoes covered in wet grass.

Now comes the worst part of any engineering design problem – 'Where do you start?'

Perhaps you should start with the framework of the mower because this is the part that would support all the other components. You have a good understanding of statics, which is the field of engineering concerned with supporting loads, and you could design a tubular steel frame without too much of a problem if you know the loads and their distribution. The trouble, however, is that you do not know the load that needs to be carried and you cannot base your design on the company's existing products as all their current ride-on mowers are shaped more like small versions of farm tractors. You could calculate the load on the basis of an average driver weight but you do not yet know how much the engine will weigh because its power, and hence its size, has not been established. Furthermore, if the mower is to be driven fast around the park over bumpy ground then the effective dynamic loads will be much greater than the static load. It is therefore probably not a good idea to start with the frame design unless you are willing to involve a great deal of guesswork. This would run the risk of producing at one extreme a frame that would break easily because it is too flimsy and at the other extreme a frame that is unnecessarily strong and hence too expensive or heavy.

Time to think again!

Perhaps you should start the design by selecting a suitable engine so that the total static weight of the mower could be calculated. You have a good basic knowledge of thermodynamics and you understand how an internal combustion engine works. The trouble here, however, is that you cannot easily specify the power required from the engine. So far you have not determined the maximum speed required of the mower, the maximum angle of slope it must be able to climb or the speed at which it can cut grass, let alone considered the question of whether the exhaust heat can dry the grass. In fact this last feature might be a good place to start because the whole point of a mower is that it cuts grass.

First of all you could decide on the diameter of the rotating blades by specifying that they must not protrude to the side of the mower beyond the wheels. This would give you the width of the cut. A few measurements in a field would then allow you to work out the volume and mass of grass that is cut for every metre that the mower moves forward. Lastly you could find the forward speed of your company's other ride-on mowers when cutting in order to calculate the mass of grass which is cut *per second*. From this you can eventually work out two more pieces of key information.

- Using your knowledge of fluid mechanics you could calculate the flow rate of air which is needed to sweep the grass cuttings into the collection bag or hopper as fast as they are being produced.

- Using your knowledge of thermodynamics you could calculate the rate at which heat must be supplied to the wet grass to evaporate most of the surface water from the cuttings by the time they reach the hopper.

At last you are starting to get somewhere because the first point will allow you to calculate the size of fan that is required and the power that is needed to drive it. The second point will allow you to calculate the rate at which waste heat from the engine must be supplied to the wet grass. Knowing the waste power and the typical efficiency of this type of engine you can then calculate the overall power that is needed if the engine is to meet this specification to dry the grass cuttings as they are produced.

Once you have the overall power of the engine and the portion of that power that it will take to drive the fan you can calculate the power that is available for the mowing process and for driving the mower's wheels. These two facts will allow you to use your knowledge of dynamics to estimate the performance of the mower as a vehicle: the acceleration with and without the blades cutting, the maximum speed up an incline and the maximum driving speed. Of course this relies on being able to estimate the overall mass of the mower and driver, which brings us back to the starting point where we did not know either of these two things. It is time to put the thinking cap back on, and perhaps leave it on, because this apparently straightforward design problem is turning out to be a sort of closed loop that is difficult to break into.

What can we learn from this brief look into the future? There are certainly two important conclusions to be drawn.

- The engineering design process, which is what most engineering is all about, can be very convoluted. While it relies heavily on calculation, there is often a need to make educated guesses to start the calculations. To crack problems like the one above of the new mower you will need to combine technical knowledge with practical experience, a flair for creativity and the confidence to make those educated guesses. The engineering courses that this textbook supports must therefore be seen as only the start of a much longer-term learning process that will continue throughout your professional career.
- A good engineer needs to think of all the subjects that are studied on an undergraduate course in modular chunks as being part of a single body of technical knowledge that will form the foundation on which a career can be built. At the introductory level of this book it is best to keep the distinction between the various topics otherwise it can become confusing to the student; it is difficult enough coming to terms with some of the concepts and equations in each topic without trying to master them all at the same time. The lawn mower example, however, shows that you must be able to understand and integrate all the topics, even though you may not have to become an expert in all of them, if you want to be a proficient engineer.

1.2 Units

The introduction is now over and it is almost time to plunge into the detailed treatment of the individual topics. Before we do that, however, we must look at the subject of the units that are used, not only in this book but also throughout the vast majority of the world's engineering industry.

Every engineering student is familiar with the fact that it is not good enough to calculate something like the diameter of a steel support rod and just give the answer as a number. The full answer must include the units that have been used in the calculation, such as millimetres or metres, otherwise there could be enormous confusion when somebody else used the answer in the next step of a large calculation or actually went ahead and built the support rod. However, there is much more to the question of units than simply remembering to quote them along with the numerical part of the answer. The really important thing to remember is to base all calculations on units which fit together in a single system. The system that is used in this book and throughout engineering is the International System of Units, more correctly known by its French name of *Système Internationale* which is abbreviated to SI.

SI units developed from an earlier system based on the metre, the kilogram and the second and hence is known as the MKS system from the initial letters of those three units. These three still form the basis of the SI because length, mass and time are the most important fundamental measurements that need to be specified in order to define most of the system. Most of the other units in the system for quantities such as force, energy and power can be derived from just these three building blocks. The exceptions are the units for temperature, electrical current and light intensity, which were developed much later and represent the major difference between the SI and the MKS system. One of these exceptions that is of concern to us for this book is the unit for measuring temperature, the **kelvin** (**K**). This is named after Lord Kelvin, a Scottish scientist and engineer, who spent most of his career studying temperature and heat in some form or another. The Kelvin is actually equal to the more familiar **degree Celsius** (**°C**), but the scale starts at what is called **absolute zero** rather than with the zero at the freezing point of water. The connection is that 0°C = 273K.

The SI is therefore based on the **second**, a unit which goes back to early Middle East civilization and has been universally adopted for centuries, plus two French units, the **metre** and the **kilogram**, which are much more recent and have their origin in the French Revolution. That was a time of great upheaval and terror for many people. It was also, however, a time when there were great advances in science and engineering because the revolutionaries' idealistic principles were based on the rule of reason rather than on inheritance and privilege. One of the good things to come out of the period, especially under the guidance of Napoleon, was that the old system of measurements was scrapped. Up to that time all countries had systems of measurement for length, volume and mass that were based on some famous ruler. In England it was the length between a king's nose and the fingertip of his outstretched arm that served as the standard measure of the **yard**, which could then be subdivided into three **feet**. This was all very arbitrary and would soon have caused a great deal of trouble as the nations of Western Europe were poised to start supplying the world with their manufactured goods. Napoleon's great contribution was to do away with any unit that was based on royalty and get his scientists to look for a logical alternative. What they chose was to base their unit of length on the distance from one of the earth's poles to the equator. This was a distance which could be calculated by astronomers in any country around the world and so it could serve as a universal standard. They then split this distance into ten million subdivisions called *measures*. In French this is the **metre** (**m**). From here it was straightforward to come up with a reproducible measure for volume, using the cubic metre (m^3), and this allowed the unit of mass to be defined as the **gram**, with one million grams being the mass of one cubic metre of pure water. In practice the gram is very small and so in drawing up a system of units for calculation purposes, later scientists used the **kilogram** (**kg**) which is equal to one thousand grams.

It is worth noting that the French Revolutionaries did not always get things right scientifically; for example, they took it into their heads to do away with the system of having twelve months in a year, twenty-four hours in a day and so on, replacing it with a system of recording time based on multiples of ten. It did not catch on, largely because most people at the time did not have watches and relied on the moon for the months and church clocks for the hours. The old unit of time, the **second** (**s**), therefore ultimately survived the French Revolution and was eventually used as the standard in the SI.

The idea of developing a system of units came some time later when scientists and engineers started to calculate such things as force, energy and power. In many cases units did not exist for what was being measured and in all cases there was the problem of getting a useful and reliable answer that would make sense to another engineer wanting to build some device based on the outcome of the calculation. What was needed was a set of units where people could be confident that if they used the correct units as inputs to a calculation then the answer would come out in the units they required. This is the logic behind developing a system of units based on the three fundamental units of the metre, the kilogram and the second. Below is a table of the units that you will be using to tackle the questions that are at the end of every chapter. Thanks to the SI, you can be confident that if you use the correct units for the data that you feed into your calculations then your answer will come out in the correct units.

Before leaving this topic it is perhaps interesting to note how fate or luck often seems to play tricks on anyone who tries to follow the rule of pure reason and logic. The unit of the metre, based so logically on the circumference of the earth, turns out to be very close in size to the ancient unit of the yard, based quite arbitrarily on the size of one particular king. Even more remarkable is that the mass of one cubic metre of water, one thousand kilograms, used as the standard for Napoleon's system, turns out to be almost exactly equal to the ancient standard of mass used in England at the time, the ton. For that reason a mass of one thousand kilograms is often referred to as one **tonne**.

1.3 Units used in this book

After that little diversion into the backwater of scientific history, it just remains to tabulate all the units that will be used in this book. This table will meet all your needs for our purposes, although it is far from a complete coverage of all the units that are within the SI. For example, there is no need for us to consider the units that relate specifically to electricity, such as the ampere or the volt. The table gives the quantity that is being measured, such as force, and then gives the appropriate unit, in this case the newton. The newton is an example of a **derived unit**; if we were to write it out in full based on just the three fundamental units of metre, kilogram and second then it would be the **kilogram metre per second squared** (**kg m/s^2**). Clearly this would be very difficult and time consuming to say and so by convention the name of a famous scientist who is associated with that quantity is used. Sir Isaac Newton is famous for his pioneering work on force and so his name has been adopted for the unit of force, as a sort of shorthand to save having to write out the full version. Other names to watch out for are joule, pascal and watt, which are used for energy, pressure and power respectively.

Some units are not of a very useful size for all purposes. For example, the unit of pressure or stress, the **pascal (Pa)**, is tiny. It is equivalent to the weight of one average eating apple spread out over an engineering drawing board. For practical purposes it is necessary to come up with a large number of pascals under a single name. The unit often employed is the **bar**, which is equal to one hundred thousand pascals (1 bar = 1×10^5 Pa). The name this time comes from the same word that is used as the basis of 'barometer' and the unit of one bar, again by one of those amazing coincidences, is almost exactly equal to the typical value of the atmospheric pressure. Normally, if it is necessary to change the size of a unit, the conventional prefixes of **micro**, **milli**, **kilo** and **mega** are used, as follows:

Prefix	*Symbol*	*Factor*
micro	μ	one millionth
milli	m	one thousandth
kilo	k	one thousand times
mega	M	one million times

So, for example, 1 km = 1000 m (one kilometre equals one thousand metres).

Table of SU units

Physical quantity	*Name of unit*	*Abbreviation*	*Formula*
Basic units			
Length	metre	m	–
Mass	kilogram	kg	–
Time	second	s	–
Temperature	kelvin	K	–
Derived units			
Area	square metre	m^2	–
Volume	cubic metre	m^3	–
Density	kilograms per cubic metre	kg/m^3	–
Velocity	metres per second	m/s	–
Angular velocity	radians per second	rad/s	–
Acceleration	metres per second squared	m/s^2	–
Angular acceleration	radians per second squared	rad/s^2	–
Force	newton	N	$kg\ m/s^2$
Pressure and stress	pascal	Pa	N/m^2
Dynamic viscosity	pascal second	Pa s	Ns/m^2
Work, energy and quantity of heat	joule	J	N m
Power	watt	W	J/s
Entropy	joule per kelvin	J/K	–
Specific heat	joule per kilogram kelvin	J/kgK	–
Thermal conductivity	watts per metre kelvin	W/mK	–
Supplementary unit			
Angle	radian	rad	–

2 Thermodynamics

Summary

Thermodynamics is an essential part of the study of mechanical engineering. It involves knowledge basic to the functioning of prime movers such as petrol and diesel engines, steam turbines and gas turbines. It covers significant operating parameters of this equipment in terms of fuel consumption and power output.

In industrial and domestic heating systems, refrigeration, air conditioning and in thermal insulation in buildings and equipment, the understanding of basic thermodynamic principles allows effective systems to be developed and applied.

In almost all manufacturing industry there are processes which involve the use of heat energy.

This chapter imparts the fundamentals of thermodynamics in the major fields and then applies them in a wide range of situations. The basis is therefore laid for further study and for the understanding of related processes in plant and equipment not covered here.

Objectives

By the end of this chapter, the reader should be able to:

- understand the principle of specific and latent heat and apply it to gases and vapours;
- appreciate the processes which can be applied to a gas and the corresponding heat and work energy transfers involved;
- relate the gas processes to power cycles theoretical and actual;
- understand the processes relating to steam and apply them in steam power plant;
- apply the vapour processes to refrigeration plant, and establish refrigeration plant operating parameters;
- understand the principles of heat transfer by conduction through plane walls and pipework.

2.1 Heat energy

This chapter introduces heat energy by looking at the specific heat and latent heat of solids and gases. This provides the base knowledge required for many ordinary estimations of heat energy quantities in heating and cooling, such as are involved in many industrial processes, and in the production of steam from ice and water. The special case of the specific heats of gases is covered, which is important in later chapters, and an introduction is made in relating heat energy to power.

Specific heat

The *specific heat* of a substance is the heat energy required to raise the temperature of unit mass of the substance by one degree. In terms of the quantities involved, the specific heat of a substance is the heat energy required to raise the temperature of 1 kg of the material by 1°C (or K, since they have the same interval on the temperature scale). The units of specific heat are therefore J/kgK.

Different substances have different specific heats, for instance copper is 390 J/kgK and cast iron is 500 J/kgK. In practice this means that if you wish to increase the temperature of a lump of iron it would require more heat energy to do it than if it was a lump of copper of the same mass.

Alternatively, you could say the iron 'soaks up' more heat energy for a given rise in temperature.

- Remember that heat energy is measured in joules or kilojoules (1000 joules).
- The only difference between the kelvin and the centigrade temperature scales is where they start from. Kelvin starts at –273 (absolute zero) and centigrade starts at 0. A degree change is the same for each.

Key points

- The specific heat of a substance varies depending on its temperature, but the difference is very small and can be neglected.
- You can find the specific heat of a substance by applying a known quantity of heat energy to a known mass of the substance and recording the temperature rise. This would need to be done under laboratory conditions to achieve an accurate answer.
- A body will give out the same quantity of heat energy in cooling as it requires in heating up over the same temperature range.

Specific heats of common substances can be found on data sheets and in reference books, and the values used in the calculations here are realistic.

The equation for calculating heat energy required to heat a solid is therefore the mass to be heated multiplied by the specific heat of the substance, c, available in tables, multiplied by the number of degrees rise in temperature, δT.

$$Q = m.c.\delta T$$

Putting in the units,

$$\text{kJ} = \text{kg} \times \frac{\text{kJ}}{\text{kg.K}} \times \text{K}$$

Note that on the right-hand side, the kg and K terms cancel to leave kJ. It is useful to do a units check on all formulas you use.

Example 2.1.1

The boiler in a canteen contains 6 kg of water at 20°C. How much heat energy is required to raise the temperature of the water to 100°C? Specific heat of water = 4190 J/kgK.

$$Q = m.c.\,\delta T$$

$$Q = 6 \times 4190 \times (100 - 20)$$

$$Q = 2\,011\,200\,\text{J} = \underline{2011.2\,\text{kJ}}$$

Example 2.1.2

How many kilograms of copper can be raised from 15°C to 60°C by the absorption of 80 kJ of heat energy? Specific heat of copper = 390 kJ/kgK.

$$Q = m.c.\,\delta T$$

$$80\,000 = m \times 390 \times (60 - 15)$$

$$m = \frac{80\,000}{390 \times 45} = \underline{4.56\,\text{kg}}$$

Problems 2.1.1

(1) Calculate the heat energy required to raise the temperature of 30 kg of copper from 12°C to 70°C. Assume the specific heat of copper to be 390 J/kgK.

(2) A body of mass 1000 kg absorbs 90 000 kJ of heat energy. If the temperature of the body rises by 180°C, calculate the specific heat of the material of the body.

Power

It is not always useful to know only how much heat energy is needed to raise the temperature of a body. For instance, if you are boiling a kettle, you are more interested in how long it will be before you can make the tea. The quantity of heat energy needed has to be related to the *power* available, in this case the rating of the heating element of the kettle, and if you have a typical kettle of, say, 2 kW, it means that in 1 second it provides 2000 joules of heat energy.

Remember that power is the rate at which the energy is delivered, i.e. work, or heat energy delivered, divided by time taken.

Let us say the kettle contains 2 kg of water and is at a room temperature of 18°C, and the kettle is 2 kW. Specific heat of water = 4.2 kJ/kgK.

$$Q = m.c.\,\delta T$$

$$Q = 2 \times 4.2 \times (100 - 18) = 688.8\,\text{kJ}$$

This is the heat energy required to raise the temperature of the water to 100°C. The kettle is producing 2 kW, i.e. 2 kJ/s. Therefore,

$$\text{time} = \frac{688.8}{2} = 344.4\,\text{sec} = \underline{5.74\,\text{min}} \text{ to boil}$$

The specific heats of gases

Solids have a value of specific heat which varies only slightly with temperature. On the other hand, gases can have many different values of specific heat depending on what happens to it while it is being heated or cooled. The two values which are used are the specific heat at constant pressure, c_p, and the specific heat at constant volume, c_v. See Figure 2.1.1.

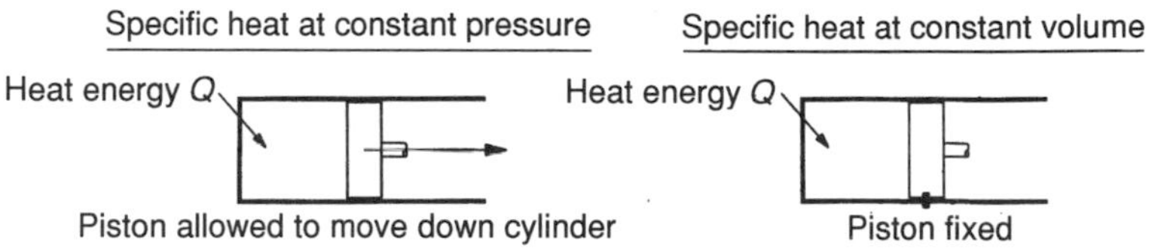

Specific heat at constant pressure — Heat energy Q — Piston allowed to move down cylinder

Specific heat at constant volume — Heat energy Q — Piston fixed

Figure 2.1.1 *Specific heats of gases*

Specific heat at constant pressure, c_p. This is the quantity of heat energy supplied to raise 1 kg of the gas through 1°C or K, while the gas is at constant pressure.

Think of 1 kg of gas trapped in a cylinder. As heat energy is added, the pressure will rise. If the piston is allowed to move down the cylinder to prevent the rise in pressure, the amount of heat energy supplied to raise the temperature of the gas by 1°C is the specific heat of the gas at constant pressure.

Specific heat at constant volume, c_v. This is the quantity of heat energy supplied to raise 1 kg of the gas through 1°C or K, while the gas is at constant volume.

Thinking again of the gas in the cylinder, in this case the heat energy is supplied while the piston is fixed, i.e. the volume is constant. The amount of heat energy added for the temperature to rise 1°C is the specific heat at constant volume, c_v.

The specific heat of a gas at constant pressure is always a higher value than the specific heat at constant volume, because when the gas is receiving heat it must be allowed to expand to prevent a rise in pressure, and, while expanding, the gas is doing work in driving the piston down the cylinder. Extra heat energy must be supplied equivalent to the work done.

Note that this is an example of the equivalence of heat energy and work energy.

Example 2.1.3

1.5 kg of gas at 20°C is contained in a cylinder and heated to 75°C while the volume remains constant. Calculate the heat energy supplied if c_v = 700 J/kgK.

$$Q = m.c_v.\delta T$$

$$Q = 1.5 \times 700 \times (75 - 20) = 57\,750\,\text{J} = \underline{57.75\,\text{kJ}}$$

Example 2.1.4

A gas with a specific heat at constant pressure, c_p = 900 J/kgK, is supplied with 80 kJ of heat energy. If the mass of the gas is 2 kg and its initial temperature is 10°C, find the final temperature of the gas if it is heated at constant pressure.

$$Q = m.c_p.\delta T$$

$$80\,000 = 2 \times 900 \times \delta T$$

$$\delta T = \frac{80\,000}{1800} = 44.44°\text{C}$$

Final temperature = 10 + 44.44 = $\underline{54.44°\text{C}}$

Latent heat

Latent means 'hidden', and is used in this connection because, despite the addition of heat energy, no rise in temperature occurs.

This phenomenon occurs when a solid is turning into a liquid and when a liquid is turning into a gas, i.e. whenever there is a 'change in state'.

In the first case, the heat energy supplied is called the latent heat of fusion, and in the second case it is called the latent heat of vaporization.

The best example to use is water. A lump of ice at, say, –5°C, will need to receive heat energy (sometimes called sensible heat because in this case the temperature does change) to reach 0°C. It will then need latent heat to change state, or melt, during which time its temperature will stay at 0°C.

Further sensible heat energy will then be needed to raise its temperature to boiling point, followed by more latent heat (of vaporization) to change it into steam. See Figure 2.1.2.

Just as each substance has its own value of specific heat, so each substance has a value of latent heat of fusion and latent heat of vaporization.

The latent heat of fusion of ice is 335 kJ/kg, that is, it needs 335 kJ for each kg to change it from ice to water. Note that there is no temperature term in the unit because, as we have already seen, no temperature

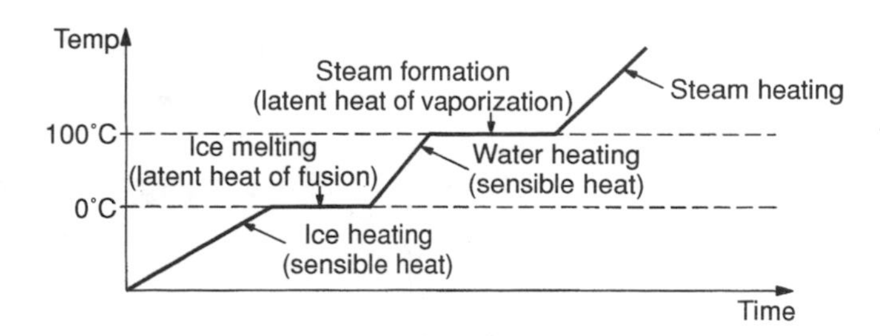

Figure 2.1.2 *Sensible and latent heat*

change occurs. Compare this with the unit for specific heat (kJ/kgK).

The latent heat of vaporization of water at atmospheric pressure is 2256.7 kJ/kg. If instead of boiling the water, we are condensing it, we would need to extract (in a condenser) 2256.7 kJ/kg. The values vary with pressure.

The significance of this theory cannot be underestimated, since it relates directly to steam plant and refrigeration plant, both of which we look at later.

Example 2.1.5

Calculate the heat energy required to change 4 kg of ice at –10°C to steam at atmospheric pressure.

Specific heat of ice = 2.04 kJ/kgK

Latent heat of fusion of ice = 335 kJ/kg

Specific heat of water = 4.2 kJ/kgK

Latent heat of vaporization of water = 2256.7 kJ/kg

Referring to Figure 2.1.2 we can see that all we have to do is add four values together, i.e. the heat energy to raise the temperature of the ice to 0°C, to turn the ice into water, to raise the water to 100°C, and to change the water at 100°C into steam.

To heat the ice,

$$Q_1 = m.c_{\text{ice}}\,\delta T$$

$$Q_1 = 4 \times 2.04 \times 10 = \underline{81.6\,\text{kJ}}$$

To change the ice at 0°C into water at 0°C,

$$Q_2 = m \times \text{latent heat of fusion}_{\text{ice}}$$

$$Q_2 = 4 \times 335 = \underline{1340\,\text{kJ}}$$

To heat the water to 100°C,

$$Q_3 = m.c_{\text{water}}\,\delta T$$

$$Q_3 = 4 \times 4.2 \times 100 = \underline{1680\,\text{kJ}}$$

To change the water into steam at 100°C,

$$Q_4 = m \times \text{latent heat of vaporization}_{\text{water}}$$

$$Q_4 = 4 \times 2256.7 = \underline{9026.8\,\text{kJ}}$$

$$\begin{aligned}\text{Total heat energy} &= Q_1 + Q_2 + Q_3 + Q_4\\ &= 81.6 + 1340 + 1680 + 9026.8\\ &= \underline{12\,128.4\,\text{kJ}}.\end{aligned}$$

In this example, we could provide the 12 128.4 kJ very quickly with a large kW heater, or much more slowly with a small kW heater. Neglecting losses, the result would be the same, i.e. steam would be produced.

As an exercise, and referring to the earlier example of the kettle, find how long it would take to produce the steam in Example 2.1.5 if you used a 2 kW heater and then a 7 kW heater, neglecting losses.

You will notice that we have dealt here mainly with water, since this is by far the most important substance with which engineers must deal. The theory concerning the heating of water to produce steam is the same as for the heating of liquid refrigerant in a refrigeration plant.

Problems 2.1.2

(1) Calculate the quantity of heat energy which must be transferred to 2.25 kg of brass to raise its temperature from 20°C to 240°C if the specific heat of brass is 394 J/kgK.

(2) Find the change in temperature produced by 10 kJ of heat energy added to 500 g of copper. Specific heat of copper = 0.39 kJ/kgK.

(3) Explain why, for a gas, the specific heat at constant volume has a different value from the specific heat at constant pressure.

An ideal gas is contained in a cylinder fitted with a piston. Initially the temperature of the gas is 15°C. If the mass of the gas is 0.035 kg, calculate the quantity of heat energy required to raise the temperature of the gas to 150°C when:

(a) the piston is fixed;
(b) the piston moves and the pressure is constant.

For the gas, c_v = 676 J/kgK and c_p = 952 J/kgK.

(4) 2 kg of ice at −10°C is heated until it has melted. How much heat energy has been used? Specific heat of ice = 2.1×10^3 J/kgK. Latent heat of fusion of ice = 335 kJ/kg.

(5) Calculate the heat energy required to melt 6 kg of lead at 20°C.

Latent heat of fusion of lead = 24.7 kJ/kg.
Specific heat of lead = 126 J/kgK
Melting point of lead = 327°C.

2.2 Perfect gases, gas laws, gas processes

The expansion and compression of gases, such as air and combustion gases, is a very important subject in the study of the operation of compressors, all types of reciprocating engines, gas turbines, and in pneumatic systems.

We need to be able to predict how the volume, temperature and pressure of a gas inter-relate in a process in order to design systems in which gas is the working medium. The cycle of operations of your car engine is a good example of the practical application of this study.

In this chapter, therefore, we look at the different ways in which a gas can be expanded and compressed, what defines the 'system', the significance of reversibility and the First Law of Thermodynamics. In order to be able to calculate changes in properties during a process, gas law expressions and equations are introduced.

Boyle's and Charles' laws

The gases we deal with are assumed to be '*perfect gases*', i.e. theoretically ideal gases which strictly follow Boyle's and Charles' laws. What are these laws?

Boyle's law. This says that if the temperature of a gas is kept constant, the product of its pressure and its volume will always be the same. Hence,

$$p \times V = \text{constant}$$

or,

$$p_1 V_1 = p_2 V_2 = p_3 V_3, \text{ etc.}$$

This means that if you have a quantity of gas and you change its pressure and therefore its volume, as long as the temperature is kept constant (this would require heating or cooling), you will always get the same answer if you multiply the pressure by the volume.

Charles' law. This says that if you keep the pressure of a gas constant, the value of its volume divided by its temperature will always be the same. Hence,

$$\frac{V}{T} = \text{constant}$$

or,

$$\frac{V_1}{T_1} = \frac{V_2}{T_2} = \frac{V_3}{T_3}, \text{ etc.}$$

These laws can be remembered separately, but from an engineer's point of view they are better combined to give a single very useful expression. This is,

$$\frac{p_1 V_1}{T_1} = \text{constant}$$

or,

$$\frac{p_1 V_1}{T_1} = \frac{p_2 V_2}{T_2} = \frac{p_3 V_3}{T_3}, \text{ etc.}$$

It is important to remember that this expression is always valid, no matter what the process, and when doing calculations it should always be the first consideration.

Key points

- There are some instances, as we shall see later, where the pressure must be in N/m^2, but for the time being, bar can be used. 1 bar = $10^5\,N/m^2$.
- Volume is always m^3.
- Temperature is always absolute, i.e. K. The best practice is to convert all temperatures to K immediately by adding 273 to your centigrade temperatures. There is an exception to this when you have a temperature difference, since this gives the same value in °C or K.

Units

p = pressure (bar) or $\left(\dfrac{N}{m^2}\right)$

V = volume (m^3)

T = absolute temperature (K) (°C + 273)

Example 2.2.1

A perfect gas in an engine cylinder at the start of compression has a volume of 0.01 m^3, a temperature of 20°C and a pressure of 2 bar.

The piston rises to compress the gas, and at top dead centre the volume is 0.004 m^3 and the pressure is 15 bar. Find the temperature. See Figure 2.2.1.

$p_1 = 2$ bar $\quad p_2 = 15$ bar

$V_1 = 0.01\,m^3 \quad V_2 = 0.004\,m^3$

$T_1 = 20 + 273 = 293\,K \quad T_2 = ?$

$$\frac{p_1 V_1}{T_1} = \frac{p_2 V_2}{T_2}$$

$$\frac{2 \times 0.01}{293} = \frac{15 \times 0.004}{T_2}$$

$$T_2 = \frac{15 \times 0.004 \times 293}{2 \times 0.01} = \underline{879\,K}$$

The geometry of the cylinder and piston is fixed, so the only way in which the value of temperature (or pressure) at the end of compression can be different for the same initial conditions is if there are different rates of heat energy transfer across the cylinder walls, i.e. how much cooling there is of the engine.

Figure 2.2.1 *p/V diagram. Example 2.2.1*

The pressure/volume (*p/V*) diagram

It is important to examine the pressure/volume diagram for a process, because this gives a visual appreciation of what is happening, and also, some formulae are derived from the diagrams. *p/V* diagrams of actual engine cycles are used to calculate power and to adjust timing.

The *p/V* diagram for Example 2.2.1 is shown in Figure 2.2.1. The volume axis is drawn vertically. To aid understanding, the piston and cylinder are shown beneath the diagram in this case. The direction of the arrow indicates if the process is an expansion or a compression, and the curve would normally be labelled to show the type of process occurring.

Reversibility

We draw the p/V diagrams for our processes assuming that the processes are *reversible*. Put simply, this means that the process can be reversed so that no evidence exists that the process happened in the first place. The best analogy is of a frictionless pendulum swinging back and forth without loss of height.

However, this is not what happens in practice, because of temperature differences, pressure differences, and turbulence within the fluid during the process, and because of friction, all of which are 'irreversibilities'.

The only way in which a reversible process could be achieved is to allow equilibrium to be reached after each of an infinite number of stages during the process, i.e. extremely slowly to let things settle down. This, unfortunately, would take an infinite time.

Strictly speaking, the p/V diagrams should be dotted to show that we are dealing with irreversible processes, but for convenience solid lines are usually used. We assume that at the end states our gas is in equilibrium.

The significance of the concept of reversibility becomes apparent in later studies of thermodynamics.

The first law of thermodynamics

This says that energy cannot be created or destroyed and the total quantity of energy available is constant. Heat energy and mechanical energy are convertible.

As a crude illustration we can consider the closed process of work input to rotate a flywheel which is then brought to a stop by use of a friction brake. The kinetic energy of rotation of the flywheel is converted into heat energy at the brake.

> 'If any system is taken through a closed thermodynamic process, the net work energy transfer and the net heat energy transfer are directly proportional'

A corollary of the first law says that a property exists such that a change in its value is equal to the difference between the net work transfer and the net heat transfer in a closed system.

This property is the internal energy, U, of the fluid, which is the energy the fluid has before the process begins and which can change during the process.

Internal energy

Internal energy is the intrinsic energy of the fluid, i.e. the energy it contains because of the movement of its molecules.

Joule's law states that the internal energy of a gas depends only upon its temperature, and is independent of changes in pressure and volume. We can therefore assume that if the temperature of a gas increases, its internal energy increases and if the temperature falls, the value of internal energy falls.

We will always be dealing with a change in internal energy, and we can show by applying the non-flow energy equation (see below, 'The non-flow energy equation') to a constant volume process that the change in internal energy is given by,

$$U_2 - U_1 = m.c_v(T_2 - T_1)$$

where c_v is the specific heat of the gas at constant volume and m is the mass.

This expression is true for all the processes which can be applied to a gas, and will be used later.

The non-flow energy equation

This is a very important expression, which we use later in non-flow processes

$$Q = W + (U_2 - U_1)$$

In words, the heat energy supplied is equal to the work done plus the change in internal energy.

This can be thought of as an expression of the first law with the internal energy change taken into account.

The system

To study thermodynamics properly, we must know what we are dealing with and where the boundaries are, so that our system is defined.

The gas in an engine cylinder forms a *closed system* bounded by the cylinder walls and the piston head. The processes we have been looking at have occurred without mass flow of the gas across these boundaries.

- These are called *non-flow processes.*

On the other hand, if we move the boundary to encompass the complete engine so as to include the inlet and exhaust, there is a mass flow into and out of the system.

- This is called a *steady flow process.*

Steady flow processes also occur in gas turbines, boilers, nozzles and condensers, wherever there is an equal mass flow in and out across the boundary of the system. It is traditional to show the boundary of a system on a diagram by a broken line.

The characteristic gas equation

In the section on gas processes, we saw that

$$\frac{p.V}{T} = \text{constant}$$

This is true for any gas, and we can use it for any process.

If we take a specific volume of gas, i.e. the volume which 1 kg occupies, we give it the designation, v, and the result of our expression is a constant called the *gas constant*, R.

Putting in the units,

$$\frac{p.v}{T} = \frac{\text{N}}{\text{m}^2} \times \frac{\text{m}^3}{\text{kg}} \times \frac{1}{\text{K}} = \frac{\text{N.m}}{\text{kgK}} = \frac{\text{J}}{\underline{\text{kgK}}} = R$$

Reverting to our use of V for non-specific volume, and using m for the corresponding mass, the equation can be arranged to give,

$$\underline{p.V = m.R.T}$$

This is called the *characteristic gas equation.*

Each gas has a particular value of gas constant. Air, for instance, has a gas constant R = 287 J/kgK. The equation is often used to find the mass of a gas when other properties (p, V and T) are known, but can be transposed to find any of the variables. It must be used at only one point in the process. It is often necessary to use it during a gas process calculation, alongside the gas laws expressions.

Example 2.2.2

An air receiver contains 40 kg of air at 20 bar, temperature 15°C. 5 kg of air leaks from the receiver until the pressure is 15 bar and the temperature falls to 10°C. Find the volume of the receiver and the mass of air lost.

R = 287 J/kgK

$p.V = m.R.T$

Using the initial conditions to find the volume of the receiver,

$$V = \frac{m.R.T}{p} = \frac{40 \times 287 \times (15 + 273)}{20 \times 10^5} = \underline{1.65\ \text{m}^3}$$

Hence, after leakage,

$$m = \frac{p.V}{R.T} = \frac{15 \times 10^5 \times 1.65}{287 \times (10 + 273)} = \underline{30.47\ \text{kg}}$$

Mass used = 40 − 30.47 = $\underline{9.53\ \text{kg}}$

Gauge pressure and absolute pressure

As the name implies, *gauge pressure* is the pressure recorded by a gauge attached to the tank or receiver. An empty tank would record a gauge pressure of zero, but in fact atmospheric pressure would be acting.

The pressures used in thermodynamics are *absolute pressures*, i.e. gauge pressure + atmospheric pressure.

To change a gauge to an absolute pressure, add atmospheric pressure, which is 1.013 bar = $1.013 \times 10^5\,N/m^2$ = $101.3\,kN/m^2$.

Problems 2.2.1

(1) An air bottle has a volume of $1\,m^3$. Find the mass of air in the bottle when the contents are at 20 bar and 15°C. $R = 287\,J/kgK$.

(2) An air receiver contains 25 kg of air at a pressure of $2000\,kN/m^2$, gauge, temperature 20°C. Find the volume of the receiver. $R = 287\,J/kgK$. Atmospheric pressure = 1 bar.

(3) A starting-air receiver for a large diesel engine has an internal volume of $1.5\,m^3$. Calculate:

- (a) The mass of air in the receiver when the contents are at 25 bar, 15°C;
- (b) The starting-air pressure in the receiver if 2 kg of air are used and the temperature remains constant.

$R = 287\,J/kgK$.

The gas processes

We have already seen that for any process we can use the combined Boyle's and Charles' laws expression,

$$\frac{p_1 V_1}{T_2} = \frac{p_2 V_2}{T_2}$$

If we apply this to a constant pressure, constant volume or isothermal (constant temperature) process, all we need do is strike out the constant term before we start calculating.

The use of other expressions will depend upon the particular process.

We will consider the following non-flow processes, e.g. gas in an engine cylinder.

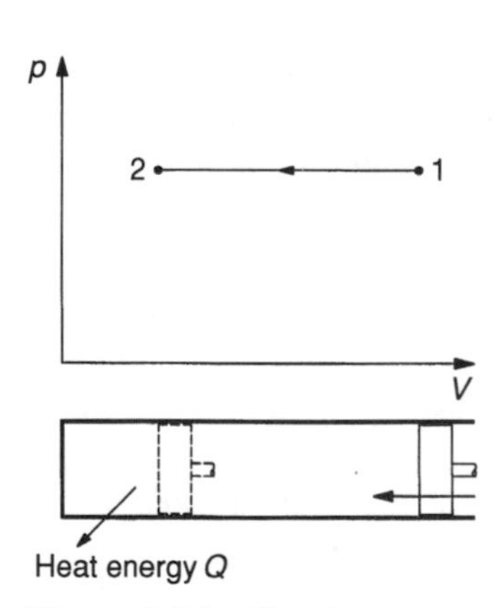

Figure 2.2.2 *Constant pressure process*

Constant pressure

The gas is held at constant pressure as the volume changes. This would require the addition or extraction of heat energy. For instance, if the piston is moved up the cylinder, the heat energy produced would need to be taken away if the pressure was to remain constant (Figure 2.2.2).

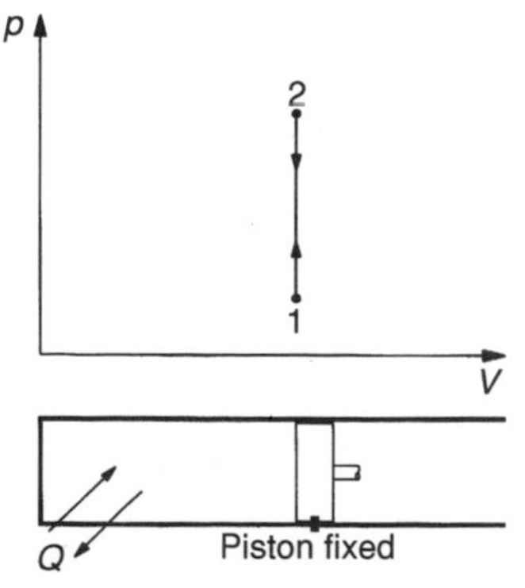

Figure 2.2.3 *Constant volume process*

Constant volume

The volume remains constant, i.e. the piston is fixed. Clearly, the only process which can occur is heating or cooling of the gas (Figure 2.2.3).

- It is important to remember that these are the only two processes which are straight lines on the p/V diagram.

Adiabatic compression and expansion

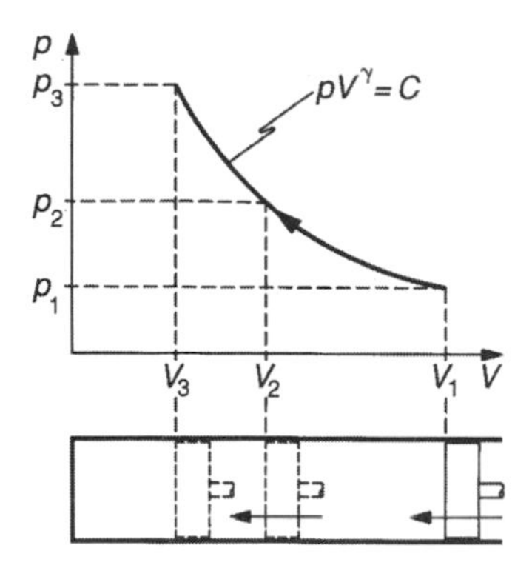

Figure 2.2.4 *Adiabatic process*

During an adiabatic process, no heat transfer occurs to or from the gas during the process. This would require the perfect insulation of the cylinder, which is not possible, and it is worth noting that even the insulation itself will absorb some heat energy (Figure 2.2.4).

The index of expansion, γ, for a reversible adiabatic process is the ratio of the specific heats of the gas, i.e.,

$$\gamma = \frac{c_p}{c_v}$$

The equation of the curve is

$$p.V^{\gamma} = \text{constant}$$

$$p_1 V_1^{\gamma} = p_2 V_2^{\gamma} = p_3 V_3^{\gamma}, \text{ etc.}$$

- Note that a reversible adiabatic process is known as an isentropic process, i.e. constant entropy. Entropy is discussed later.

Polytropic expansion and compression

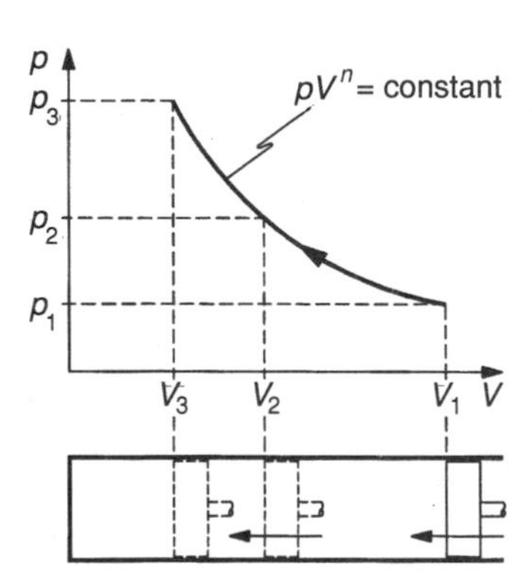

Figure 2.2.5 *Polytropic process*

This is the practical process in which the temperature, pressure and volume of the gas all change. All gas processes in the real world are polytropic – think of the gas expanding and compressing in an engine cylinder (Figure 2.2.5).

The equation of the curve is

$$p.V^n = c$$

$$p_1 V_1^n = p_2 V_2^n = p_3 V_3^n, \text{ etc.}$$

where 'n' is the index of polytropic expansion or compression.

Isothermal process

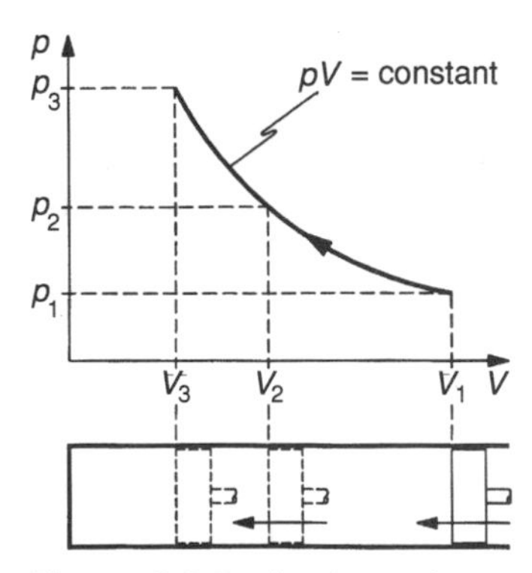

Figure 2.2.6 *Isothermal process*

In this case, which is Boyle's law, the temperature of the gas remains constant during the process. Like the adiabatic process, this cannot be achieved in practice (Figure 2.2.6).

The equation of the curve is

$$p.V = C$$

$$p_1 V_1 = p_2 V_2 = p_3 V_3, \text{ etc.}$$

Key points

- Before calculating, determine the type of process.
- Always sketch the p/V diagram for the process.
- Remember that polytropic, adiabatic and isothermal processes are all curves on the p/V diagram. A curve must therefore be labelled to show which it is.
- The adiabatic curve is steepest and, as the index lowers, so the p/V curve flattens (see Figure 2.2.7).
- The three curves, i.e. polytropic, adiabatic and isothermal, are all of the general form $pV^n = C$, where in the isothermal case, $n = 1$, and in the adiabatic case, $n = \gamma$. The value of n in the polytropic case lies between 1 and γ.
- Compression and expansion processes are calculated in exactly the same way, but remember that on the p/V diagram they must be indicated by the direction of the arrow.

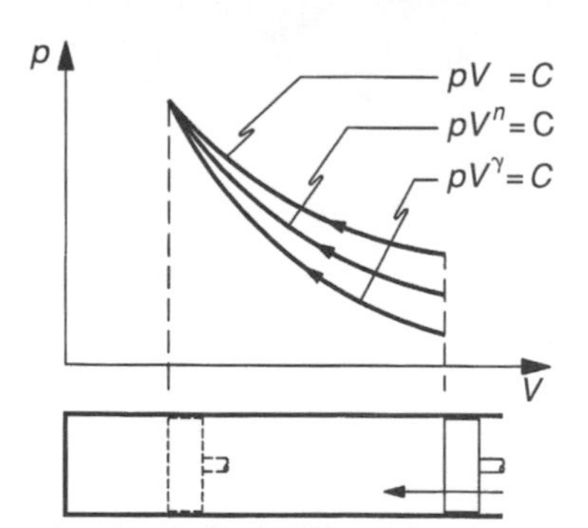

Figure 2.2.7 *Compression curves*

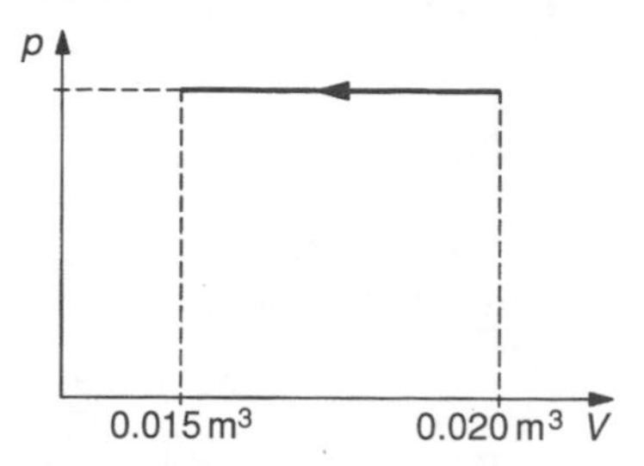

Figure 2.2.8 *Example 2.2.3*

By combining the equation $\dfrac{p_1V_1}{T} = \dfrac{p_2V_2}{T_2}$

with $\underline{p_1V_1^n = p_2V_2^n}$

we can produce another important expression,

$$\frac{T_2}{T_1} = \left(\frac{p_2}{p_1}\right)^{\frac{n-1}{n}} = \left(\frac{V_1}{V_2}\right)^{n-1} \quad \text{for a polytropic process}$$

This can be applied to the adiabatic process to give,

$$\frac{T_2}{T_1} = \left(\frac{p_2}{p_1}\right)^{\frac{\gamma-1}{n}} = \left(\frac{V_1}{V_2}\right)^{\gamma-1}$$

Maths in action

In determining values of pressure, volume and temperature using the gas laws formulae, it is necessary to transpose equations which include powers. We have the formula,

$$p_1V_1^n = p_2V_2^n$$

This gives,

$$p_1 = \frac{p_2V_2^n}{V_1^n} = p_2\left(\frac{V_2}{V_1}\right)^n$$

and

$$V_1 = \left(\frac{p_2V_2^n}{p_1}\right)^{\frac{1}{n}} = \sqrt[n]{\frac{p_2V_2^n}{p_1}}$$

Example 2.2.3

Gas of volume 0.02 m^3 is cooled until its volume is 0.015 m^3 while its pressure remains constant. If the initial temperature is 50°C, find the final temperature (Figure 2.2.8).

$$\frac{p_1V_1}{T} = \frac{p_2V_2}{T_2}$$

The pressure terms are omitted.

$$\frac{0.02}{(50+273)} = \frac{0.015}{T_2}$$

$$T_2 = \frac{0.015 \times 323}{0.02} = \underline{242.25\,\text{K}}$$

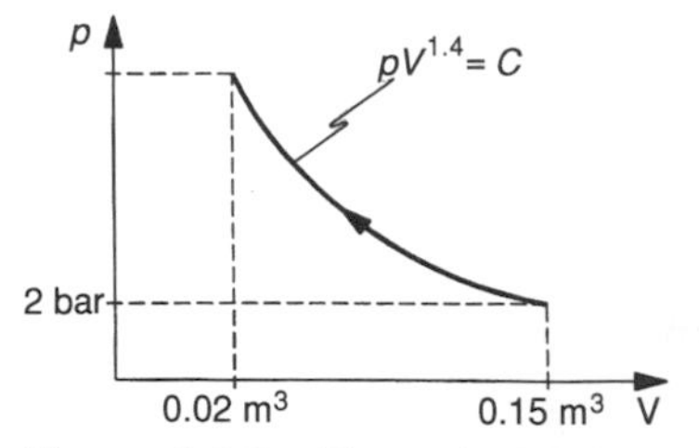

Figure 2.2.9 *Example 2.2.4*

Example 2.2.4

Air of volume 0.15 m³ is compressed adiabatically from a pressure of 2 bar, temperature 12°C, to a volume of 0.02 m³. If $\gamma = 1.4$, find the final pressure and temperature (Figure 2.2.9).

$$p_1 V_1^{\gamma} = p_2 V_2^{\gamma}$$

$$2 \times 0.15^{1.4} = p_2 \times 0.02^{1.4}$$

$$p_2 = 2 \times \frac{0.15^{1.4}}{0.02^{1.4}} = \underline{33.58 \text{ bar}}$$

$$\frac{p_1 V_1}{T_1} = \frac{p_2 V_2}{T_2}$$

$$\frac{2 \times 0.15}{(12 + 273)} = \frac{33.58 \times 0.02}{T_2}$$

$$\underline{T_2 = 638 \text{ K}}$$

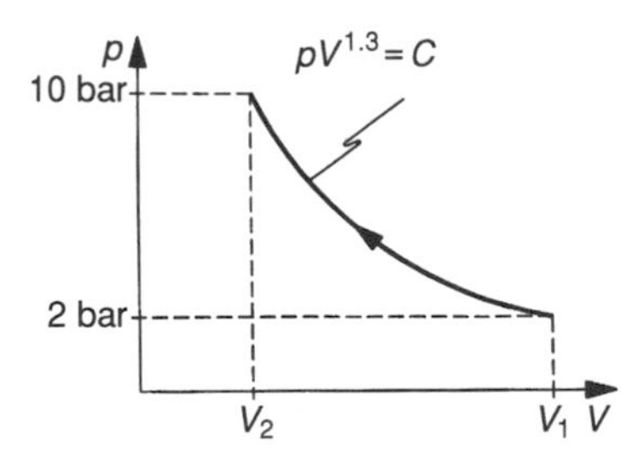

Figure 2.2.10 *Example 2.2.5*

Example 2.2.5

Gas at a temperature of 20°C, pressure 2 bar, is compressed to 10 bar. The compression is polytropic, $n = 1.3$. Find the final temperature (Figure 2.2.10).

In this case we must use the relationship

$$\frac{T_2}{T_1} = \left(\frac{P_2}{P_1}\right)^{\frac{n-1}{n}}$$

$$\frac{T_2}{(273 + 20)} = \left(\frac{10}{2}\right)^{\frac{1.3-1}{1.3}}$$

$$T_2 = 293 \times 5^{0.23}$$

$$\underline{T_2 = 424.3 \text{ K}}$$

Maths in action

Using logs

It may be that properties are known, but the index of compression or expansion is not.

Let us assume in a compression process we know p_1 and V_1 and p_2 and V_2, and we want to find the equation of the curve.

$$p_1 V_1^n = p_2 V_2^n$$

Rearranging,

$$\frac{p_1}{p_2} = \left(\frac{V_2}{V_1}\right)^n$$

Taking logs of both sides,

$$\log\left(\frac{p_1}{p_2}\right) = n.\log\left(\frac{V_2}{V_1}\right)$$

$$n = \frac{\log(p_1/p_2)}{\log(V_2/V_1)}$$

Problems 2.2.2

(In each case sketch the p/V diagram.)

(1) A quantity of gas is heated from 15°C to 50°C at constant volume. If the initial pressure of the gas is 3 bar, find the new pressure.

(2) Gas of volume 0.5 m^3 is heated at constant pressure until its volume is 0.9 m^3. If the initial temperature of thc gas is 40°C, find the final temperature.

(3) 2 m^3 of gas is compressed polytropically ($n = 1.35$) from a pressure of 2 bar to a pressure of 25 bar. Calculate:

(a) final volume;
(b) final temperature if the initial temperature is 25°C.

(4) Gas at 20°C, volume 1 m^3, is compressed polytropically from a pressure of 2 bar to a pressure of 10 bar. If the index of compression is $n = 1.25$, find:

(a) final volume;
(b) final temperature.

(5) Gas of volume 0.2 m^3, pressure 5 bar, temperature 30°C, is compressed until its volume is 0.04 m^3. If the compression occurs in a perfectly insulated cylinder, calculate the final temperature and pressure. $\gamma = 1.4$.

(6) Gas at 35°C, volume 0.75 m^3, is compressed polytropically to the law $pV^{1.25} = C$, from a pressure of 5 bar to a pressure of 15 bar. Find:

(a) final volume;
(b) final temperature of the gas.

(7) Polytropic expansion takes place in an engine cylinder. At the beginning of expansion, the pressure and volume are 1750 kN/m^2 and 0.05 m^3, and at the end expansion they are 122.5 kN/m^2 and 0.375 m^3. Find the index of polytropic expansion.

(8) Gas at a pressure of 0.95 bar, volume 0.2 m^3 and temperature 17°C, is compressed until the pressure is 2.75 bar and volume 0.085 m^3. Calculate the compression index and the final temperature.
(9) A gas initially at 12 bar, 200°C and volume 100 cm^3, is expanded in a cylinder. The volumetric expansion is 6 and the index of expansion is 1.35. Calculate the final volume, pressure and temperature.
(10) A gas is compressed in a cylinder from 1 bar and 35°C at the beginning of the stroke to 40 bar at TDC. The volume of gas in the cylinder at the start of compression is 800 cm^3. If $n = 1.3$, find:

(a) the clearance volume;
(b) temperature at the end of compression.

2.3 Work done and heat energy supplied

We have looked at gas processes and how to calculate pressures, temperatures and volumes after a process has occurred. This is very important in, for instance, engine design, where important parameters are piston and cylinder size and combustion temperature.

We can now go on to find work and heat energy transfer occurring during a process, which prepares us for the next chapter in which these elements are put together to form complete engine cycles.

Using the *p/V* diagram to find work

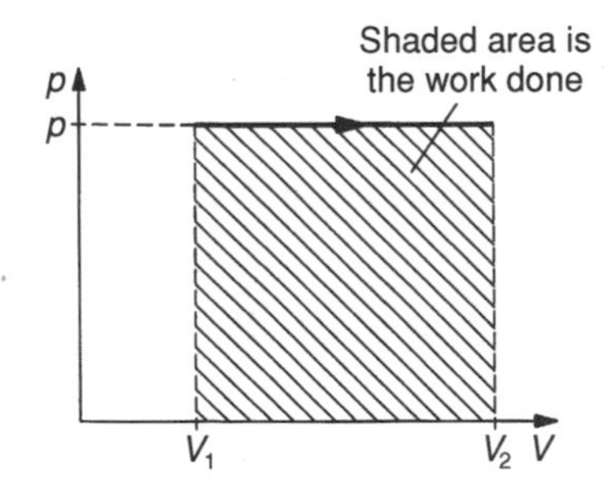

Figure 2.3.1 *Constant pressure process*

We have seen the usefulness of the p/V diagram in visualizing the gas processes and showing the variables. They are also important in allowing us to find expressions for the work done during a process. To demonstrate this, we will look at the constant pressure process.

Figure 2.3.1 shows the p/V diagram as the piston moves down the cylinder with no change in pressure. The area of the rectangle under the constant pressure line is its height multiplied by its breadth, i.e. $p(V_2 - V_1)$.

Putting in the units:

$$p = \frac{\text{N}}{\text{m}^2}$$

$$V = \text{m}^3$$

$$\text{Area of rectangle} = p(V_2 - V_1) = \frac{\text{N}}{\text{m}^2} \times \text{m}^3 = \text{N.m}$$

$$1\ (\text{N.m}) = 1\ (\text{joule})$$

A joule (J) is the unit of work.

- Hence, by finding the area under the p/V diagram, we have found the work done during the process.

We now have an expression for the work done during a constant pressure process, and by looking at the other processes on the p/V

diagram and finding the area under each, we will have all the expressions for work done.

Unfortunately, the polytropic, adiabatic and isothermal processes are, as we have seen, curves on the p/V diagram, and to find these areas we must use calculus.

Expressions for work done

Constant volume

$$W = 0$$

This must be so because on the p/V diagram the process is a straight vertical line which has no area beneath it, and because the piston does not move.

Constant pressure

$$W = p.\delta V$$

Adiabatic

$$W = \left(\frac{p_1 V_1 - p_2 V_2}{\gamma - 1}\right)$$

Polytropic

$$W = \left(\frac{p_1 V_1 - p_2 V_2}{n - 1}\right)$$

Isothermal

$$W = p_1 V_1 \ln\left(\frac{V_2}{V_1}\right)$$

Note: ln is shorthand for $\log_e$.

- Using these expressions, we can find work done during any gas process.

In reality, almost all practical processes are polytropic, i.e. $pV^n = c$. The adiabatic and isothermal processes are impossible to achieve, but useful as 'ideal' processes.

In an engine, the non-flow expansions and compressions are all polytropic, meaning that heat energy is transferred through the cylinder walls, and pressure, volume and temperature all change.

The same is true in flow processes such as gas turbines where the same equations can be used across inlet and outlet conditions.

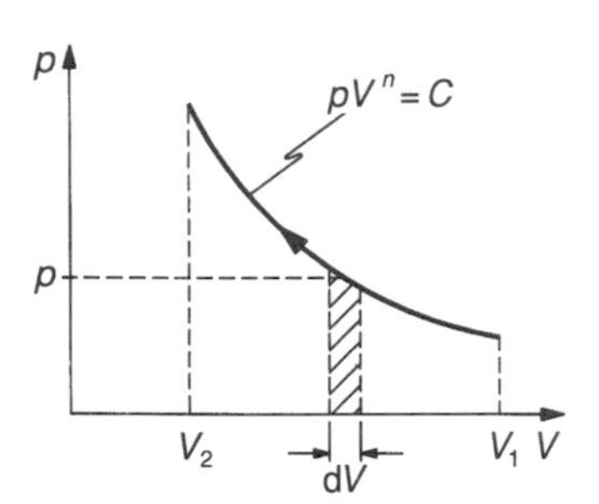

Figure 2.3.2 *Polytropic work done*

Maths in action

Finding areas under curves is a common task in mathematics and an extremely useful facility for engineers. We can find areas by using Simpson's rule and other similar methods, or simply by plotting on graph paper and counting the number of squares. But the best method – if the equation of the curve is known – is to use calculus, which leads to an equation which gives a precise answer.

In this example, we are going to find the area under a curve of the form pV^n = Constant. This, as we have seen, is the relationship between p and V when a polytropic process occurs, in other words what is happening in almost all engines during expansion and compression of the gases. It gives the expression for work done during a polytropic process.

Figure 2.3.2 shows the p/V diagram for a polytropic compression. We carry out the usual procedure for integration which is to consider a strip of the area we want to find, and add up all the strips in the area to give an expression for the total area.

$$\text{Area of strip} = p.\mathrm{d}V$$

$$\text{Total area} = \int_{V_1}^{V_2} p.\mathrm{d}V$$

$$\text{From } p.V^n = c,\ p = \frac{c}{V^n},$$

$$\text{Total area} = \int_{V_1}^{V_2} \frac{c}{V^n}\,.\,\mathrm{d}V$$

Taking c outside and putting V^n at the top,

$$\text{Area} = c\int_{V_1}^{V_2} V^{-n}\,\mathrm{d}V$$

Integrating,

$$A = c\left[\frac{V^{1-n}}{1-n}\right]_{V_1}^{V_2}$$

substituting V_1 and V_2,

$$A = \frac{c}{1-n}\left[V_2^{1-n}1 - V_1^{1-n}\right]$$

$$= \frac{1}{1-n}\left[\frac{c}{V_2^n}V_2 - \frac{c}{V_1^n}V_1\right]$$

substituting $p_1 = \dfrac{c}{V_1^n}$ and $p_2 = \dfrac{c}{V_2^n}$,

$$A = \frac{1}{1-n}\left[p_2 V_2 - p_1 V_1\right]$$

It is more convenient to write $(1 - n)$ as $(n - 1)$, giving,

$$A = \frac{\left[p_1 V_1 - p_2 V_2\right]}{n-1} = \text{polytropic work done}$$

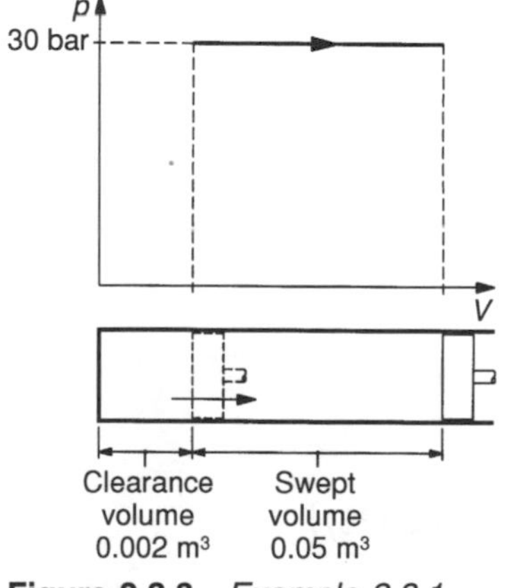

Figure 2.3.3 *Example 2.3.1*

Example 2.3.1

A piston moves down the cylinder in an engine from top dead centre to bottom dead centre, while the pressure in the cylinder remains constant at 30 bar. The clearance volume is $0.002\,m^3$ and the swept volume is $0.05\,m^3$. What is the work done?

The process is shown in Figure 2.3.3.

$P = 30$ bar $= 30 \times 10^5\,N/m^2$

V_1 = clearance volume $= 0.002\,m^3$

V_2 = swept volume + clearance volume

$= 0.002 + 0.05 = 0.052\,m^3$

In a constant pressure process,

Work done $= p(V_2 - V_1)$

$30 \times 10^5 \times (0.052 - 0.002) = 150\,000$ J

$= 150$ kJ

Key point

In the expressions for work done, pressure must be in N/m^2 or kN/m^2, giving answers in J and kJ respectively.

As an example of how we could calculate a value of power produced, if we assume in this example (forgetting for the moment that a constant pressure expansion in an engine cylinder is in practice highly unlikely), that the piston is forced down the cylinder every revolution – as it would be in a two-stroke engine – it would mean that if the engine was turning at 300 rpm then $300 \times 150 = 45\,000$ J of work would be done each minute.

This is 750 J/s = 750 W, the power produced in the cylinder.

Example 2.3.2

The bore and stroke of the cylinder of a diesel generator are 146 mm and 280 mm respectively, and the clearance volume is 6% of the swept volume. If the air before compression is at 1 bar, find the work done in compression if the process is:

(a) polytropic, $n = 1.3$;
(b) isothermal.

$$\text{Swept volume} = \frac{\pi \times (\text{bore})^2}{4} \times \text{stroke}$$
$$= \frac{\pi \times 0.146^2}{4} \times 0.28$$
$$= 0.00469\,\text{m}^3$$

$$\text{Clearance volume} = 0.06 \times \text{swept volume}$$
$$= 0.06 \times 0.00469 = 0.000\,28\,\text{m}^3$$

$$\text{Initial volume} = \text{swept volume} + \text{clearance volume}$$
$$= 0.00469 + 0.000\,28$$
$$= 0.00497\,\text{m}^3$$

Polytropic compression

$$P_1 V_1^n = p_2 V_2^n, \quad 1 \times 0.004\,97^{1.3} = p_2 \times 0.000\,28^{1.3}$$

$$p_2 = 1 \times \left(\frac{0.00497}{0.00028}\right)^{1.3} = 42.1\ \text{bar}$$

$$W = \frac{p_1 V_1 - p_2 V_2}{n - 1}$$

$$= \frac{10^5\left[(1 \times 0.00497) - (42.1 \times 0.00028)\right]}{1.3 - 1}$$

$$= 2273\,\text{J} = \underline{2.273\,\text{kJ}}$$

Isothermal compression

$$W = p_1 V \ln\left(\frac{V_2}{V_1}\right)$$

$$= 1 \times 10^5 \times 0.004\,97 \times \ln\left(\frac{0.00028}{0.00497}\right)$$

$$= 1430\,\text{J} = \underline{1.43\,\text{kJ}}$$

Note that least work is done when the compression is isothermal, and in the case of an air compressor for instance, isothermal compression is desirable in order to reduce the power input necessary from its engine or motor.

An approximation to isothermal compression is achieved by multi-stage compression with intercooling.

Alternative expressions for work done

The characteristic gas equation gives us the relationship,

$pV = m.R.T$

You will see that pV occurs in the expressions for work done. We can substitute $m.R.T$ instead. The expressions for work done then become,

Constant volume

$W = 0$

Constant pressure

$W = m.R(T_1 - T_2)$

Adiabatic

$$W = \frac{m.R(T_1 - T_2)}{\gamma - 1}$$

Polytropic

$$W = \frac{m.R(T_1 - T_2)}{n - 1}$$

Isothermal

$$W = m.R.T_1 \ln\left(\frac{V_2}{V_1}\right)$$

These expressions are necessary if the values given in a problem do not allow the use of the 'pV' equations, or simply if they are preferred in cases when either could be used.

Key point

The work done is negative, because we have done work *on* the gas. Work done *by* the gas is positive.

Example 2.3.3

1.5 kg of gas is compressed isothermally at a temperature of 20°C, from a volume of 0.5 m^3 to a volume of 0.3 m^3. If R for the gas is 290 J/kgK, find the work done.

We cannot use the '$p.V$' expression for the work done, because the initial pressure is unknown. We could find the pressure using $p.V = m.R.T$ and then use it, but this is a case where the '$m.R.T$' version can be used directly.

$$W = m.R.T_1 \ln\left(\frac{V_2}{V_1}\right)$$

$$W = 1.5 \times 290 \times (20 + 273) \ln\left(\frac{0.3}{0.5}\right)$$

$$W = -65\ 107\ \text{J} = \underline{-65.2\ \text{kJ}}$$

Heat energy transferred during a process

If we have a non-flow process occurring, say, within an engine cylinder, then, inevitably, some heat energy is transferred to or from the gas through the cylinder walls. Usually, this heat energy is transferred out of the cylinder walls to the engine cooling water.

The quantity of heat energy transferred affects the values of pressure and temperature achieved by the gas within the cylinder. Only in an adiabatic process is there no transfer of heat energy to or from the gas, and this situation is impossible to achieve in practice.

In order to find how much heat energy has been transferred during a non-flow process, we use the non-flow energy equation (NFEE) (see page 17, 'The non-flow energy equation'.

$$Q = W + (U_2 - U_1)$$

In words, this means that the heat energy transferred through the cylinder walls is the work done during the process added to the change in internal energy during the process.

Remember that the internal energy of the gas is the energy it has by virtue of its temperature, and that if the temperature of the gas increases, its internal energy increases, and vice versa.

Using the NFEE is straightforward, in that we have already seen the equations for work done, and the change in internal energy for any process (see page 16, 'Internal energy') is given by

$$U_2 - U_1 = m.c_v(T_2 - T_1)$$

We can calculate each and add them together.

Applying the NFEE to each process gives,

Constant volume

$Q = W + (U_2 - U_1)$. $W = 0$ in a constant volume process.

$$Q = (U_2 - U_1)$$

$$\underline{Q = m.c_v\,(T_2 - T_1)}$$

Constant pressure

$Q = W + (U_2 - U_1)$. $W = p.\delta V$ in a constant pressure process.

$$\underline{Q = p(V_2 - V_1) + m.c_v\,(T_2 - T_1)}$$

It is useful to digress here to establish an important expression concerning the gas constant, R, and values of c_p and c_v.

We know that the heat energy supplied in the constant pressure process is

$$Q = m.c_p(T_2 - T_1)$$

therefore,

$$m.c_p(T_2 - T_1) = p(V_2 - V_1) + m.c_v(T_2 - T_1)$$

Substituting from $p.V = m.R.T$), $p(V_2 - V_1) = m.R(T_2 - T_1)$, and dividing by $(T_2 - T_1)$,

$c_p = R + c_v$ for unit mass

$\underline{R = c_p - c_v}$

Adiabatic

$Q = W + (U_2 - U_1)$

$\underline{Q = 0.}$

This is the definition of an adiabatic process.

Note: if W and δU were calculated and put in the formula, the answer would be 0.

Polytropic

$Q = W + (U_2 - U_1)$

$$Q = \frac{(p_1 V_1 - p_2 V_2)}{n - 1} + m.c_v (T_2 - T_1)$$

Isothermal

$Q = W + (U_2 - U_1)$. No change in temperature, therefore no change in U.

$$Q = p_1 V_1 \ln \left(\frac{V_2}{V_1}\right)$$

Note that the '$m.R.T$' versions of the work done expressions could be used instead.

Example 2.3.4

0.113 m^3 of air at 8.25 bar is expanded in a cylinder until the volume is 0.331 m^3. Calculate the final pressure and the work done if the expansion is polytropic, $n = 1.4$. If the temperature before expansion is 500°C, and $c_v = 245$ J/kgK, find the heat energy transferred during the process. $R = 810$ J/kgK.

The process is shown in Figure 2.3.4

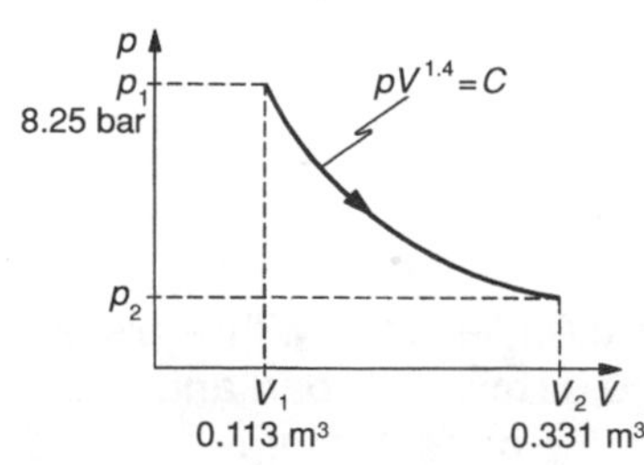

Figure 2.3.4 *Example 2.3.4*

$p_1 V_1^n = p_2 V_2^n$

$8.25 \times 0.113^{1.4} = p_2 \times 0.331^{1.4}$

$$p_2 = 8.25 \times \frac{0.113^{1.4}}{0.331^{1.4}} = 8.25 \times \left(\frac{0.113}{0.331}\right)^{1.4} = \underline{1.83 \text{ bar}}$$

$$W = \frac{p_1 V_1 - p_2 V_2}{n - 1}$$

$$= \frac{(8.25 \times 10^5 \times 0.113) - (1.83 \times 10^5 \times 0.331)}{1.4 - 1}$$

$$W = 81\ 630\,\text{J} = \underline{81.63\,\text{kJ}}$$

To find the heat energy transferred we use the non-flow energy equation,

$$Q = W + (U_2 - U_1)$$

In this case, $W = 81.63\,\text{kJ}$, and $(U_2 - U_1) = m.c_v\,(T_2 - T_1)$.

We need to find T_2 and m, the mass of gas in the cylinder.

$$\frac{p_1 V_1}{T_1} = \frac{p_2 V_2}{T_2}$$

$$\frac{8.25 \times 0.113}{(500 + 273)} = \frac{1.83 \times 0.331}{T_2}$$

$$T_2 = \frac{1.83 \times 0.331 \times 773}{8.25 \times 0.113} = \underline{502.3\,\text{K}}$$

$$p_1 V_1 = m.R.T_1$$

$$m = \frac{p_1 V_1}{RT_1} = \frac{8.25 \times 10^5 \times 0.113}{810 \times 773} = \underline{0.149\,\text{kg}}$$

$$U_2 - U_1 = m.c_v\,(T_2 - T_1)$$

$$= 0.149 \times 245(502.3 - 773)$$

$$= -9881.9\,\text{J} = \underline{-9.9\,\text{kJ}}$$

$$Q = W + (U_2 - U_1) = 81.63 + (-9.9) = \underline{71.7\,\text{kJ}}$$

For this process to occur with these conditions, 71.7 kJ of heat energy must be supplied to the gas.

Key point

The heat energy transfer is positive because heat energy has been supplied to the gas. Heat energy rejected is negative.

Problems 2.3.1

(In all cases, sketch the p/V diagram).

(1) An ideal gas is contained in a cylinder fitted with a piston. Initially the V, p and T are $0.02\,\text{m}^3$, 1.4 bar and 15°C respectively. Calculate the quantity of heat energy required to raise the temperature of the gas to 150°C when:

(a) the piston is fixed;
(b) the piston moves and the pressure is constant.

$$c_v = 676\,\text{J/kgK}$$

$$c_p = 952\,\text{J/kgK}$$

(2) $0.25\,m^3$ of gas is compressed isothermally from a pressure of 1 bar until its volume is $0.031\,m^3$. Calculate the heat energy transfer.

(3) $1\,m^3$ of gas at 10 bar is expanded polytropically with an index of $n = 1.322$ until its volume is doubled. The final pressure is 4 bar. Calculate the work done during the process.

(4) Air expands adiabatically, $\gamma = 1.4$, from 12 bar, 50°C to 4 bar. Calculate the final temperature and the work done per kg of air if the expansion takes place in a cylinder fitted with a piston. $R = 287$ J/kgK.

(5) A perfect gas of mass 0.2 kg at 3.1 bar and 30°C is compressed according to the law $pV^{1.3} = C$ until the pressure is 23.54 bar, temperature 210°C.

Find the initial volume, the final volume, the work done and the heat energy transferred during the process.

$$c_v = 700\,\text{J/kgK}$$

$$c_p = 980\,\text{J/kgK}$$

(6) A perfect gas of volume $0.085\,m^3$ pressure 27.6 bar and temperature 1800°C is expanded to 3.3 bar and at this pressure its volume is $0.45\,m^3$. If the index of expansion is 1.27, find:

(a) The final temperature;
(b) The work done during the process.

(7) In a marine diesel cylinder, $0.03\,m^3$ of gas is expanded with an index of 1.4 from 3.5 bar to $0.09\,m^3$. Calculate the final pressure and the work done during the process.

(8) An engine of 250 mm bore, 300 mm stroke, compression ratio 11 to 1 air takes in air at 1 bar, 20°C. If the compression is polytropic with an index of $n = 1.31$, find the pressure at the end of compression, the work done and the heat energy transferred during the process. Note: compression ratio = V_1/V_2.

$$R = 287\,\text{J/kgK},$$

$$c_p = 1005\,\text{J/kgK}.$$

2.4 Internal combustion engines

We have seen the non-flow processes of constant volume, constant pressure, adiabatic, polytropic and isothermal expansion and compression, and have found how to calculate work done and heat energy transferred during these operations. In this chapter we apply this knowledge to theoretical and practical engine cycles, looking at the processes which make up the cycle, the calculation of brake and indicated power, mean effective pressure, efficiency and fuel consumption, and the relationship between theoretical and actual. Also, we examine how power and other parameters can be found in practice.

The second law of thermodynamics

It is useful at this stage to consider the heat engine cycle in general terms.

The *thermal efficiency* will always be given by the ratio of what we get out of the engine in terms of work, to the amount of heat energy supplied. From the first law we know that in a closed cycle, the change in internal energy is zero, and the net work energy transfer equals the net heat energy transfer. Therefore the work done is the difference between the heat energy supplied and the heat energy rejected, hence,

$$\text{Thermal efficiency} = \frac{\text{work done}}{\text{heat energy supplied}}$$

$$= \frac{\text{heat supplied} - \text{heat rejected}}{\text{heat supplied}}$$

$$\text{Thermal efficiency} = 1 - \frac{\text{heat rejected}}{\text{heat supplied}} = 1 - \frac{Q_R}{Q_s}$$

This expression can be applied to all the *ideal cycles* we look at next.

Clearly, the efficiency can only be 100% if no heat energy is rejected. We know, and can demonstrate, that it is not possible to construct a heat engine which can operate without rejecting heat energy.

The *second law of thermodynamics* encapsulates this:

> 'It is impossible for a heat engine to produce work if it exchanges heat energy from a single heat reservoir'

In other words, there must be a 'cold reservoir' to which heat energy is rejected.

The second law has other versions, e.g.

> 'Wherever a temperature difference occurs, motive power can be produced'

> 'If a system is taken through a cycle and produces work, it must be exchanging heat with two reservoirs at different temperatures'

The establishment of the second law is due mainly to Sadi Carnot in the nineteenth century.

The Carnot cycle

Carnot proposed a cycle which would give the maximum possible efficiency between temperature limits.

Figure 2.4.1 shows this cycle which consists of an isothermal expansion ($pV = C$) of the gas in the cylinder from point 1 to point 2 as heat energy is supplied, followed by an adiabatic expansion ($pV^\gamma = C$) to point 3. Between 2 and 3 the gas has cooled. The piston moves up the

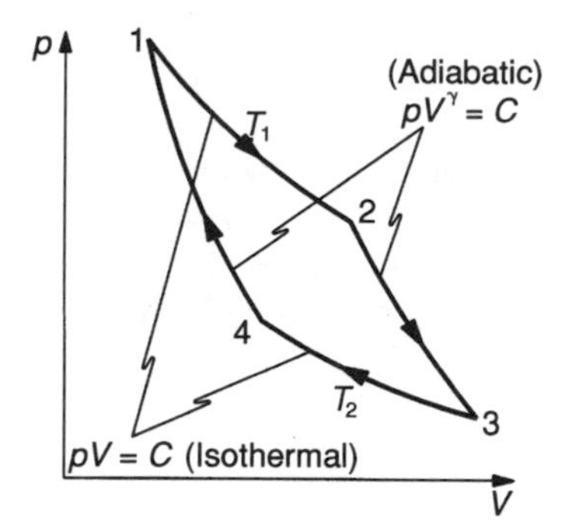

Figure 2.4.1 *Carnot cycle*

cylinder between 3 and 4 compressing the gas isothermally as heat energy is rejected, and between 4 and 1 the gas is compressed adiabatically.

All the processes are reversible, and heat energy is supplied and rejected at constant temperature. The cycle is therefore impossible to create in practice.

A lengthy proof, using the non-flow energy equation and the expressions for work done substituted into the expression for efficiency we have just derived, gives an expression for the efficiency of the cycle, i.e. the Carnot efficiency.

$$\text{Carnot efficiency, } \eta = 1 - \frac{T_2}{T_1}$$

T_2 is the lowest and T_1 the highest of only two temperatures involved in the cycle.

The cycle cannot be created in practice, since we have reversible processes and must supply and reject heat at constant temperature. It does, however, supply a means of rating the effectiveness of a cycle or plant by allowing us to calculate a maximum theoretical efficiency based on maximum and minimum temperatures, even if the cycle is not operating on the Carnot cycle.

Example 2.4.1

A diesel engine works between a maximum temperature of 600°C and a minimum temperature of 65°C. What is the Carnot efficiency of the plant?

$$\text{Carnot efficiency} = 1 - \frac{T_2}{T_1}$$

$$= 1 - \frac{(65 + 273)}{(600 + 273)} = 1 - \frac{338}{873}$$

$$= 0.613 = \underline{61.3\%}$$

The air standard cycles

In what are known as air standard cycles, or ideal cycles, the constant volume, constant pressure and adiabatic processes are put together to form theoretical engine cycles which we can show on the p/V diagram. The actual p/V diagram is different from what is possible in practice, because, for instance, we assume that the gas is air throughout the cycle when in fact it may be combustion gas. We also assume that valves can open and close simultaneously and that expansion and compressions are adiabatic.

Air standard cycles are reference cycles which give an approximation to the performance of internal combustion engines.

Constant volume (Otto) cycle

This is the basis of the petrol engine cycle.

Figure 2.4.2 shows the cycle, made up of an adiabatic compression, 1–2 (piston rises to compress the air in the cylinder), heat energy added at constant volume, 2–3 (the fuel burns), adiabatic expansion, 3–4 (the hot gases drive the piston down the cylinder), and heat energy rejected at constant volume, 4–1 (exhaust).

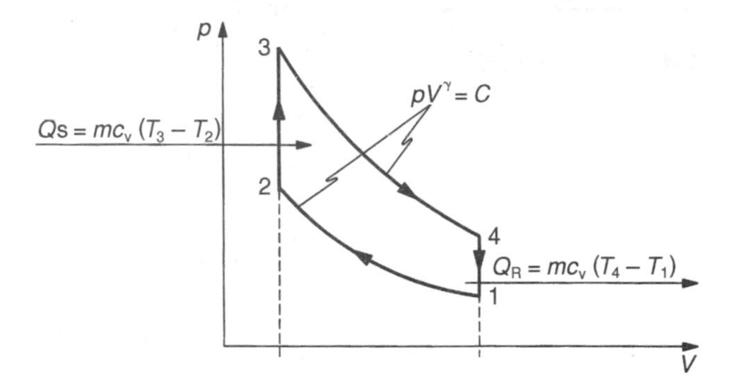

Figure 2.4.2 *Constant volume (Otto) cycle*

What can we do with this? We can calculate from our knowledge so far:

- the pressures, volumes and temperatures around the cycle;
- the work done during each of the processes and therefore the net work done;
- the heat energy transferred during each process;
- the ideal – or air standard – efficiency of the cycle using the expression we derived earlier in this chapter.

Indicated mean effective pressure, P_{mi}

The air standard efficiency of the cycle is a useful indicator of actual performance, but limited because it is not easy to decide in practice where heat energy transfers begin and end.

The ideal cycle diagram – and an actual indicator diagram, which we see later – can also provide a value of *indicated mean effective pressure*, P_{mi}, which is another useful comparator. This is found by dividing the height of the diagram by its length to produce a rectangle of the same area. See Figure 2.4.3, in which the rectangle is shown hatched.

$$P_{mi} = \frac{\text{area of diagram}}{\text{length}} = \frac{A}{L}$$

The area of the diagram is work, joules = N.m. The length of the diagram is volume, i.e. m^3.

$$P_{mi} = \frac{\text{N.m}}{\text{m}^3} = \frac{\text{N}}{\text{m}^2}$$

Key points

- The indicated mean effective pressure can be thought of as the pressure acting on the piston over full stroke which would give the same work output. Generally speaking, the higher this is the better.
- Since the brake power of an engine is usually quoted and easier to record, instead of an indicated mean effective pressure, a value of *brake mean effective pressure*, P_{mb}, is more often used as the comparator. It is found by using the expression for indicated power with values of brake power and brake mean effective pressure substituted. This is covered later in the chapter.

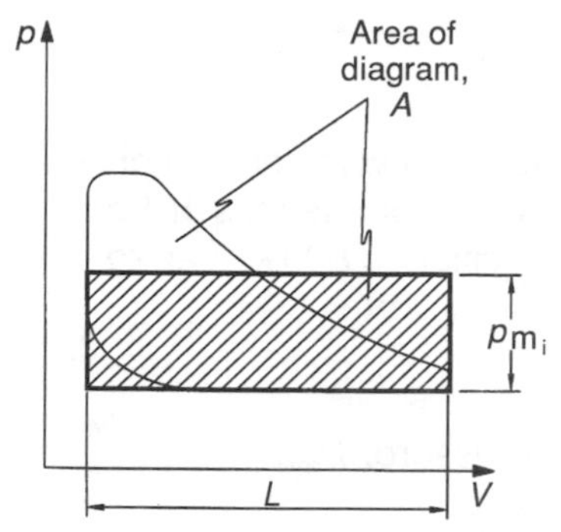

Figure 2.4.3 *Mean effective pressure*

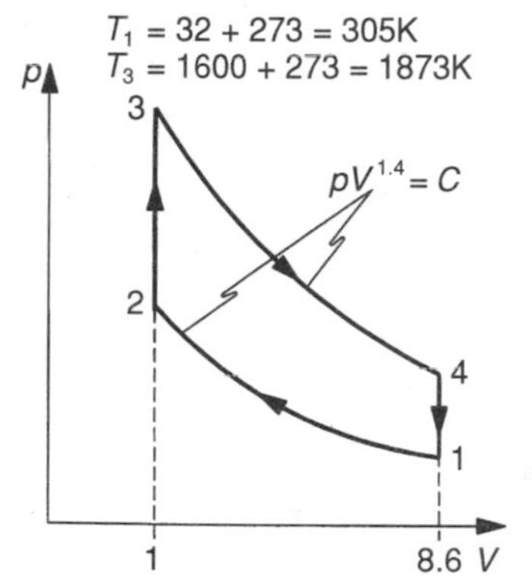

Figure 2.4.4 *Example 2.4.2*

Example 2.4.2

The ratio of compression of an engine working on the constant volume cycle is 8.6:1. At the beginning of compression the temperature is 32°C and at the end of heat supply the temperature is 1600°C. If the index of compression and expansion is 1.4, find:

(a) the temperature at the end of compression;
(b) the temperature at the end of expansion;
(c) the air standard efficiency of the cycle.

Figure 2.4.4 shows the cycle.

Note:

- the compression ratio is a ratio of volumes, V_1/V_2, not a ratio of pressures;
- the dimensionless ratio values of 8.6 and 1 are used directly in the equations;
- the heat energy transfer in a constant volume process is $(m.c_v.\delta T)$.

$$\frac{T_2}{T_1} = \left(\frac{V_1}{V_2}\right)^{\gamma-1}, \quad \frac{T_2}{305} = \left(\frac{8.6}{1}\right)^{1.4-1}$$

$$T_2 = 305 \times 8.6^{0.4}$$

$$= \underline{721.3\,\text{K}} \text{ temperature at end of compression.}$$

$$\frac{T_4}{T_3} = \left(\frac{V_3}{V_4}\right)^{\gamma-1} \quad \frac{T_4}{1873} = \left(\frac{1}{8.6}\right)^{0.4},$$

$$T_4 = 1873 \times 0.42$$

$$= \underline{792\,\text{K}} \text{ temperature at end of expansion.}$$

Air standard efficiency, η

$$= 1 - \frac{\text{heat rejected}}{\text{heat supplied}}$$

$$= 1 - \frac{m.c_v\,(T_4 - T_1)}{m.c_v\,(T_3 - T_2)} \quad m \text{ and } c_v \text{ cancel}$$

$$= 1 - \frac{(792 - 305)}{(1873 - 721.3)} = 1 - \frac{487}{1151.7}$$

$$= 0.577 = \underline{57.7\%} \text{ air standard efficiency}$$

Example 2.4.3

In an air standard (Otto) constant volume cycle, the compression ratio is 8 to 1, and the compression commences at 1 bar, 27°C. The constant volume heat addition is 800 kJ per kg of air. Calculate:

(a) the thermal efficiency;
(b) the indicated mean effective pressure, P_{mi}.

$c_v = 718$ J/kgK

$\gamma = 1.4$

Figure 2.4.5 shows the cycle.

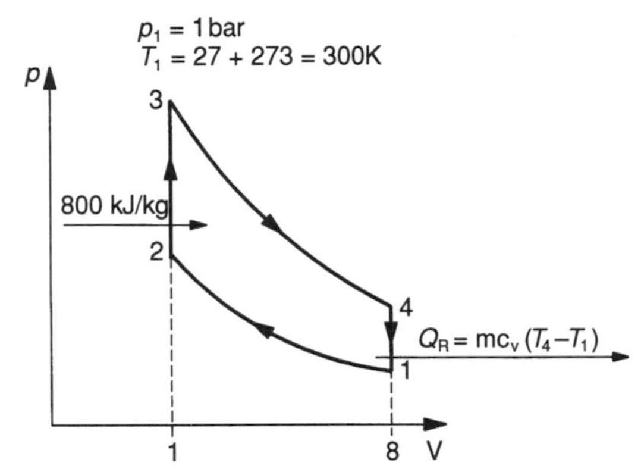

Figure 2.4.5 *Example 2.4.3*

$$\gamma = \frac{c_p}{c_v},\ c_p = \gamma \times c_v = 1.4 \times 718$$

$= \underline{1005\ \text{J/kgK}}$ (see page 20, 'Adiabatic compression and expansion')

$$R = c_p - c_v,\ R = 1005 - 718 = \underline{287\ \text{J/kgK}}$$

$$p_1 V_1 = \underline{m.R.T_1},\ V_1 = \frac{mRT_1}{p_1} = \frac{1 \times 287 \times 300}{1 \times 10^5}$$

$$= \underline{0.861\ \text{m}^3}$$

$$\frac{V_1}{V_2} = 8,\quad V_2 = \frac{0.861}{8} = \underline{0.1076\ \text{m}^3}$$

$$\text{Swept volume} = V_2 - V_1 = 0.861 - 0.1076 = \underline{0.7534\ \text{m}^3}$$

$$\frac{T_2}{T_1} = \left(\frac{V_1}{V_2}\right)^{\gamma - 1},\ \frac{T_2}{300} = \left(\frac{8}{1}\right)^{1.4-1},$$

$$T_2 = 300 \times 8^{0.4} = \underline{689.2\ \text{K}}$$

$$Q_{1-2} = m.c_v\,(T_3 - T_2)$$

$$T_3 = 689.2 + \frac{800 \times 10^3}{718} = \underline{1803.4\,\text{K}}\ (m = 1\,\text{kg})$$

$$\frac{T_4}{T_3} = \left(\frac{V_3}{V_4}\right)^{1.4-1} \qquad T_4 = 1803.4\left(\frac{1}{8}\right)^{0.4} = \underline{785\,\text{K}}$$

$$\text{Heat energy rejected} = Q_{4-1} = m.c_v\,(T_4 - T_1)$$

$$= 718(785 - 300) = \underline{348.2\,\text{kJ}}$$

$$\text{Work} = \text{heat supplied} - \text{heat rejected}$$

$$= 800 - 348.2 = \underline{451.8\,\text{kJ}}$$

$$\text{Air standard efficiency} = \eta = 1 - \frac{\text{heat rejected}}{\text{heat supplied}}$$

$$= 1 - \frac{451.8}{800} = 1 - 0.5647$$

$$= 0.435 = \underline{43.5\%}$$

$$\text{Mean effective pressure, } P_{mi} = \frac{\text{area of diagram}}{\text{length}}$$

$$= \frac{\text{work}}{\text{swept volume}}$$

$$= \frac{451.8}{0.7534} = \underline{599.7\,\text{kN/m}^2}$$

Constant pressure (diesel) cycle

This is the basis of some diesel engine cycles.

Figure 2.4.6 shows the cycle which consists of adiabatic compression, 1–2, constant pressure heat addition, 2–3, adiabatic expansion, 3–4 and exhaust at constant volume, 4–1

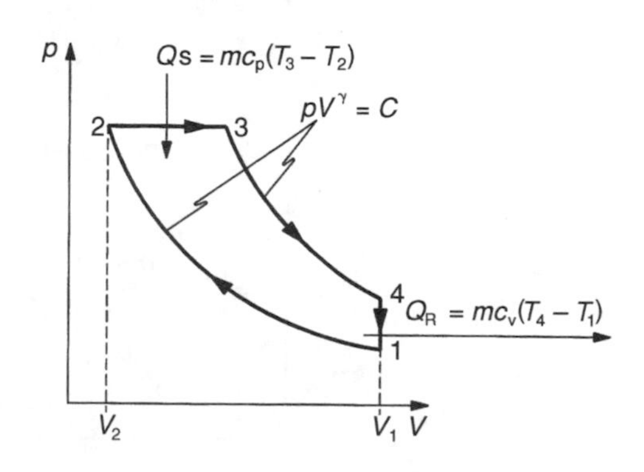

Figure 2.4.6 *Constant pressure (diesel) cycle*

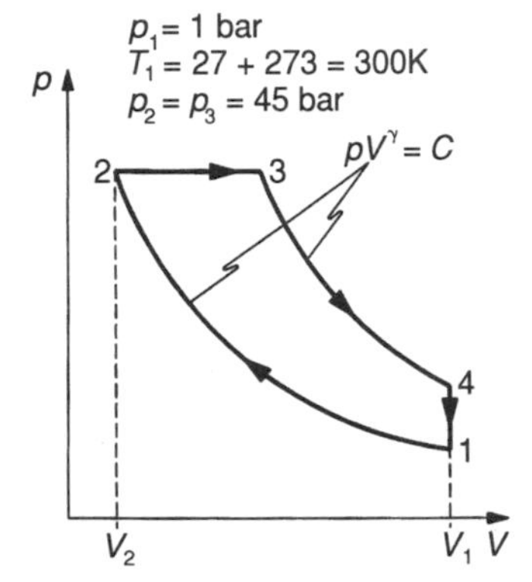

Figure 2.4.7 *Example 2.4.4*

Example 2.4.4

In an air standard diesel cycle, the compression commences at 1 bar, 27°C. Maximum pressure is 45 bar and the volume doubles during the constant pressure process. Calculate the air standard efficiency.

For air;

$\gamma = 1.4$, $c_p = 1005$ J/kgK. $c_v = 718$ J/kgK

Figure 2.4.7 shows the cycle.

$$R = c_p - c_v = 1005 - 718 = \underline{287\text{ J/kgK}}$$

Using a mass of 1 kg,

$$p_1 V_1 = m.R.T_1$$

$$V_1 = \frac{m.R.T_1}{P_1} = \frac{1 \times 287 \times (27 + 273)}{1 \times 10^5} = \underline{0.861\text{ m}^3}$$

$$\frac{T_2}{T_1} = \left(\frac{p_2}{p_1}\right)^{\frac{\gamma-1}{\gamma}} \quad T_2 = T_1\left(\frac{p_2}{p_1}\right)^{\frac{\gamma-1}{\gamma}}$$

$$T_2 = 300\left(\frac{45}{1}\right)^{0.286} = \underline{891\text{ K}}$$

$$\frac{p_1 V_1}{T_1} = \frac{p_2 V_2}{T_2}$$

$$V_2 = \frac{p_1 V_1 T_2}{p_2 T_1} = \frac{1 \times 0.861 \times 891}{45 \times 300} = \underline{0.0568\text{ m}^3}$$

$$\frac{\not{p}_2 V_2}{T_2} = \frac{\not{p}_3 V_3}{T_3}$$

$$T_3 = \frac{V_3 T_2}{V_2} = \frac{(0.0568 \times 2) \times 891}{0.0568} = \underline{1782\text{ K}}$$

$$T_4 = T_3\left(\frac{V_3}{V_4}\right)^{\gamma-1} = 1780\left(\frac{2 \times 0.0568}{0.861}\right)^{0.4} = \underline{791.7\text{ K}}$$

$$\text{Air standard efficiency} = 1 - \frac{\text{heat rejected}}{\text{heat supplied}}$$

$$= 1 - \frac{\not{m}.c_v\,(T_4 - T_1)}{\not{m}.c_p\,(T_3 - T_2)} = 1 - \frac{718(791.7 - 300)}{1005(1780 - 891)}$$

$= 0.605 = \underline{60.5\%}$ air standard efficiency.

Mixed pressure (dual combustion) cycle

This is the basis of most diesel engine cycles.

Figure 2.4.8 shows the cycle, consisting of heat addition at constant volume *and* at constant pressure, two adiabatics and constant volume heat rejection. Note the expressions for heat addition and rejection on the diagram.

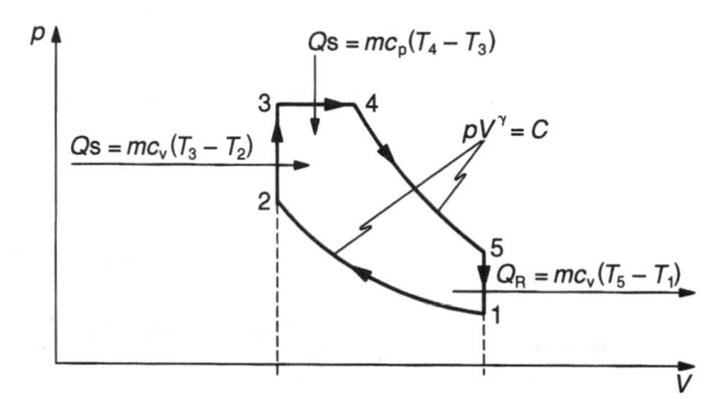

Figure 2.4.8 *Mixed pressure (dual combustion) cycle*

Example 2.4.5

A compression ignition engine working on the ideal dual combustion cycle has a compression ratio of 16:1. The pressure and temperature at the beginning of compression are 98 kN/m^2 and 30°C respectively. The pressure and temperature at the completion of heat supplied are 60 bar and 1300°C. Calculate the thermal efficiency of the cycle.

c_v = 717 J/kgK

c_p = 1004 J/kgK

Figure 2.4.9 shows the cycle.

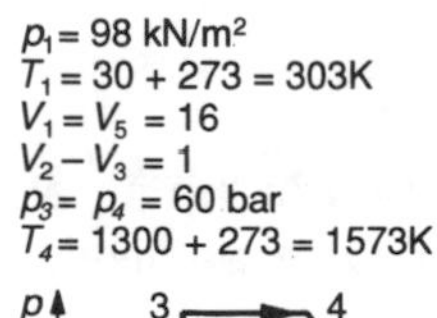

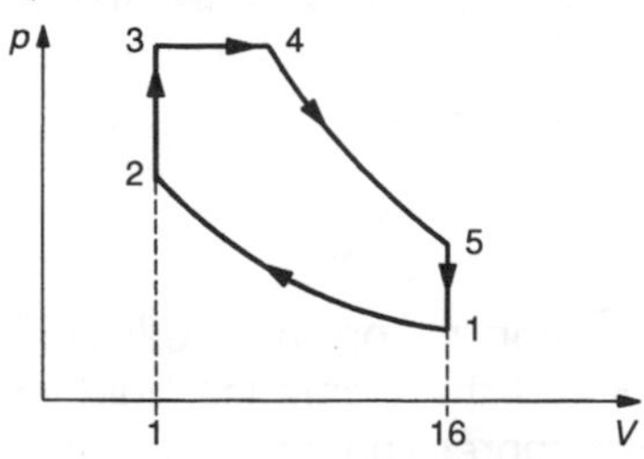

Figure 2.4.9 *Example 2.4.5*

$$\gamma = \frac{c_p}{c_v} = \frac{1004}{717} = 1.4$$

$$p_2 = p_1\left(\frac{V_1}{V_2}\right)^{\gamma} = 0.98\left(\frac{16}{1}\right)^{1.4} = \underline{47.53\text{ bar}}$$

$$T_2 = T_1\left(\frac{V_1}{V_2}\right)^{\gamma-1} = 303\left(\frac{16}{1}\right)^{0.4} = \underline{918\text{ K}}$$

$$\frac{p_2 \not{V}_2}{T_2} = \frac{p_3 V_3}{T_3},\; T_3 = \frac{60 \times 918}{47.53} = \underline{1159\text{ K}}$$

$$\frac{\not{p}_3 V_3}{T_3} = \frac{p_4 V_4}{T_4},\; V_4 = \frac{V_3 \times 1573}{1159} = \underline{1.358 V_3}$$

$$p_4 V_4^{\gamma} = p_5 V_5^{\gamma}$$

$$p_5 = p_4 \left(\frac{V_4}{V_5}\right)^{\gamma} = 60 \left(\frac{1.358 V_3}{16 V_3}\right)^{1.4} = \underline{1.899 \text{ bar}}$$

$$T_5 = T_4 \left(\frac{V_4}{V_5}\right)^{\gamma - 1} = 1573 \left(\frac{1.358\ V_c}{16\ V_c}\right)^{0.4} = \underline{586.5 \text{ K}}$$

Heat energy supplied = $m.c_v\,(T_3 - T_2) + m.c_p\,(T_4 - T_3)$

$= 0.717(1159 - 918) + 1.004(1573 - 1159)$, using a mass of 1 kg

$= 173 + 416 = 589$ kJ/kg

Heat energy rejected $= m.c_v\,(T_5 - T_1)$

$= 0.717\,(586.5 - 303)$

$= \underline{203.3 \text{ kJ/kg}}$

$$\text{Air standard efficiency} = 1 - \frac{\text{heat rejected}}{\text{heat supplied}}$$

$$= 1 - \frac{203.3}{589} = 0.6548$$

$$= \underline{65.48\%}$$

Problems 2.4.1

(1) A petrol engine working on the Otto cycle has a compression ratio of 9:1, and at the beginning of compression the temperature is 32°C. After heat energy supply at constant volume, the temperature is 1700°C. The index of compression and expansion is 1.4. Calculate:

(a) temperature at the end of compression;
(b) temperature at the end of expansion;
(c) air standard efficiency of the cycle.

(2) In a diesel cycle the pressure and temperature of the air at the start of compression are 1 bar and 57°C respectively. The volume compression ratio is 16 and the energy added at constant pressure is 1250 kJ/kg. Calculate:

(a) theoretical cycle efficiency;
(b) mean effective pressure.

(3) The swept volume of an engine working on the ideal dual combustion cycle is 0.1068 m^3 and the clearance volume is 8900 cm^3. At the beginning of compression the pressure is 1 bar, and temperature is 42°C. If the temperature after

expansion is 450°C, the maximum temperature 1500°C and the maximum pressure 45 bar, calculate the air standard efficiency of the cycle.

$\gamma = 1.4,$

$c_v = 0.715$ J/kgK

(4) A compression ignition engine cycle is represented by compression according to the law $pV^{1.35} = C$, 1160 kJ/kg of heat energy supplied at constant pressure, expansion according to the law $pV^{1.3} = C$ back to the initial volume at bottom dead centre, and completed by heat rejection at constant volume. The initial conditions are 1 bar, 43°C, and the compression ratio is 13:1.

Assuming air to be the working fluid throughout, determine the heat transfer per kg during:

(a) the compression process;
(b) the expansion process;
(c) the constant volume process.

$c_p = 1005$ J/kgK

$c_v = 718$ J/kgK

(5) In an engine operating on the ideal dual combustion cycle the compression ratio is 13.5:1. The maximum cycle pressure and temperature are 44 bar and 1350°C respectively. If the initial pressure and temperature are 1 bar and 27°C, calculate the thermal efficiency of the cycle and the mean effective pressure of the cycle.

$c_p = 1.005$ kJ/kgK

$c_v = 0.718$ kJ/kgK

The indicator diagram

A real-life p/V diagram is called an *indicator diagram*, which shows exactly what is happening inside the cylinder of the engine.

This plot is useful because it allows us to find the work which the engine is doing and therefore its power, and it also enables us to see the effect of the timing of inlet, exhaust and fuel burning, so that adjustments can be made to improve cycle efficiency.

In the case of a large slow-speed engine, like a marine diesel engine which typically rotates at about 100 rpm, an indicator diagram can be produced by screwing a device called an engine indicator onto a special cock on the cylinder head of the engine. The indicator records the pressure change in the cylinder and the volume change (which is proportional to crank angle), and plots these on p/V axes using a needle acting on pressure sensitive paper wrapped around a drum. This produces what is known as an 'indicator card'.

Figure 2.4.10 shows the indicator. The spring in the indicator can be changed to suit the maximum cylinder pressure, so that a reasonable plot can be obtained.

Such a mechanical device is not satisfactory for higher-speed engines, but the same result can be plotted electronically.

Brake mean effective pressure, P_{mb}

It was explained (see page 36) that a value of *brake mean effective pressure*, P_{mb}, is used as a comparator between engines, because it is easier to find than indicated mean effective pressure, P_{mi}.

Brake mean effective pressure is calculated from the indicated power formula with brake power and P_{mb} substituted,

$$\text{bp} = P_{mb} \times A \times L \times n$$

Fuel consumption

The fuel consumption of an engine is of great importance, and is affected by detail engine design. The figure most often used to express it is a *specific fuel consumption* (*sfc*) based on the number of kg of fuel burned per second for a unit of power output, i.e. the kg of fuel burned per second for each brake kW.

$$\text{sfc} = \frac{\text{kg fuel burned per sec}}{\text{brake power in kW}}$$

putting in the units,

$$\text{sfc} = \frac{\text{kg}}{\text{s}} \times \frac{\text{s}}{\text{kJ}} = \frac{\text{kg}}{\text{kJ}}$$

An alternative is to express the fuel consumption for each unit of power, e.g for 1 kWh, brake or indicated. A kilowatt hour is a power of 1 kW delivered for 1 hour.

We then have,

Brake specific fuel consumption,

$$\text{bsfc} = \frac{\text{kg fuel burned per hour}}{\text{bp}} = \text{kg/bkWh}$$

and

Indicated specific fuel consumption,

$$\text{isfc} = \frac{\text{kg fuel burned per hour}}{\text{ip}} = \text{kg/ikWh}$$

These values are also often quoted in grammes, i.e. g/kWh.

Brake and indicated thermal efficiency

The thermal efficiency of the engine can be found by considering, as for all values of efficiency, what we get out for what we put in. In this case we get out a value of brake power and we put in heat energy from the fuel burned. The amount of heat energy we put in is the kg of fuel burned per second multiplied by the calorific value of the fuel, CV in kJ/kg.

If we are using the brake power, the efficiency we get is called the brake thermal efficiency, η_b.

$$\eta_b = \frac{\text{brake power}}{\text{kg fuel per sec} \times \text{CV}}$$

which gives units,

$$\eta_b = \text{kW} \times \frac{\text{s}}{\text{kg}} \times \frac{\text{kg}}{\text{kJ}} = 1$$

This can be a decimal 0–1, or a percentage.

Indicated thermal efficiency is found in a similar way, i.e.,

$$\eta_i = \frac{\text{indicated power}}{\text{kg of fuel per sec} \times \text{CV}}$$

Example 2.4.6

An indicator diagram taken from a large diesel engine has an area of 400 mm^2 and length 50 mm. The indicator spring is such that the scale of the pressure axis is 1 mm = 1 bar. If the cylinder diameter and stroke are both 250 mm and the engine is 4-stroke running at 6 rev/s, find the indicated power if the engine has six cylinders.

$$\text{Mean effective pressure} = P_{mi}$$

$$= \frac{\text{area of diagram}}{\text{length of diagram}} \times \text{spring rate}$$

$$= \frac{400}{50} \times 1 \times 10^5$$

$$= \underline{8 \times 10^5 \text{ N/m}^2}$$

$$\text{Indicated power} = P_{mi}\, A.L.n$$

$$= 8 \times 10^5 \times \frac{\pi \times 0.25^2}{4} \times 0.25 \times \frac{6}{2}$$

$$= 29\,452 \text{ W per cylinder}$$

$$\text{Indicated power} = \text{ip per cylinder} \times \text{number of cylinders}$$

$$= 176\,715 \text{ W} = \underline{176.7 \text{ kW}}$$

Example 2.4.7

The area of an indicator diagram taken off a 4-cylinder, 4-stroke engine when running at 5.5 rev/s is 390 mm^2, the length is 70 mm, and the scale of the indicator spring is 1 mm = 0.8 bar. The diameter of the cylinders is 150 mm and the stroke is 200 mm. Calculate the indicated power of the engine assuming all cylinders develop equal power.

$$P_m = \frac{A}{L} \times \text{spring rate} = \frac{390}{70} \times 0.8$$

$$= 4.46\,\text{bar} = 4.46 \times 10^5\,\text{N/m}^2$$

$$\text{ip} = P_m A.L.n \times \text{number of cylinders}$$

$$= 4.46 \times 10^5 \times \frac{\pi \times 0.15^2}{4} \times 0.2 \times \frac{5.5}{2} \times 4$$

$$= 17\,339\,\text{W} = \underline{17.34\,\text{kW}}$$

Example 2.4.8

During a test, a 2-cylinder, 2-stroke diesel engine operating at 2.75 rev/s records at the dynamometer a brake load of 2.7 kN acting at a radius of 1.6 m. The bore of the cylinder is 0.35 m and the stroke is 0.5 m. If the indicated mean effective pressure is 3 bar, calculate:

(a) the indicated power;
(b) the brake power;
(c) the mechanical efficiency.

$$\text{ip} = P_{mi}\,A.L.n = 3 \times 10^5 \times \frac{\pi \times 0.35^2}{4} \times 0.5 \times 2.75 \times 2$$

$$= 79\,374\,\text{W} = \underline{79.37\,\text{kW}}$$

$$\text{bp} = T\omega = (2.7 \times 1.6) \times 2.75 \times 2\pi$$

$$= \underline{74.64\,\text{kW}}$$

Note: Torque = force × radius.

$$\eta_m = \frac{\text{bp}}{\text{ip}} = \frac{74.64}{79.37} = 0.94 = \underline{94\%}$$

Example 2.4.9

A marine 4-stroke diesel engine develops a brake power of 3200 kW at 6.67 rev/s with a mechanical efficiency of 90% and a fuel consumption of 660 kg/hour. The engine has eight cylinders of 400 mm bore and 540 mm stroke. Calculate:

(a) the indicated mean effective pressure;
(b) the brake thermal efficiency.

The calorific value of the fuel = 41.86 MJ/kg.

$$\eta_m = \frac{\text{bp}}{\text{ip}} \quad \text{ip} = \frac{\text{bp}}{\eta_m} = \frac{3200}{0.9} = \underline{3555.6\,\text{kW}}$$

$$\text{ip} = P_m A.L.n$$

$$\frac{3555.6}{8} = P_{mi} \times \frac{\pi \times 0.4^2}{4} \times 0.54 \times \frac{6.67}{2}$$

$$P_{mi} = \frac{3555.6 \times 4 \times 2}{8 \times \pi \times 0.4^2 \times 0.54 \times 6.67} = 1963.9\,\text{kN/m}^2$$

$$= \underline{19.63\text{ bar}}$$

$$\text{Brake thermal efficiency} = \frac{\text{brake power}}{\text{kg fuel/s} \times \text{CV}}$$

$$= \frac{3200}{\frac{660}{3600} \times 41.86 \times 10^3}$$

$$= 0.417 = \underline{41.7\%}$$

Example 2.4.10

A 6-cylinder 4-stroke internal combustion engine is run on test and the following data was noted:

Compression ratio = 8.2:1
Speed = 3700 rpm
Brake torque = 0.204 kN.m
Bore = 90 mm
Fuel consumption = 26 kg/h
Stroke = 110 mm
Calorific value of fuel = 42 MJ/kg
Indicated mean effective pressure = 7.82 bar

Calculate:

(a) the mechanical efficiency;
(b) the brake thermal efficiency;
(c) the brake specific fuel consumption.

$$\text{ip} = P_{mi}A.L.n \times \text{number of cylinders}$$

$$= 782 \times \pi \times \frac{0.09^2}{4} \times 0.11 \times \frac{3700}{120} \times 6 = 101.2\,\text{kW}$$

$$\text{bp} = T.\omega = 0.204 \times \frac{3700 \times 2\pi}{60} = 79\,\text{kW}$$

$$\eta_m = \frac{\text{bp}}{\text{ip}} = \frac{79}{101.2} = 0.781 = \underline{78.1\%}$$

$$\eta_b = \frac{\text{bp}}{\text{kg fuel/s} \times \text{CV}} = \frac{79}{\frac{26}{3600} \times 42 \times 10^3}$$

$$= 0.26 = \underline{26\%}$$

$$\text{Brake specific fuel consumption} = \frac{\text{kg fuel/h}}{\text{brake power}} = \frac{26}{79}$$

$$= \underline{0.329\,\text{kg/kWh}}$$

Volumetric efficiency

The *volumetric efficiency* of an engine – or a reciprocating compressor – is a measure of the effectiveness of the engine in 'breathing in' a fresh supply of air.

Under perfect circumstances, when the piston starts to move from top dead centre down the cylinder, fresh air is immediately drawn in. However, above the piston at TDC there is a residual pressure which remains in the cylinder until the piston has moved down the cylinder a sufficient distance to relieve it and create a pressure slightly below atmospheric. Only then will a fresh charge of air be drawn in.

A further difficulty is the heating of the air in the hot inlet manifold, which also reduces the mass of air entering the cylinder.

The ratio of the swept volume of the engine to the volume of air actually drawn in is called the volumetric efficiency, η_v.

$$\eta_v = \frac{\text{volume of charge induced at reference temperature and pressure}}{\text{piston swept volume}}$$

The reference temperature and pressure are usually the inlet conditions.

Example 2.4.11

A 4-stroke, 6 cylinder engine has a fuel consumption of 26 kg/h and an air/fuel ratio of 21:1. The engine operates at 3700 rpm and has a bore of 90 mm, stroke 110 mm. Calculate the volumetric efficiency referred to the inlet conditions of 1 bar, 15°C. $R = 287$ J/kgK.

Using the characteristic gas equation, $p_1 V_1 = m.R.T_1$

Volume of air induced/minute

$$= \frac{m.R.T_1}{p_1} = \frac{\frac{(26 \times 21)}{60} \times 287 \times (15 + 273)}{1 \times 10^5}$$

$= 7.52\,\text{m}^3/\text{min}$

$$\text{Swept volume} = \frac{\pi \times 0.09^2}{4} \times 0.11 = 7 \times 10^{-4}\ \text{m}^3/\text{rev}$$

$$= 7\times10^{-4} \times \frac{3700}{2} \times 6 = 7.76\,\text{m}^3/\text{min}$$

$$\eta_v = \frac{\text{volume induced at reference}}{\text{swept volume}} = \frac{7.52}{7.76} = 0.97 = \underline{97\%}$$

If, given a volume, you need to change it to a different set of conditions, use can be made of

$$\frac{p_1 V_1}{T_1} = \frac{p_2 V_2}{T_2}$$

Case study

Marine diesels

Diesel engines are produced by many manufacturers, in a range of power outputs, for very many applications.

The largest diesel engines are to be found in ships, and these operate on the 2-stroke cycle, which makes them quite unusual. The piston is bolted to a piston rod which at its lower end attaches to a crosshead running in vertical guides, i.e. a crosshead bearing. A connecting rod then transmits the thrust to the crank to turn the crankshaft. The arrangement is the same as on old triple expansion steam engine, from which they were derived. They have the further peculiarity of being able to run in both directions by movement of the camshaft. This provides astern movement without the expense of what would be a very large gearbox.

These very large engines are the first choice for most merchant ships because of their economy and ability to operate on low quality fuel. A typical installation on a container ship, for instance, would be a 6-cylinder turbocharged engine producing 20 000 kW at a speed of about 100 rpm. The engine is connected directly to a fixed-pitch propeller.

Most diesel engines are now turbocharged. Exhaust gas from the engine drives a gas turbine connected to a fan compressor which forces air into the cylinder at a raised pressure. This has the main advantage of charging the cylinder with a greater mass of air (the mass is proportional to the pressure, from $pV = m.R.T$), thereby allowing more fuel to be burned, so for the same size cylinder more power can be produced. An added advantage in the case of a 2-stroke engine is that by pressurizing the air into the cylinder, the exhaust gas is more effectively removed or 'scavenged' before the next cycle begins

One of the main problems with large slow-speed engines is the headroom necessary to accommodate them, and in a vessel such as a car ferry, they are not usually fitted because they would limit car deck space. Instead, medium-speed engines are used which are 4-stroke and are of the more usual trunk-piston configuration, the same as a car engine and almost all other engines too.

One of the latest engines, developed for fast ferries, has the following particulars:

Power output	8200 kW
Operating cycle	4-stroke
Number of cylinders	20, in 'V' configuration
Bore	265 mm
Stroke	315 mm
Operating speed	1150 rpm
Dimensions	7.4 m long × 1.9 m wide × 3.3 m high
Weight	43 tonnes (43 000 kg)
Mean effective pressure	24.6 bar
Specific fuel consumption	195 kg/kWh
Time between overhauls	24 000 hours

The engine has a single large turbocharger at one end. Clearly, this is a sizeable engine, and typically a large ferry would need two or three of them. Most cruise ships also have these 'medium-speed' diesel engines.

Many manufacturers produce a single engine design in which the number of cylinders in the complete engine can be varied to suit the required output. This simplifies spares and maintenance requirements and means that the engine builder can tailor an engine of a standard design to meet different requirements.

The details below illustrate this for an engine type now in production. Note the number of variations which can be obtained and therefore the range of power outputs available:

Operating cycle	2-stroke
Bore	350 mm
Stroke	1400 mm
Number of cylinders	4, 5, 6, 7, 8, 9, 10, 11 or 12
Power output	2900–8900 kW
Mean effective pressure	19 bar
Fuel consumption	180 g/kWh

Review exercise problems 2.4.2

(1) An indicator diagram taken from one cylinder of a 6-cylinder 2-stroke engine has an area of 2850 mm^2 and length 75 mm when running at 2 rev/s. The indicator spring rate is 1 mm = 0.2 bar. Given that the cylinder bore is 550 mm and the stroke is 850 mm, calculate the indicated power of the engine, assuming each cylinder develops the same power.

(2) A 6-cylinder, 4-stroke diesel engine has a bore of 150 mm and a stroke of 120 mm. The indicated mean effective pressure is 9 bar, the engine runs at 300 rpm, and the mechanical efficiency is 0.85. Calculate the indicated power and the brake power.

(3) A single cylinder 4-stroke engine is attached to a dynamometer which provides a braking load of 362 N. The radius at which the brake acts is 800 mm. If at this load the engine has a speed of 318 rpm, find the brake power.

(4) A single cylinder 4-stroke oil engine has a cylinder diameter of 180 mm and stroke 300 mm. During a test, the following results were recorded,

Area of indicator = 500 mm^2
Length of indicator card = 70 mm
Card scale (spring rate), 1 mm = 0.8 bar
Brake load = 354 N
Brake load radius = 780 mm
Engine speed = 5 rev/s
Fuel consumption = 3.2 kg/h
Calorific value of fuel = 43.5 MJ/kg

Calculate:

(a) the indicated power;
(b) the brake power;
(c) the brake thermal efficiency.

(5) A 3-cylinder, 4-stroke engine has a bore of 76 mm and a stroke of 125 mm. It develops 12 kW at the output shaft when running at 1500 rpm. If the mechanical efficiency is 85% and it burns 3.2 kg of oil per hour of calorific value 42 000 kJ/kg, find the indicated mean effective pressure, assuming all cylinders produce the same power, and the brake thermal efficiency.

(6) A 4-cylinder, 4-stroke engine of 78 mm bore and 105 mm stroke develops an indicated power of 47.5 kW at 4400 rpm. The air/fuel ratio is 21 kg air/kg fuel, the fuel consumption is 13.6 kg/h and the calorific value of the fuel is 41.8 MJ/kg. Calculate for the engine:

(a) the indicated mean effective pressure;
(b) the indicated thermal efficiency;
(c) the volumetric efficiency referred to inlet conditions of 1 bar, 15°C.

R = 287 J/kgK

(7) A 6-cylinder, 4-stroke diesel engine has a bore of 210 mm and a stroke of 315 mm. At 750 rpm, the brake mean effective pressure is 4.89 bar and the specific brake fuel consumption is 0.195 kg/kWh. The air to fuel mass flow ratio is 28 to 1 and the atmospheric conditions are 0.95 bar, 17°C. Calculate the volumetric efficiency.

(8) (a) Derive an expression for the brake thermal efficiency of an engine.
(b) If an engine has an indicated power of 45 kW and a mechanical efficiency of 85%, determine its brake thermal efficiency if it consumes 0.156 kg/min of fuel of calorific value 42 MJ/kg.

2.5 The steady flow energy equation (SFEE)

In machinery such as turbines, boilers and nozzles, there is flow of the working fluid at a constant rate into and out of the system. This is also true of the internal combustion engine if we place the system boundary outside the cylinders to encompass the inlet and outlet valves.

In this chapter, we look at cases of steady flow and apply appropriate expressions, beginning by looking at all the elements involved in the steady flow process and then calculating values of work done and heat energy transferred. We are then in a position to find power and efficiency. Also in this chapter, we look at the effect of change of entropy during a steady flow process.

Figure 2.5.1 represents a steady flow system. To arrive at the steady flow energy equation, we need to include all the energies involved. These are:

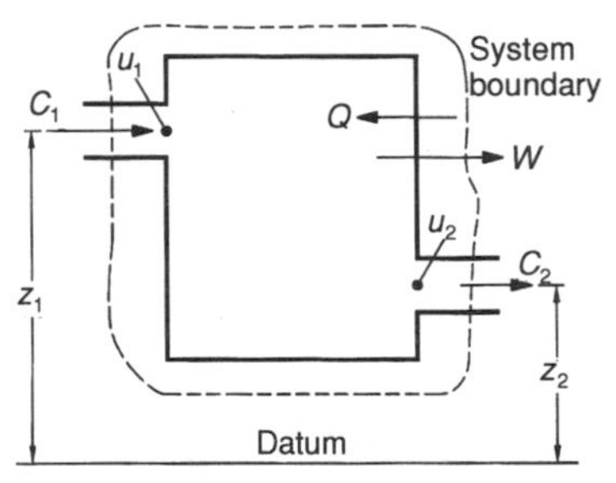

Figure 2.5.1 *Steady flow system*

- Kinetic energy
- Potential energy
- Internal energy
- Work energy
- Heat energy

Kinetic energy (due to motion), potential energy (due to height above a datum), and internal energy (the intrinsic energy contained by a fluid because of its temperature), are well known, and we also know that work can be done and heat energy transferred to or from the system.

In order to enter and leave the system, the fluid must do work on the system and on the surroundings. That is, in order to enter the system, the fluid outside must expend energy to push in, and in order to leave the system the fluid inside must expend energy to push out.

This is calculated by multiplying the pressure by the volume flow rate of fluid entering the system and leaving the system, but instead of using volume flow rate in m^3/s, it is convenient to use specific volume, i.e. the volume which 1 kg of the fluid occupies, and multiply this by the mass flow rate of the fluid in kg/s. This is so we can bracket quantities and multiply them all by a mass flow rate.

Hence,

$$\text{Volume flow rate} = \text{mass flow rate} \times \text{specific volume}$$

$$= \frac{\text{m}^3}{\text{kg}} \times \frac{\text{kg}}{\text{s}} = \frac{\text{m}^3}{\text{s}} = \dot{m}.v$$

$$\text{Pressure} \times \text{volume flow rate} = \frac{\text{N}}{\text{m}^2} \times \frac{\text{m}^3}{\text{s}} = \frac{\text{N.m}}{\text{s}} = \frac{\text{joules}}{\text{s}}$$

$$= \text{W} = \dot{m}.p.v$$

We can now say that the total energy entering the system is the same as the total energy leaving the system, and produce an equation, assuming Q is supplied to the fluid and W is done by the fluid, and remembering that in this case, Q and W are per second,

$$Q + \dot{m}\left(u_1 + z_1 g + \frac{c_1^2}{2}\right) + \dot{m}p_1 v_1 = W + \dot{m}\left(u_2 + z_2 g + \frac{c_2^2}{2}\right) + \dot{m}p_2 v_2$$

The u, z and c terms represent the internal, potential and kinetic energies respectively.

$$Q - W = \dot{m}\left[(u_2 + p_2 v_2) - (u_1 + p_1 v_1) + \frac{c_2^2 - c_1^2}{2} + (z_2 g - z_1 g)\right]$$

writing

$$h_1 = (u_1 + p_1 v_1), \text{ and } h_2 = (u_2 + p_2 v_2)$$

and neglecting potential energy terms,

$$Q - W = \dot{m}\left[(h_2 - h_1) + \frac{c_2^2 - c_1^2}{2}\right]$$

This is the form of the steady flow energy equation which is suitable for most cases.

Where:

Q = rate of heat energy transfer to or from the system, kJ/s = (kW)

W = rate of work energy transfer to or from the system, kJ/s = (kW)

$\dot{m}$ = mass flow rate of the fluid $\left(\frac{\text{kg}}{\text{s}}\right)$

h = specific enthalpy $\left(\frac{\text{kJ}}{\text{kg}}\right)$

c = velocity of fluid $\left(\frac{\text{m}}{\text{s}}\right)$.

Note that in this case, Q and W are per second.

Putting in the units,

$$\frac{\text{J}}{\text{s}} = \left[\frac{\text{kg}}{\text{s}}\left(\frac{\text{J}}{\text{kg}} + \frac{\text{m}^2}{\text{s}^2}\right)\right] = \left[\frac{\text{J}}{\text{s}} + \frac{\text{kg.m}^2}{\text{s.s}^2}\right] = \left[\frac{\text{J}}{\text{s}} + \left(\frac{\text{kg.m}}{\text{s}^2}\right)\frac{\text{m}}{\text{s}}\right]$$

$$= \left[\frac{\text{J}}{\text{s}} + \frac{\text{N.m}}{\text{s}}\right] = \frac{\text{J}}{\text{s}}$$

Note: 1 newton = $\frac{1\,\text{kg.m}}{\text{s}^2}$, from force = mass × acceleration.

Key points

- Heat energy is rejected and is therefore negative.
- There is an enthalpy drop through the turbine, so this is also negative.
- Quantities have been changed to joules so that velocities can be entered in m/s.

Enthalpy

We have given the symbol h, *specific enthalpy*, for the sum of the internal energy u, and the product of pressure and volume, pv, i.e.

$$h = u + pv$$

h is the specific enthalpy of the fluid, a property which is found in tables of properties, e.g. for steam and refrigerants, and can be thought of as a figure for the 'total energy' of a fluid.

For a perfect gas, the specific enthalpy change is calculated using,

$$h_2 - h_1 = c_p.(T_2 - T_1)$$

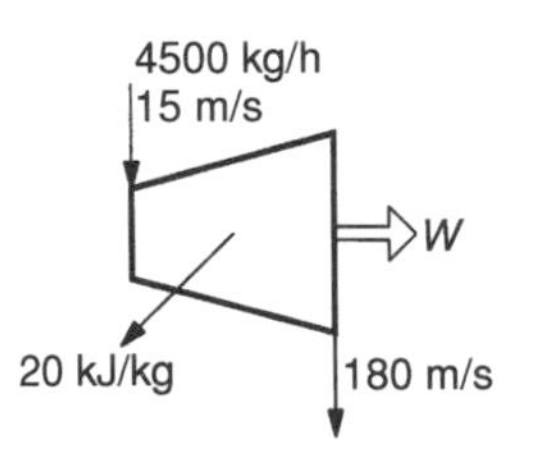

Figure 2.5.2 *Example 2.5.1*

Example 2.5.1

Gas enters a turbine with a velocity of 15 m/s at a rate of 4500 kg/h and is discharged with a velocity of 180 m/s. If the turbine loses 20 kJ to the surroundings for every kg of gas flow, calculate the power developed if the enthalpy drop is 420 kJ/kg. Figure 2.5.2 represents the turbine.

$$Q - W = m\left[(h_2 - h_1) + \frac{c_2^2 - c_1^2}{2}\right]$$

$$\text{Mass flow of gas} = \frac{4500}{3600} = 1.25\text{ kg/s}$$

$$(-20\,000 \times 1.25) - W = 1.25\left[-420\,000 + \frac{180^2 - 15^2}{2}\right]$$

$$-25\,000 - W = -525\,000 + 20\,109$$

$$-W = -525\,000 + 20\,109 + 25\,000 = -497\,981\text{ J}$$

$$\underline{W = 480\text{ kW}}$$

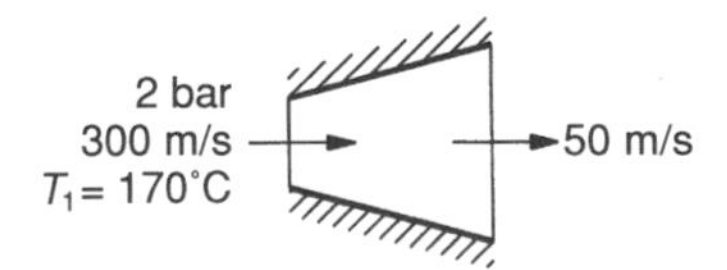

Figure 2.5.3 *Example 2.5.2*

Example 2.5.2

Air is delivered to a diffuser at 2 bar, 170°C. The air velocity is reduced from 300 m/s at inlet to 50 m/s at outlet. Assuming adiabatic flow, find the air pressure at diffuser outlet.

$$c_p = 1005\text{ J/kgK}$$

$$\gamma = 1.4$$

See Figure 2.5.3.

Key points

- A diffuser has no moving parts, so there is no work done, and because the flow is assumed to be adiabatic, there is no heat energy transfer.
- A diffuser is the opposite of a nozzle, it being divergent, producing a pressure rise between inlet and exit.
- To find the final pressure, we have made use of an expression we have used so far only in non-flow processes, and applied it across the inlet and exit points in this steady flow system.

$$Q - W = m\left[(h_2 - h_1) + \frac{c_2^2 - c_1^2}{2}\right]$$

$$0 - 0 = m\left[(h_2 - h_1) + \frac{50^2 - 300^2}{2}\right]$$

$$-(h_2 - h_1) = \frac{50^2 - 300^2}{2} = -43\,750$$

$$h_1 - h_2 = c_p\,(T_1 - T_2) = -43\,750 \quad (\text{from } \delta h = m.c_p\,\delta T)$$

$$1005(170 - T_2) = -43\,750,\ T_2 = 213.5°\text{C}$$

$$\frac{T_2}{T_1} = \left(\frac{p_2}{p_1}\right)^{\frac{\gamma-1}{\gamma}},\ \frac{213.5 + 273}{170 + 273} = \left(\frac{p_2}{2}\right)^{\frac{1.4-1}{1.4}},$$

$$\frac{486.5}{443} = \left(\frac{p_2}{2}\right)^{0.286},\ p_2 = \underline{2.78\text{ bar}}$$

Applications of the SFEE

In each case we will begin by writing down the SFEE. We can then take out non-relevant properties.

Steam or gas turbine

$$Q - W = \dot{m}\left[(h_2 - h_1) + \frac{c_2^2 - c_1^2}{2}\right]$$

Usually, we can neglect the heat loss from the turbine because the fluid flows through very quickly, giving insufficient time for heat energy transfer to occur. Also, because the inlet and outlet velocities are similar, the kinetic energy term can be neglected.

This leaves us with an extremely useful equation,

$$-W = \dot{m}(h_2 - h_1)$$

This is written as,

$$W = \dot{m}(h_1 - h_2)$$

in order to give a positive value of work when point 1 is turbine inlet.

For a steam turbine, the values of h are obtained from steam tables. For a gas turbine, the enthalpy difference is obtained from,

$$\dot{m}(h_1 - h_2) = \dot{m}.c_p\,(T_1 - T_2)$$

This expression is true for any process.

Putting in units,

$$\dot{m}(h_1 - h_2) = \frac{\text{kg}}{\text{s}} \times \frac{\text{kJ}}{\text{kg}} = \frac{\text{kJ}}{\text{s}} = \text{kW}$$

$$\dot{m}.c_p\ (T_1 - T_2) = \frac{\text{kg}}{\text{s}} \times \frac{\text{kJ}}{\text{kg.K}} \times \text{K} = \frac{\text{kJ}}{\text{s}} = \text{kW}$$

Example 2.5.3

A steam turbine receives steam with an enthalpy of 3467 kJ/kg. At the outlet from the turbine, the enthalpy of the steam is 2570 kJ/kg. If the mass flow rate of steam is 2 kg/s, find the power of the turbine.

$$W = \dot{m}(h_1 - h_2) = 2(3476 - 2570) = \underline{1812\,\text{kW}}$$

Example 2.5.4

The temperature of the gas entering a turbine is 750°C. If the gas leaves the turbine at a temperature of 500°C, and the mass flow rate of the gas is 3.5 kg/s, calculate the power developed by the turbine.

$$c_p = 980\ \text{J/kgK}$$

$$W = \dot{m}.c_p\ (T_1 - T_2) = 3.5 \times 980 \times (750 - 500)$$

$$= 857\,500\ \text{J} = \underline{857.5\,\text{kJ}}$$

Compressor

The same argument concerning heat energy loss and fluid velocity applies to the compressor. We are thinking here of the rotary, or axial compressor, since for a reciprocating compressor we are more likely to be considering the non-flow processes occurring in the cylinders.

The SFEE becomes

$$\underline{-W = (h_2 - h_1)}$$

Key point

Power input is negative because work is done on the system.

Example 2.5.5

An axial flow gas compressor takes in gas with a specific enthalpy of 200 kJ/kg, and discharges it to a receiver with a specific enthalpy of 1500 kJ/kg. If the mass flow rate of the gas is 5 kg/s, find the power required.

$$-W = \dot{m}(h_2 - h_1) = 5(1500 - 200) = \underline{6500\,\text{kW}}$$

Boiler

$$Q - W = \dot{m}\left[(h_2 - h_1) + \frac{c_2^2 - c_1^2}{2}\right]$$

A boiler does no work (there are no moving parts), and velocities into and out of the boiler are low. The SFEE can therefore be written,

$$\underline{Q = \dot{m}(h_2 - h_1)}$$

Example 2.5.6

A boiler receives feedwater with an enthalpy of 505 kJ/kg and produces steam with an enthalpy of 2676 kJ/kg. Neglecting losses, find the heat energy supplied to the boiler if the mass flow rate is 2 kg/s.

$$Q = \dot{m}(h_2 - h_1) = 2(2676 - 505) = \underline{4342\,\text{kW}}$$

Key point

h_2 will be smaller than h_1, giving a negative Q, indicating heat energy given out by the system.

Condenser

The condenser cools the vapour – steam or refrigerant – to produce a liquid, usually by passing the vapour over tubes circulated with a cooler liquid, or air. As in the case of the boiler, there are no moving parts, and velocities are low.

The SFEE becomes,

$$Q = \dot{m}(h_2 - h_1).$$

Throttle

A throttle is used in steam plant for pressure reduction, for example in a steam reducing valve to lower high pressure steam coming from a boiler to make it suitable for heating purposes. Throttling involves passing the steam through a restricting orifice, or a partially open valve, thereby introducing friction to reduce the pressure.

There is very little time for heat energy transfer, velocities before and after the orifice are similar, and there is no work done.

The SFEE becomes,

$$0 = m(h_2 - h_1)$$

hence,

$$\underline{h_2 = h_1}$$

- The enthalpy before and after throttling is the same. This is made use of in the separating and throttling calorimeter which we look at in the steam section.

Nozzle

Nozzles are used in gas and steam turbines, and in many other applications to increase fluid velocity.

Because fluid velocity is high, we can assume that no heat energy transfer takes place through the nozzle, and there is no work done because there are no moving parts.

The SFEE becomes,

$$0 = \dot{m}\left[(h_2 - h_1) + \frac{c_2^2 - c_1^2}{2}\right]$$

If we assume a negligible inlet velocity, which is often the case, we can transpose to give,

$$(h_1 - h_2) = \frac{c_2^2}{2}$$

$$\underline{c_2 = \sqrt{2(h_1 - h_2)}}$$

This is an expression for the velocity of the fluid leaving the nozzle.

Putting in units,

$$\frac{m}{s} = \sqrt{\frac{J}{kg}} = \sqrt{\frac{N.m}{kg}} = \sqrt{\frac{kg.m.m}{s^2.kg}} = \sqrt{\frac{m^2}{s^2}} = \frac{m}{s}$$

Example 2.5.7

Air enters a nozzle at 400°C with negligible velocity at a rate of 1.2 kg/s. At the exit from the nozzle, the temperature is 110°C. Find the velocity at the nozzle exit and the nozzle exit area if the specific volume of the air at the exit is 1.3 m³/kg.

c_p = 1005 J/kgK

Figure 2.5.4 shows the nozzle.

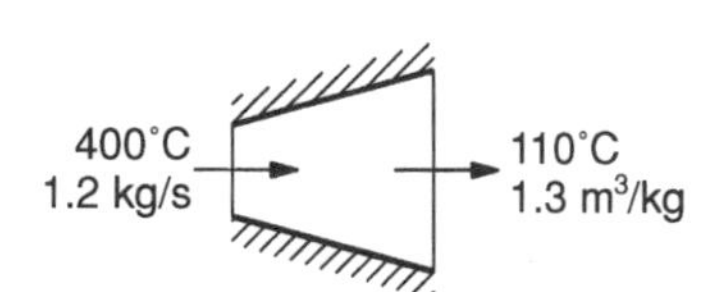

Figure 2.5.4 *Example 2.5.7*

$$c = \sqrt{2(h_1 - h_2)} = \sqrt{2.c_p\,(T_1 - T_2)}$$

$$= \sqrt{2 \times 1005(400 - 110)} = \underline{763.5\text{ m/s}}$$

Volume flow rate = velocity × area

$$\text{Area} = \frac{\text{volume flow}}{\text{velocity}} = \frac{\text{mass flow} \times \text{specific volume}}{\text{velocity}}$$

$$= \frac{1.2 \times 1.3}{763.5} = \underline{0.002\text{ m}^2}$$

Further use is made of the SFEE in the chapter on steam.

Problems 2.5.1

(1) A steam turbine receives steam with a specific enthalpy of 3117 kJ/kg at a rate of 5 kg/s. At the turbine outlet, the specific enthalpy of the steam is 2851 kJ/kg. Find the turbine power in kW.

(2) A gas turbine develops 500 kW when the mass flow through the turbine is 4 kg/s. Neglecting heat energy loss from the turbine and gas velocity changes, determine the specific enthalpy drop through the turbine.

(3) Gas enters a nozzle with negligible velocity and discharges at 100 m/s. Determine the enthalpy drop through the nozzle.

(4) A fluid flows through a turbine at the rate of 30 kg/min. Across the turbine the specific enthalpy drop of the fluid is 580 kJ/kg and the turbine loses 2500 kJ/min from the turbine casing. Find the power produced neglecting velocity changes.

(5) An air compressor takes in air at 20°C and discharges it at 35°C when the gas flow rate is 2 kg/s. Find the power absorbed by the compressor if it loses 50 kJ/min to the surroundings.

$c_p = 1005$ J/kgK

(6) Gas enters a nozzle with a velocity of 200 m/s, temperature 50°C. At the outlet from the nozzle, the gas temperature is 30°C. Determine the exit velocity of the gas, assuming no heat energy losses.

$C_p = 950$ J/kgK

(7) A nozzle receives steam with enthalpy 2900 kJ/kg at the rate of 10 kg/min, and at the outlet from the nozzle the velocity is 1050 m/s. If the inlet steam velocity is zero and there are no heat losses, find the specific enthalpy of the steam at the exit and the outlet area of the nozzle if at this point the specific volume of the steam is 20 m^3/kg.

(8) A steam turbine receives steam with a velocity of 28 m/s, specific enthalpy 3000 kJ/kg at a rate of 3500 kg per hour. The steam leaves the turbine with a specific enthalpy of 2200 kJ/kg at 180 m/s. Calculate the turbine output, neglecting losses.

Isentropic efficiency

A complication is introduced when finding the enthalpy after a process if we consider change of *entropy, s*. So far we have mentioned entropy only in saying that a reversible adiabatic process is isentropic, i.e. no change in entropy. Since a process can be neither reversible nor adiabatic, we can never have a process in which there is no change of entropy; the entropy after the process is always greater than at the start of the process.

Entropy can be defined as that property such that if we plot it against absolute temperature, the area under the process curve is the heat energy transferred.

From this,

$$\delta s = \frac{\delta Q}{T}$$

The units of specific entropy are therefore J/kgK, or, more usually, kJ/kgK.

From an engineer's point of view it is sufficient to say at this stage that the closer to zero change in entropy the process can be, the more effective the process. For instance, if we have expansion of gas or steam in a turbine, more of the heat energy is converted into work at the output shaft if the change in entropy is small.

For a turbine or compressor, the change of entropy is indicated by the 'isentropic efficiency'. The ideal is 100%, meaning no change in entropy, but a typical value is 0.8–0.9.

We have already seen that the change in specific enthalpy of a perfect gas is

$$\delta h = c_p.\delta T$$

and that, from the SFEE, the power developed in, say, a turbine is

$$P = \dot{m}(h_1 - h_2)$$

- Therefore the power produced is proportional to the change in temperature through the turbine.

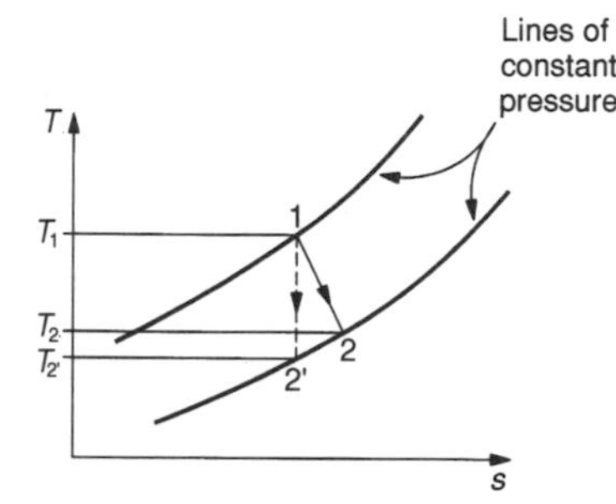

Figure 2.5.5 *T/s axes for a perfect gas*

Figure 2.5.5 shows T/s axes for a perfect gas with lines of constant pressure added.

Let us consider point 1 as the inlet condition to a gas turbine. If expansion through the turbine is isentropic, i.e. no change in entropy, the temperature drop is

$$(T_1 - T'_2)$$

T'_2 denotes the isentropic point.

If entropy increases through the turbine, the exit point is at point 2 for the same outlet pressure, and the temperature drop is

$$(T_1 - T_2)$$

T_2 denotes the exit point if the expansion is not isentropic.

Clearly,

$(T_1 - T'_2)$ is greater than $(T_1 - T_2)$

Therefore the work done and the power produced is greatest when the expansion through the turbine is isentropic.

The ratio

$$\frac{(T_1 - T_2)}{(T_1 - T'_2)} = \eta_T$$

the isentropic efficiency of the turbine.

Key points

- From the expression it can be seen that if the isentropic temperature drop is known, the actual temperature drop can be found simply by multiplying by the isentropic efficiency.
- This is a ratio of temperatures, but if it was written as a corresponding ratio of enthalpies, the same answer would result because, as we have seen, for a perfect gas, the enthalpy change is directly proportional to the temperature change.

The gas turbine

Figure 2.5.6 shows a simple *open cycle gas turbine* plant in which air is compressed between 1 and 2, fuel is added in the combustion chamber between 2 and 3 at constant pressure, and expansion of the hot gases takes place in the turbine between 3 and 4. There are only two pressures to consider, these are usually expressed as a pressure ratio across which the turbine operates. These pressures are represented by the constant pressure lines on the T/s and the p/V diagrams.

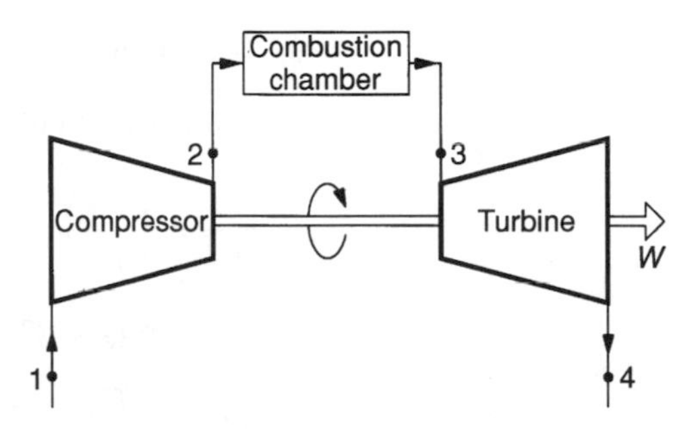

Figure 2.5.6 *Simple open cycle gas turbine*

The compressor and turbine are on a common shaft, therefore some of the work produced in the turbine is lost in driving the compressor. Usually, the mass increase of the gases after the combustion chamber because of fuel addition is neglected.

Figure 2.5.7 shows the cycle on T/s axes, assuming the compression and expansion are not isentropic, and showing the isentropic lines as dotted. Figure 2.5.7 also shows the cycle on p/V axes.

From the SFEE, we can establish the following expressions,

Work done by turbine $= W_t = (h_3 - h_4) = c_p\,(T_3 - T_4)$

Work to compressor $= W_c = (h_2 - h_1) = c_p\,(T_2 - T_1)$

Constant pressure heat addition at combustion chamber

$= (h_3 - h_2) = c_p\,(T_3 - T_2)$

$$\text{Plant efficiency} = \eta = \frac{\text{net work}}{\text{heat energy input}} = \frac{W_t - W_c}{Q_{cc}}$$

$$= \frac{(h_3 - h_4) - (h_2 - h_1)}{(h_3 - h_2)}$$

$$\eta = \frac{(T_3 - T_4) - (T_2 - T_1)}{(T_3 - T_2)}$$

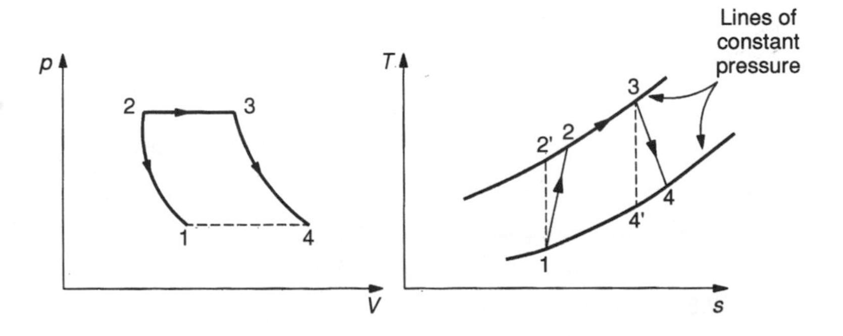

Figure 2.5.7 *Gas turbine cycle on p/V and T/s axes*

Example 2.5.8

In an open cycle gas turbine plant, the pressure ratio is 5:1 and the maximum cycle temperature is 650°C. The minimum cycle temperature is 15°C and the isentropic efficiency of both the turbine and the compressor is 0.86. Calculate the power output if the mass flow rate through the turbine is 1 kg/s.

c_p = 1005 kJ/kgK and γ = 1.4 for air and combustion gas

See Figure 2.5.8.

First, we find the isentropic temperature after compression.

$$\frac{T'_2}{T_1} = \left(\frac{p_2}{p_1}\right)^{\frac{\gamma-1}{\gamma}}, \quad \frac{T'_2}{(15+273)} = \left(\frac{5}{1}\right)^{\frac{1.4-1}{1.4}},$$

$$T'_2 = 288 \times 5^{0.286} = 456.3\,\text{K}$$

$$0.86 = \frac{T'_2 - T_1}{T_2 - T_1}, \quad 0.86 = \frac{456.3 - 288}{T_2 - 288},$$

$$T_2 = \frac{(456.3 - 288)}{0.86} + 288 = \underline{483.7\,\text{K}}$$

Following the same procedure through the turbine,

$$\frac{T'_4}{T_3} = \left(\frac{p_4}{p_3}\right)^{\frac{\gamma-1}{\gamma}}, \quad \frac{T'_4}{(650+273)} = \left(\frac{1}{5}\right)^{\frac{1.4-1}{1.4}},$$

$$T'_4 = 923 \times 0.2^{0.286} = 582.5\,\text{K}$$

$$\eta_T = \frac{T_3 - T_4}{T_3 - T'_4}, \quad 0.86 = \frac{923 - T_4}{923 - 582.5},$$

$$T_4 = 923 - 0.86(923 - 582.5) = \underline{630.17\,\text{K}}$$

Turbine power $= \dot{m}(h_3 - h_4) = \dot{m}.c_p\,(T_3 - T_4)$

$= 1 \times 1005 \times (923 - 630.17) = 294.3\,\text{kW}$

Compressor power $= \dot{m}(h_2 - h_1) = \dot{m}.c_p\,(T_2 - T_1)$

$= 1 \times 1005 \times (483.7 - 288) = 196.7\,\text{kW}$

Net power = turbine power – compressor power

$= 294.3 - 196.7 = \underline{97.6\,\text{kW}}$

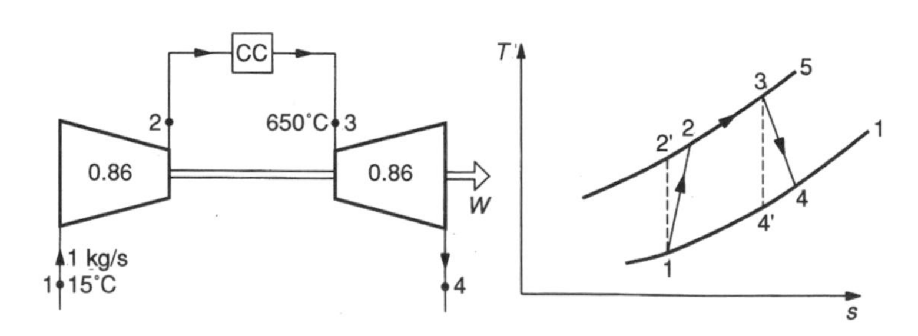

Figure 2.5.8 *Example 2.5.8*

Example 2.5.9

The compressor of a gas turbine receives air at a pressure and temperature of 1.01 bar and 20°C respectively, and delivers it to the combustion chamber at a pressure and temperature of 4.04 bar and 200°C respectively. After constant pressure heating to 680°C, the products of combustion enter the turbine, which has an isentropic efficiency of 0.84. Calculate the compressor isentropic efficiency and the thermal efficiency of the cycle.

For air, $c_p = 1.005\,\text{kJ/kgK}$, $\gamma = 1.4$
For combustion gases, $c_p = 1.15\,\text{kJ/kgK}$, $\gamma = 1.33$

See Figure 2.5.9.

$$\frac{T'_2}{T_1} = \left(\frac{p_2}{p_1}\right)^{\frac{\gamma-1}{\gamma}},\; T'_2 = 293 \times \left(\frac{4.04}{1.01}\right)^{0.286},\; T'_2 = 435.4\,\text{K}$$

$$\eta_c = \frac{T'_2 - T_1}{T_2 - T_1} = \frac{435.4 - 293}{473 - 293} = \underline{0.791}$$

$$W_c = c_p\,(T_2 - T_1) = 1.005(473 - 293) = 180.9\,\text{kJ/kg}$$

$$\frac{T'_4}{T_3} = \left(\frac{p_4}{p_3}\right)^{\frac{\gamma-1}{\gamma}},\; T'_4 = 953 \times \left(\frac{1.01}{4.04}\right)^{\frac{1.33-1}{1.33}},\; T'_4 = 676\,\text{K}$$

$$\eta_T = 0.84 = \frac{T_3 - T_4}{T_3 - T'_4}, = \frac{953 - T_4}{953 - 676}$$

$$T_4 = 953 - 0.84(953 - 676) = 720.3\,\text{K}$$

$$W_T = c_p\,(T_3 - T_4) = 1.15(953 - 720.3) = 267.6\,\text{kJ/kg}$$

$$\text{Net work} = W_T - W_c = 267.6 - 180.9 = 86.7\,\text{kJ/kg}$$

Heat energy input at combustion chamber

$$= c_p\,(T_3 - T_2) = 1.005(953 - 473) = 482.4\,\text{kJ/kg}$$

$$\text{Thermal efficiency of plant} = \frac{\text{net work}}{\text{heat energy in}}$$

$$= \frac{86.7}{482.4} = 0.18 = \underline{18\%}$$

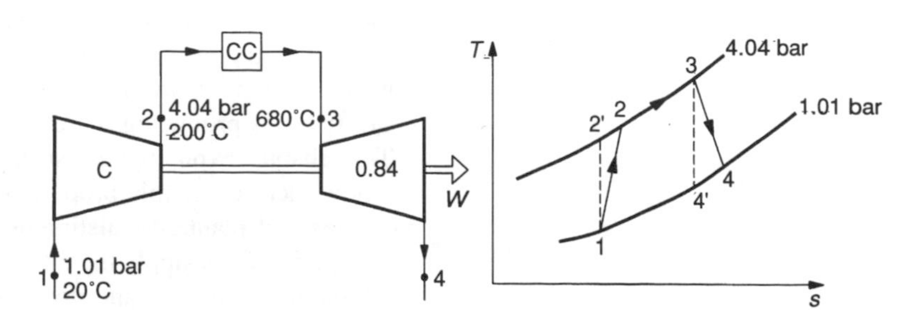

Figure 2.5.9 *Example 2.5.9*

Problems 2.5.2

In all cases, sketch the T/s diagram and a line diagram of the plant with given temperatures and pressures.

(1) In a gas turbine cycle, 4.5 kg/s of air enters a rotary compressor at a pressure and temperature of 1 bar and 18°C respectively. It is compressed through a pressure ratio of 5 to 1 with an isentropic efficiency of 0.85 and then heated to a temperature of 810°C in the combustion chamber. The air is then expanded in a gas turbine to a pressure of 1 bar with an isentropic efficiency of 0.88. Calculate the net power output of the plant and the thermal efficiency. c_p = 1.006 kJ/kgK and γ = 1.4 for both air and combustion gases.

(2) Air enters an open cycle gas turbine plant with an initial pressure and temperature of 1 bar and 15°C respectively and is compressed to a pressure of 6 bar. The combustion gas enters the turbine at a pressure and temperature of 6 bar and 727°C respectively and expands in two stages of equal pressure ratios to the initial pressure. The gases are reheated at constant pressure between the stages. If the isentropic efficiency of the compressor is 0.8, and the isentropic efficiency of the turbine is 0.85, calculate the work output per kg of air and the cycle efficiency.

(3) An open cycle gas turbine plant operates between pressures of 6 bar and 1 bar, and air enters the compressor at 20°C at a rate of 240 kg/min. After the combustion chamber, the gas temperature is 710°C. Given that the isentropic efficiencies of compressor and turbine are 0.78 and 0.837 respectively, calculate the output power and the thermal efficiency of the plant. For air and combustion gases, c_p = 1.005 J/kgK, c_v = 0.718 J/kgK.

(4) An open cycle gas turbine operates with a minimum temperature of 18°C, a pressure ratio of 4.5:1, and isentropic efficiencies of 0.85 for compressor and turbine. The fuel used has a calorific value of 42 MJ/kg, and the air/fuel ratio in the combustion chamber is 84 kg air/kg fuel. Calculate the thermal efficiency of the cycle.

(Note: equate heat addition in combustion chamber to $m.c_p.\delta T$, to find T_3, neglecting mass increase due to fuel addition.) c_p = 1005 J/kgK, δ = 1.4 for both air and combustion gases.

2.6 Steam

This section requires the use of 'steam tables', i.e. tables of steam properties.

Steam is used extensively in heating and power systems throughout industry and in power stations nuclear and conventional. It is true to say that our computers would be of no account without the existence of steam turbines producing the electricity to power them.

This chapter explains how steam is produced, the terminology used and how thermodynamic properties can be found and applied to steam processes and plant. We distinguish between different types of steam, use property tables, apply the effects of entropy change during a process and look at steam flow and non-flow processes.

Key points

- Wet steam cannot be superheated.
- Steam containing water droplets is called wet saturated steam.
- Steam not containing water droplets, but which is at saturation temperature, is called dry saturated steam.
- Steam above saturation temperature is known as superheated steam.
- The degree of wetness of saturated steam is given by its dryness fraction, x. This is a value between 0 and 1. The higher the value, the drier the steam and the more heat energy the steam will contain. Dryness fraction is the ratio of the mass of pure steam to the total mass of the steam sample, and indicates what is called the *quality* of the steam.

Types of steam

Most of us see steam only when we boil the kettle, but we can learn something even from this.

We observe that because it is able to lift the kettle lid, steam can do work. Because we can scald our hands with it, we know that it can transfer heat energy. Also, if we look closely at the steam coming out of the spout, we can see water droplets, or a water mist, within it, indicating it is what we call '*wet steam*'.

All boilers, of whatever size, are doing the same thing as the kettle, i.e. boiling water. For low grade applications such as heating, we may need only hot water. For turbines we are interested in much higher temperatures and pressures to produce steam of the right quality.

We have already said that the steam produced by a kettle is wet steam. This sort of steam is made up of water droplets and 'pure steam', i.e. steam which does not have water droplets. If the steam was taken away from the water from which it is produced, and more heat energy added, some of the droplets would change into pure steam, and the steam would be drier. Eventually all the water droplets would have changed state and the steam would be dry.

The steam produced from the boiling water is at the same temperature as the water. This is called the *saturation temperature*, t_s.

The steam cannot rise above this temperature until all the water droplets have disappeared, because all the heat energy supplied is used to change the state of the water droplets, i.e. latent heat of vaporization.

A soon as the steam has dried, and if more heat energy is supplied, the temperature of the steam will increase to produce *superheated steam*, i.e. steam above the saturation temperature of the water from which it was produced.

Using tables of steam properties – steam tables – we can find the energy of the steam. This energy is available to do work in a turbine or to be transferred for heating purposes.

Key points

To fix a value in the steam tables:

- If the steam is wet, we need to know its pressure and its dryness fraction.
- If the steam is superheated, we need to know its pressure and its temperature.

The production of steam

Figure 2.6.1 shows diagrammatically the production of steam.

(1) Boiling water, i.e. water at saturation temperature, t_s.
(2) Wet saturated steam. Steam composed of water droplets and 'pure steam'. Temperature, t_s. Low dryness fraction, x.
(3) Wet saturated steam. More water droplets have changed into 'pure' steam. Temperature, t_s. Dryness fraction, x, higher.

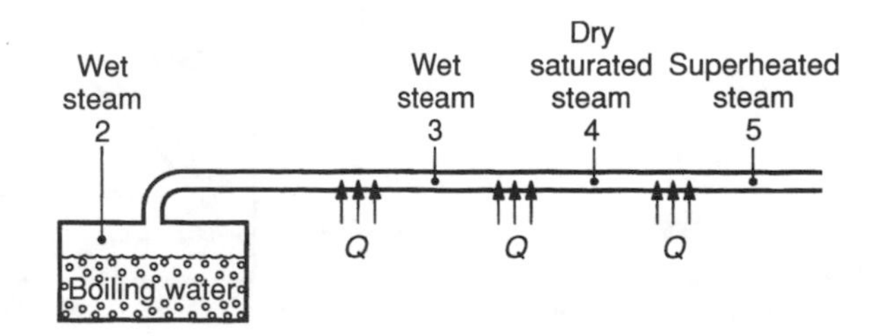

Figure 2.6.1 *Production of steam*

(4) Dry saturated steam. No water droplets. Temperature, t_s. Dryness fraction, $x = 1$.
(5) Superheated steam. Temperature above, t_s.

The steam temperature cannot be raised above saturation temperature, t_s, until all the water droplets have gone.

The temperature/enthalpy (*T/h*) diagram

Figure 2.6.2 shows a simplified diagram of temperature against enthalpy, *h*. Remember to think of enthalpy as 'total energy' made up of internal energy and pressure energy. Values of enthalpy are used to calculate heat energy transfer and work transfer, as we see later in this chapter.

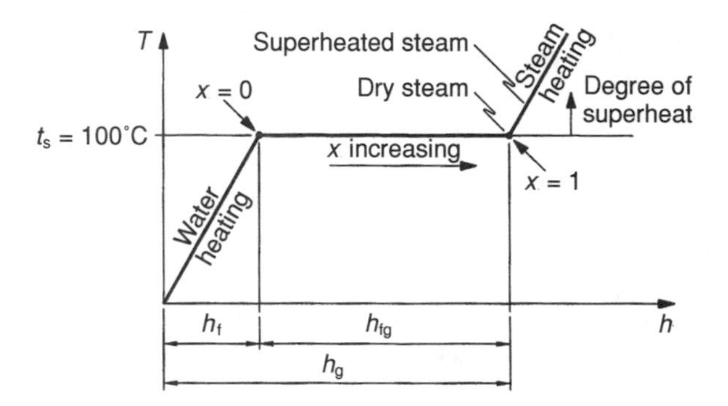

Figure 2.6.2 *Simplified T/h diagram for a pressure of 1 bar*

On the diagram:

h_f: enthalpy of water at saturation temperature, t_s (kJ/kg).

h_g: enthalpy of dry steam. (kJ/kg).

h_{fg}: $h_g - h_f$: latent heat of vaporization (kJ/kg).

All these values are found in the steam tables, which use the units given.

It is worth studying this diagram carefully because it gives a clear picture of what is happening alongside the associated steam table values, and shows dryness fraction and degree of superheat.

The diagram can be drawn for any pressure. See Figure 2.6.3 which is an accurate plot of temperature against specific enthalpy.

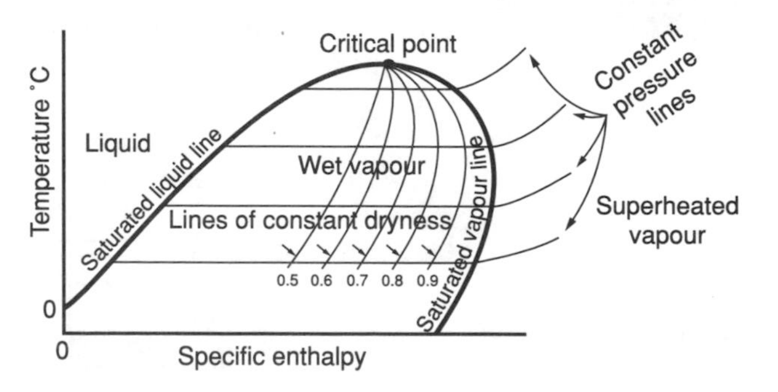

Figure 2.6.3 *Temperature/enthalpy diagram for vapour*

Note the following:

- The liquid, wet steam and superheat regions.
- The saturated liquid line.
- The saturated vapour line.
- The lines of constant pressure.

The separating and throttling calorimeter

The quality (i.e. the dryness) of wet steam can be found by using a separating and throttling calorimeter. Figure 2.6.4 shows the general arrangement of the device.

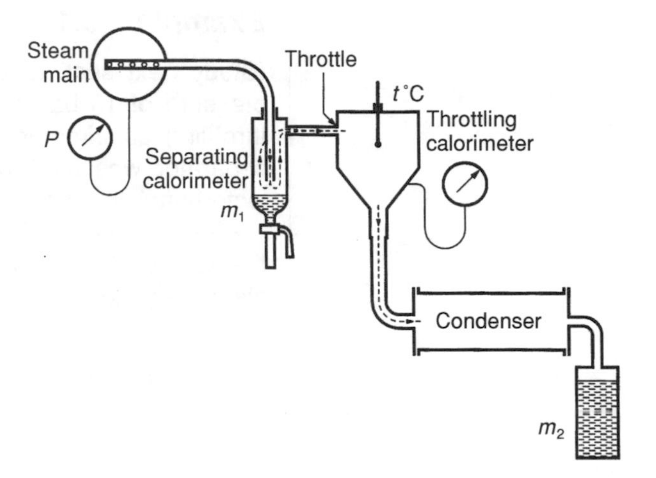

Figure 2.6.4 *Separator and throttling calorimeter*

The separator, as its name suggests, physically separates the water droplets from the steam sample. This alone would give us a good idea of the dryness of the steam, despite that the separation is not complete, because, as we have seen the dryness fraction is the ratio of the mass of pure steam to the total mass of the steam.

Having separated out the water droplets we can find their mass to give us the mass of water in the sample, m_1. The 'pure steam' is then condensed to allow its mass to be found, m_2. Then,

$$\text{Dryness fraction from separator,} \quad x = \frac{m_2}{m_1 + m_2}$$

A more accurate answer is obtained by connecting the outlet from the separator directly to a throttle and finding the dryness fraction of the partly dried steam.

In the throttling calorimeter, the steam issuing through the orifice must be superheated, or we have two dryness fractions, neither of which we can find. Throttling improves the quality of the steam, which is already high after passing through the separator, therefore superheated steam at this point is not difficult to create.

To find the enthalpy of the superheated steam, we need its temperature and its pressure.

For the throttling calorimeter,

Enthalpy before = enthalpy after throttling (see page 57, 'Applications of the SFEE')

$h_f + x.h_{fg}$ = enthalpy from superheat tables

If we call the dryness from the separator, x_1, and the dryness from the throttling calorimeter x_2, the dryness fraction of the steam sample is x, given by,

$$\underline{x = x_1 \times x_2}$$

Example 2.6.1

(Study next section for use of steam tables.) Steam at a pressure of 15 bar was tested by use of a separating and throttling calorimeter. The mass of water collected in the separator was 0.55 kg and the mass condensed and collected after the throttle was 10 kg.

After the throttle, the pressure of the steam was 1 bar and the temperature 150°C. Find the dryness fraction of the steam sample.

$$\text{Dryness from separator} = \frac{m_1}{m_1 + m_2} = \frac{10}{10 + 0.55}$$

$$= \underline{0.948} = x_1$$

For the throttle,

Enthalpy before = enthalpy after

$h_f + x.h_{fg}$ at 15 bar = h at 1 bar 150°C

$845 + (x_2 \times 1947) = 2777$

$$x_2 = \frac{2777 - 845}{1947} = \underline{0.992}$$

$$\text{Dryness of sample} = x_1 \times x_2 = 0.948 \times 0.992 = \underline{0.94}$$

The steam tables

Figure 2.6.5 shows a steam tables extract for a pressure of 2 bar.

We are concerned at this stage only with p, t_s, V_g, h_f, h_{fg} and h_g.

Specific enthalpy, h, is the total energy of 1 kg of the steam, made up of internal energy and pressure energy. It is important because it is the energy we want to use in the steam heater or turbine.

Specific volume, v, is the volume in m^3 which 1 kg of the steam will occupy. It is important because the size of the boilers and piping, etc., can be estimated using these values.

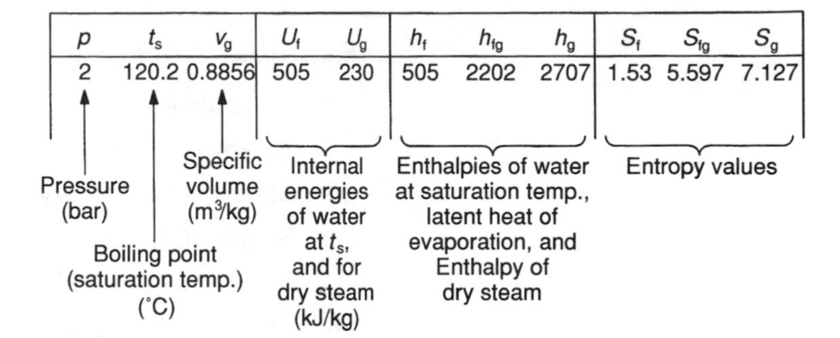

Figure 2.6.5 *Saturated water and steam extract for a pressure of 2 bar*

Pressure is p, and saturation temperature (boiling point corresponding to the pressure) is t_s.

Once again, refer to Figure 2.6.2 for an idea of these quantities.

Finding values in the saturated water and steam section

- To find the enthalpy of water, look up pressure and read off h_f.
- To find the enthalpy of dry steam, look up pressure and read off h_g.
- To find the enthalpy of wet steam, look up pressure and use $h = h_f + x.h_{fg}$.
- Similarly, to find internal energy values, use u_f, u_g, and $u_f + x.u_{fg}$.
- To find the specific volume of dry steam, use $v = v_g$.
- To find the specific volume of wet steam, use $v = x.v_g$.

Superheated steam

These tables are arranged differently from the saturated water and steam tables. Remember that dryness fraction, x, is not involved in these values, because steam cannot be superheated unless all the water droplets have gone.

The temperature of the superheated steam must be known. Figure 2.6.6 shows an extract for the details of superheated steam at 10 bar, 350°C.

The *degree of superheat* is the number of degrees above t_s. In this case, the degree of superheat is 350 – 179.9 = 170.1°C.

Note that most steam tables have a table for water between 0 and 100°C, with corresponding pressures in the second column. This is useful when finding the enthalpy of water, h_f, between 0 and 100°C.

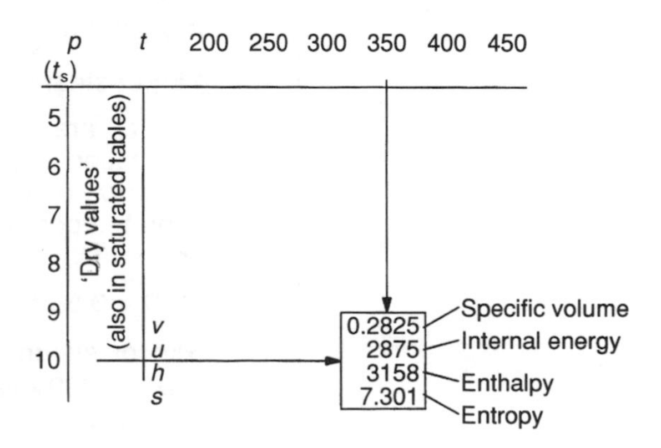

Figure 2.6.6 *Superheated steam table extract for a pressure of 10 bar, temperature 350°C*

Example 2.6.2

Find the enthalpy of:

(a) Steam at 6 bar, dry. From tables, h_g = 2757 kJ/kg

(b) Steam at 50 bar, dry. From tables, h_g = 2794 kJ/kg

(c) Steam at 12 bar, dryness, $x = 0.75$

$h = h_f + x.h_{fg} = 798 + (0.75 \times 1986) = 2287.5$ kJ/kg

(d) Steam at 70 bar, $x = 0.9$

$h = h_f + x.h_{fg} = 1267 + (0.9 \times 1505) = 2621.5$ kJ/kg

(e) Steam at 10 bar 300°C. From superheat tables, h = 3052 kJ/kg

(f) Steam at 40 bar, 400°C. From superheat tables, h = 3214 kJ/kg

(g) Water at 34°C. Refer to temperature only, h_f = 142.4 kJ/kg

(h) Water at 850°C. h_f = 355.9 kJ/kg

Example 2.6.3

(a) What volume does 1 kg of steam at 15 bar, $x = 1$, occupy? $v_g = 0.1371$ m³/kg.

(b) What volume does 5 kg of steam at 20 bar, $x = 0.75$ occupy? $v_g = 0.099\,57$ m³/kg, $v = x.v_g = 0.75 \times 0.099\,57 = 0.0747$ m³/kg.

This is the volume 1 kg will occupy. 5 kg will occupy $5 \times 0.0747 = 0.3735$ m³

Example 2.6.4

Find the internal energy of:

(a) 1 kg of dry steam at 10 bar. u_g = 2584 kJ/kg.

(b) 3 kg of water at 130°C. Refer to temperature only, u_f = 546 kJ/kg.

For 3 kg, $u = 3 \times 546 = 1638$ kJ.

(c) 1 kg of steam at 50 bar, dryness 0.8. $u = u_f + x.u_{fg}$

Values of u_{fg} are not listed in most steam tables, because of space limitations. u_{fg} is the difference between u_f and u_g, i.e. the difference between the internal energy of the water at saturation temperature and as dry steam (also at saturation temperature by definition).

$u = 1149 + 0.8(2597 - 1149) = 2307.4$ kJ/kg

(d) 5 kg of steam at 40 bar, t = 500°C. From superheat tables, u = 3099 kJ/kg. For 5 kg, $u = 5 \times 3099 = 15\,495$ kJ.

Example 2.6.5

A boiler produces steam at 50 bar, 350°C. What is the degree of superheat of the steam?

The degree of superheat is the number of degrees above saturation temperature,

i.e. degree of superheat $= t - t_s$.

In this case, degree of superheat = 350 – 263.9 = 86.1°C.

Key points

- The steam tables values are specific, i.e. for 1 kg.
- Values of enthalpy and internal energy are kJ/kg.
- For enthalpy of water between 0 and 100°C, use the page of the steam tables where temperature is in the first column. When finding the enthalpy of water, refer to the temperature only, disregarding the pressure which makes little difference.
- In the superheat tables, for convenience the saturation temperature is in brackets beneath the pressure.
- It is necessary to interpolate between values if a value is in between the values given. See page 86, 'Maths in action'.

Example 2.6.6

A vessel of volume 0.2816 m^3 contains dry steam at 14 bar. What mass of steam does the vessel contain?

Specific volume of dry steam at 14 bar = v_g = 0.1408 m^3/kg

We have 0.2816 m^3, which is twice this volume, therefore the vessel contains 2 kg.

Example 2.6.7

A steam space in a boiler drum has a volume of 0.5 m^3. If it contains steam at 10 bar, dryness = 0.75, what is the mass of steam in the drum?

Specific volume of the steam, $v = x.v_g = 0.75 \times 0.1944$

$= 0.1458\ m^3/kg$

$$\text{Mass of steam in the drum} = \frac{0.5}{0.1458} = \underline{3.43\ kg}.$$

Problems 2.6.1

(1) Find the specific enthalpy of the following:

(a) Water at 50°C.
(b) Steam at 50 bar, $x = 0.9$.
(c) Steam at 2 bar, $x = 0.76$.
(d) Steam at 10 bar, 500°C.
(e) Steam at 20 bar, 250°C.
(f) Dry saturated steam at 8 bar.

(2) If steam at a pressure of 40 bar has a temperature of 450°C, what is the degree of superheat?

(3) Find the volume of 6 kg of steam at 5 bar, $x = 0.8$.

(4) The steam drum in a boiler contains steam at 30 bar, 350°C. If the volume of the drum is 0.4525 m^3, what mass of steam does it contain?
(5) Find the internal energy of 6 kg of steam at 5 bar, dryness 0.9.
(6) Steam at 6.5 bar passes from a steam main through a separating and throttling calorimeter. The condition of the steam after throttling is 1 bar, 125°C. The mass of steam condensed after throttling is 25 kg and 1.31 kg of water is collected in the separator. Calculate the dryness of the steam.

It is important to have a good working knowledge of the steam tables, and be able to use them routinely for finding the values required.

Steam flow processes

We have already seen the application of the steady flow energy equation in gas processes.

In applying the same equation to steam, instead of using the expression

$$\delta h = m.c_p.\delta T$$

which applies only to a perfect gas, we refer to steam tables and read off values of h for wet, dry or saturated steam, as required.

Example 2.6.8

Determine the power output from a steam turbine if it receives steam at 40 bar, 350°C, and the steam leaving the turbine is at 15 bar, 200°C. The steam mass flow rate is 0.5 kg/s.

Figure 2.6.7 shows the turbine.

SFEE

$$Q - W = \dot{m}\left[(h_2 - h_1) + \frac{c_2^2 - c_1^2}{2}\right]$$

neglecting heat losses and velocity changes,

$$W = \dot{m}(h_2 - h_1)$$

From steam tables,

$$h_1 = 3094 \text{ kJ/kg}$$

$$h_2 = 2796 \text{ kJ/kg}$$

$$W = 0.5(3094 - 2796) = \underline{149 \text{ kW}}$$

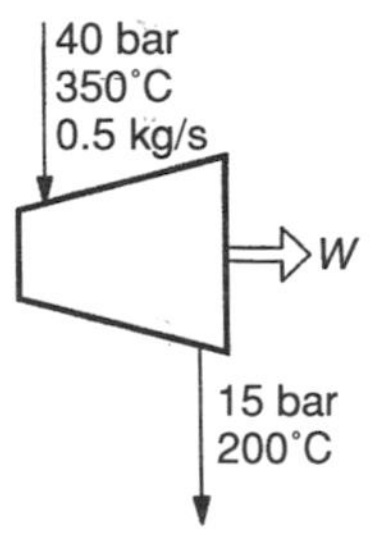

Figure 2.6.7 *Example 2.6.8*

Example 2.6.9

Steam enters a heater at 3 bar, dryness fraction, $x = 0.8$. If the drain from the heater is steam at 1 bar, $x = 0.3$, what is the heat energy transferred per kg of steam, assuming no losses?

3 bar
$x = 0.8$
Heater
Drain
1 bar
$x = 0.3$

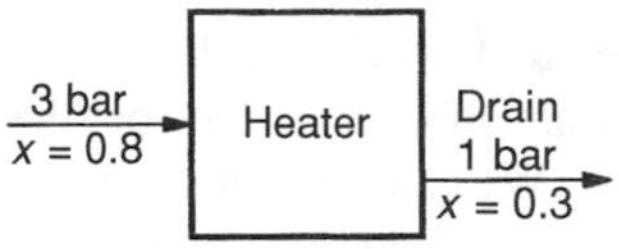

Figure 2.6.8 *Example 2.6.9*

Figure 2.6.8 shows the heater.

From the SFEE,

$$Q = \dot{m}(h_2 - h_1)$$

From steam tables,

$$h_1 = h_f + x.h_{fg} = 561 + (0.8 \times 2164) = 2292.2\,\text{kJ/kg}$$

$$h_2 = h_f + x.h_{fg} = 417 + (0.3 \times 2258) = 1094\,\text{kJ/kg}$$

$$Q = 1(2292.2 - 1094) = 1198.2\,\text{kJ/kg}$$

If the mass flow rate of steam through the heater is 0.2 kg/s, what is the heater rating in kW?

$$Q = 0.2 \times 1197.8 = \underline{239.6\,\text{kW}}$$

Problems 2.6.2

(1) Determine the power output from a steam turbine if it receives 2 kg/s of steam at 50 bar, 400°C and the steam leaves the turbine at 0.5 bar with a dryness of 0.2.

(2) In a steam turbine plant, the steam supply to the turbine is 44 bar dry saturated and the exhaust steam is 0.04 bar with a dryness of 0.69. Determine the power output from the turbine for a steam mass flow rate of 3 kg/s.

(3) A turbo-generator is supplied with superheated steam at a pressure of 30 bar and temperature 350°C. The exhaust steam has a pressure of 0.06 bar, dryness 0.88. Find:

  (a) The enthalpy drop per kg of steam;
  (b) The power developed if the steam flow rate is 0.25 kg/s.

(4) Steam enters the heating coils of an evaporator at 3 bar, 300°C and leaves as water at 25°C. How much heat energy has been given up per kg of steam to the water in the evaporator? If there are 75 kg of water at 25°C in the evaporator, to what temperature will it rise after 6 kg of steam have passed through the heating coil? Specific heat of water = 4.2 kJ/kgK.

(5) Steam enters the superheaters of a boiler at a pressure of 20 bar, dryness 0.8, and leaves at the same pressure at a temperature of 300°C. Calculate the heat energy supplied per kg of steam in the superheaters and the increase in volume of the steam through the superheaters.

Steam plant

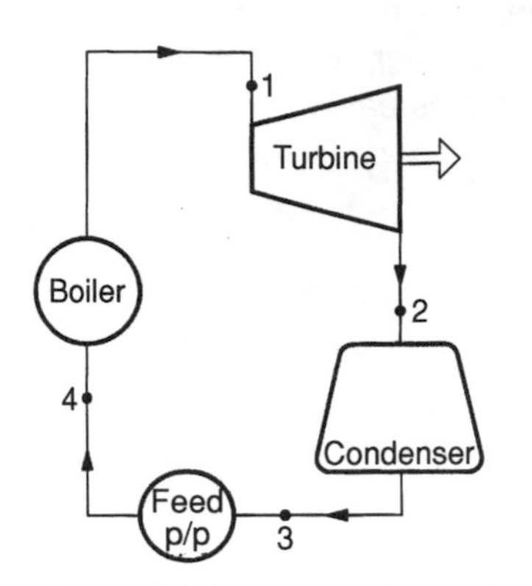

Figure 2.6.9 *Basic steam plant*

If we consider all the items in a simple steam plant, we have boiler, turbine and condenser. For calculation purposes it is usual to neglect the feed pump which supplies water to the boiler. Figure 2.6.9 shows the basic steam plant. Each of these items involves a steady flow process, and as we saw in Section 2.5, and in the foregoing examples, the SFEE can be reduced to a very simple form.

For each item then, it is a simple matter to find the work or heat energy transfers if we know the condition of the steam or the temperature of the water as the case may be. We can then find turbine power, heat energy lost to condenser cooling, heat energy supplied to the boiler and plant efficiency.

If we made the plant more sophisticated by adding feed heaters (steam heaters which pre-heat the boiler feedwater), multi-stage expansion in the turbines, de-aerators (to remove oxygen from the feed to reduce corrosion), and other refinements, we could use the simplified SFEE for these too, since they are all steady flow.

It is convenient to show steam processes on axes of properties, and a complete steam plant cycle can then be seen.

In the introduction to steam, we looked at a simplified temperature/enthalpy curve as an aid to understanding the production of steam and the values shown in the steam tables. To show steam processes in steam plant, pressure/volume, temperature/entropy, enthalpy/entropy and pressure/enthalpy diagrams can be used, but usually the T/s diagram is sufficient.

The h/s diagram is available as a chart from which values can be read instead of using the steam tables.

The p/h diagram is usually used for refrigerant plant.

Figures 2.6.10, 11, 12 and 13 show the form of these plots for any vapour. They can be produced for water by taking values from the steam tables.

For each plot note:

- The saturated liquid and saturated vapour lines.
- The liquid, vapour and superheat regions.
- The critical point is at the turning point of the saturated liquid and the saturated vapour lines, where there is a zero value of latent heat of vaporization, and therefore boiling to produce the vapour does not occur.

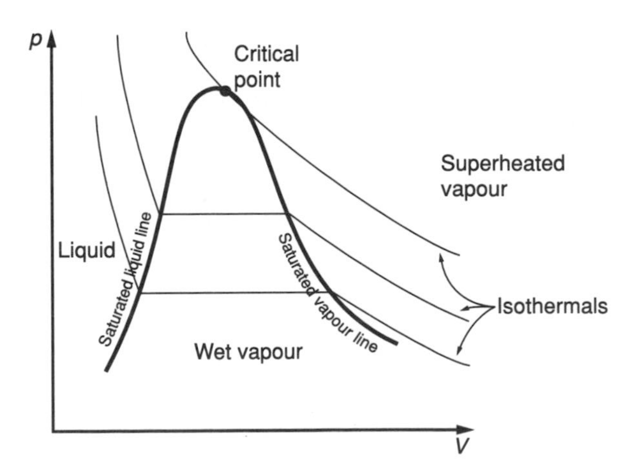

Figure 2.6.10 *p/V diagram for vapour*

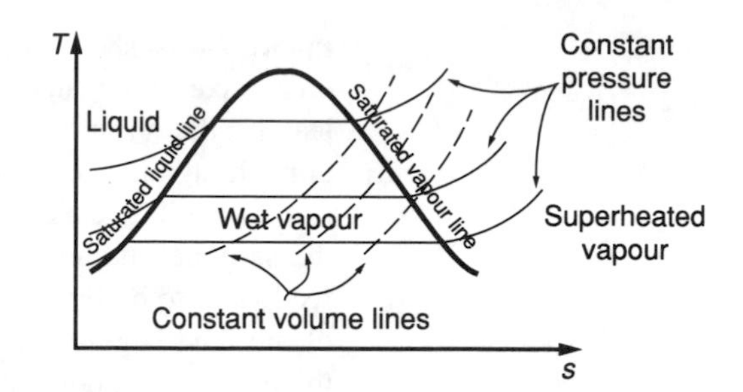

Figure 2.6.11 *T/s diagram for steam*

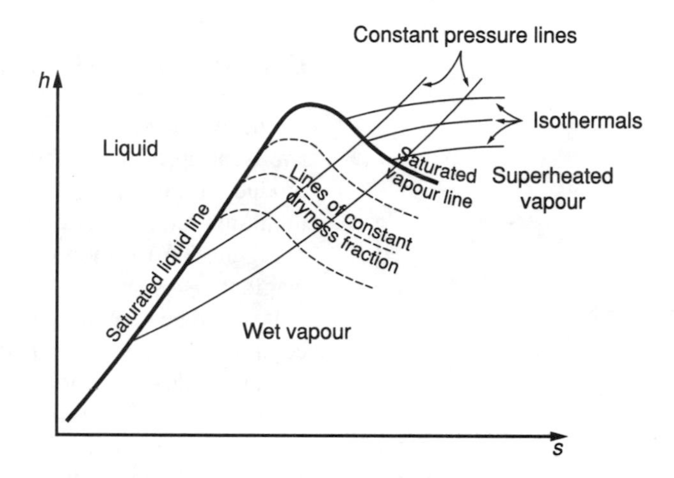

Figure 2.6.12 *h/s diagram for vapour*

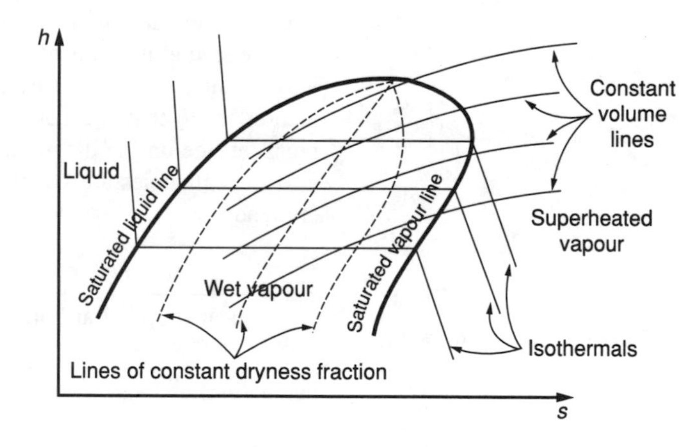

Figure 2.6.13 *p/h diagram for vapour*

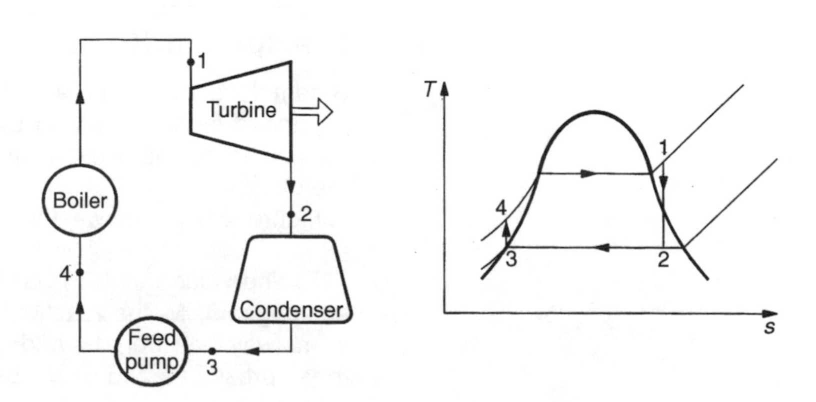

Figure 2.6.14 *Steam plant and T/s diagram*

Figure 2.6.14 shows the basic steam plant and the thermodynamic cycle on *T/s* axes, with superheated steam entering the turbine at 1, wet steam leaving the turbine, 2, and condensation to saturation temperature (i.e. only enough heat energy removed to produce condensation into water – further cooling, called sub-cooling, would take the process to the left of the saturated liquid line), 3. The feed pump raises the pressure of the feedwater to boiler pressure, and the feed enters the boiler at 4. Heating of the feed begins at this point, in a preheater or in the boiler itself if there is no economizer. Note that the boiler is a constant pressure process and follows the constant pressure line in the *T/s* diagram.

Carnot and Rankine cycles

We have looked at the Carnot cycle, consisting of two isothermals and two isentropics, as a reference cycle, which could, if it were possible to produce it, give the greatest efficiency for the available maximum and minimum temperatures occurring in the cycle.

Figure 2.6.15 shows the Carnot cycle for a vapour on the *T/s* diagram.

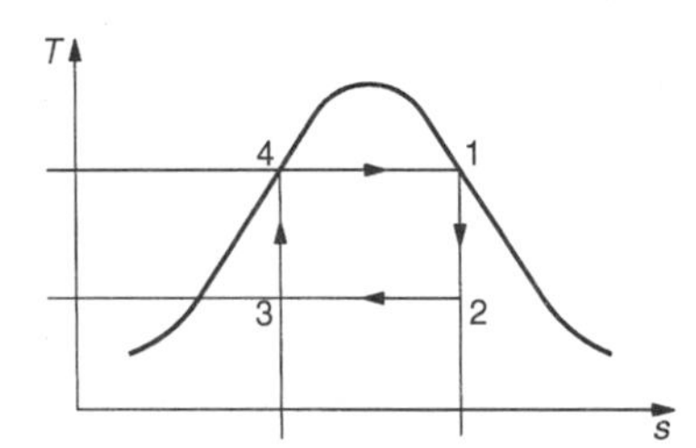

Figure 2.6.15 *Carnot cycle for vapour on T/s diagram*

It would be very difficult to stop the condensation at point 3, and then compression of a wet vapour follows, which is also not realistic. The Carnot cycle, therefore, is not suitable in this case as a reference cycle.

Instead, the modified Carnot cycle, called the Rankine cycle is used. Figure 2.6.16 shows this cycle on the *T/s* diagram. There is isentropic compression in the feedpump from 3 to 4 and isentropic expansion from 1 to 2, but condensation is extended to the saturated liquid line. From 4 to 1 the feed is at first heated to saturation temperature and then changes into steam at constant temperature, because of latent heat of vaporization. Note that the constant pressure heating follows the constant pressure line on the diagram.

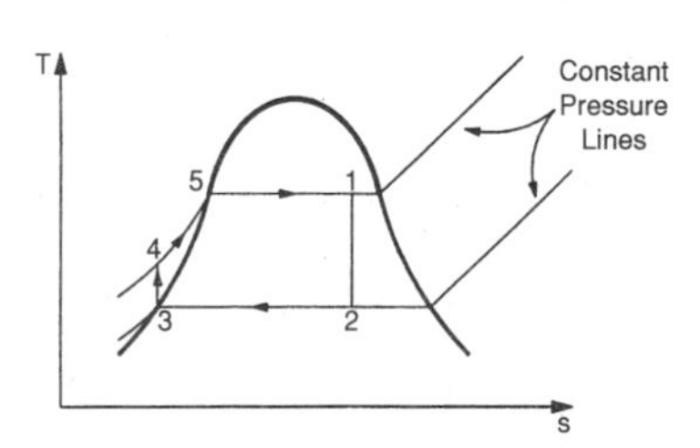

Figure 2.6.16 *Rankine cycle*

If the enthalpies are known, it is an easy matter to calculate Rankine eficiency.

$$\eta_R = \frac{\text{turbine work}}{\text{heat supplied at boiler}} = \frac{h_1 - h_2}{h_1 - h_3}$$

neglecting feedpump work.

Example 2.6.10

Steam is supplied to a turbine at 17.5 bar, 300°C and expanded isentropically to 0.07 bar. Calculate the dryness fraction at the end of expansion and the Rankine efficiency of the plant.

Figure 2.6.17 shows the plant and the process on *T/s* axes.

To find values at 17.5 bar it is necessary to interpolate – see page 86, 'Maths in action'. In this case, because 17.5 bar is midway between 15 and 20 bar, we can take values at these pressures and divide by 2.

Key point

h_4 is h_f at 007 bar, because although the feedwater has been pumped to boiler pressure to get it into the boiler, it has not been preheated to the corresponding saturation temperature, but remains with the same enthalpy (neglecting feedpump work) which it had leaving the condenser as a saturated liquid.

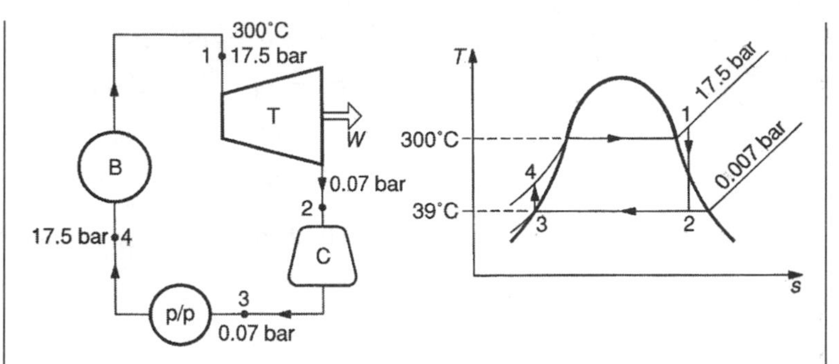

Figure 2.6.17 *Example 2.6.10*

At 17.5 bar, 300°C, $h_1 = 3032\,\text{kJ/kg}$, $s_1 = 6.8435\,\text{kJ/kgK}$.

$s_1 = s_2$ because the expansion is isentropic.

$6.8435 = s_f + x.s_{fg}$ at 0.07 bar

$$6.8435 = 0.559 + x(7.715), \quad x = \frac{6.8435 - 0.559}{7.715} = \underline{0.815}$$

$$h_2 = h_f + x_{hfg} = 163 + 0.815(2409) = 2126\,\text{kJ/kg}$$

$$\text{Rankine efficiency} = \frac{h_1 - h_2}{h_1 - h_4} = \frac{3032 - 2126}{3032 - 163} = 0.3161$$

$$= \underline{31.6\%}$$

Example 2.6.11

Steam enters a turbine at 50 bar, 500°C and exhausts into the condenser at 0.2 bar with a dryness of 0.92. The condensate leaves the condenser at 55°C. Calculate:

(a) the power developed for a steam mass flow of 1.5 kg/s;
(b) the heat energy lost to the condenser cooling water;
(c) the thermal efficiency of the plant.

Figure 2.6.18 shows the plant.

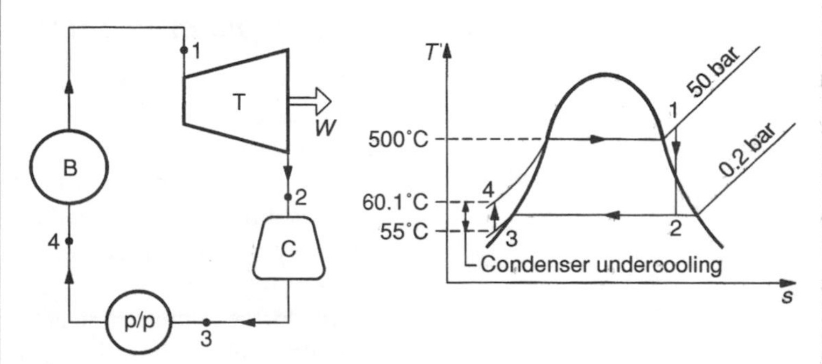

Figure 2.6.18 *Example 2.6.11*

$h_1 = 3433\,kJ/kg$

$h_2 = h_f + x.h_{fg}$ at 0.2 bar $= 251 + 0.92(2358)$

$= 2420.36\,kJ/kg$

$P = \dot{m}(h_1 - h_2) = 1.5(3433 - 2420.36) = \underline{1519\,kW}$

Heat to condenser cooling $= \dot{m}(h_2 - h_3)$

$= 1.5(2420.36 - 230.2)$

$= \underline{3285.24\,kW}$

$$\text{Thermal efficiency} = \frac{h_1 - h_2}{h_1 - h_4} = \frac{3433 - 2420.36}{3433 - 230.2}$$

$$= 0.316 = \underline{31.6\%}$$

Note:

- Condenser undercooling is $t_s - 55 = 60.1 - 55 = 5.1K$.
- h_3 can be found by referring to h_f at 55°C, neglecting the pressure.
- The thermal efficiency in this example is not the Rankine efficiency, because the expansion is not isentropic and there is condenser undercooling.
- Feedpump work is neglected, therefore $h_3 = h_4$.

Problems 2.6.3

In each case sketch the *T*/*s* diagram

(1) In a steam turbine plant, the steam supply to the turbine is 44 bar dry saturated, and the exhaust steam is at 0.04 bar after isentropic expansion. Calculate the Rankine efficiency of the cycle.

(2) Steam enters a turbine at 40 bar, 300°C and exhausts into the condenser at 0.065 bar, dryness 0.75. If there is 11K of undercooling in the condenser, calculate the thermal efficiency of the cycle.

Isentropic efficiency

When examining gas turbines, we saw that a plot of absolute temperature against entropy enabled us to show the effect of an increase of entropy on the output power of the turbine. In that case, because we were dealing with a gas, for which the enthalpy change is proportional to the temperature change, that plot was suitable (remembering also that enthalpy change through the turbine is proportional to work done and power produced).

To do the same for steam, for which we usually do not calculate enthalpy values, but simply look them up in steam tables, we plot enthalpy against entropy on the *h*/*s* diagram.

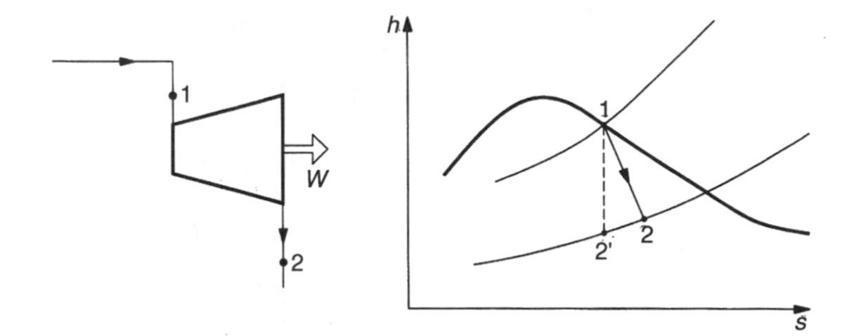

Figure 2.6.19 *Isentropic efficiency*

Figure 2.6.19 shows the h/s diagram in a form suitable for using with calculations. If point 1 is the steam condition entering the turbine, 2′ is the exit condition if the expansion is isentropic, and 2 is the exit condition if the expansion is not isentropic, the isentropic efficiency of the turbine is given by,

$$\eta_T = \frac{h_1 - h_2}{h_1 - h_2'}$$

Contrast this with the corresponding expression for expansion through a gas turbine where, for the reasons explained, we used temperatures instead of enthalpies directly.

Clearly, if the expansion is not isentropic, which in all practical cases it is not, the actual enthalpy drop through the turbine is reduced, and therefore so is the work done. By multiplying the isentropic enthalpy drop by the isentropic efficiency, the actual enthalpy drop can be found.

Example 2.6.12

Steam enters a turbine at 44 bar, dry, and expands with an isentropic efficiency of 0.85 to 0.04 bar. If the mass flow rate of steam through the turbine is 2 kg/s, calculate the power developed by the turbine.

Figure 2.6.20 shows the turbine and the process on the h/s diagram.

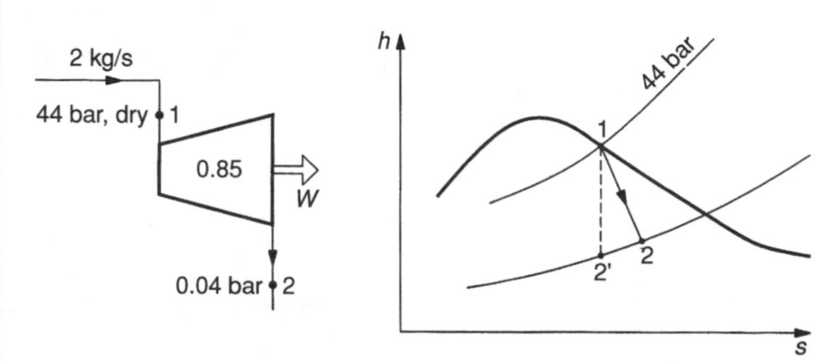

Figure 2.6.20 *Example 2.6.12*

From steam tables,

$h_1 = 2798\,\text{kJ/kg},\ s_1 = 6.029\,\text{kJ/kgK}$

At 0.04 bar, for isentropic expansion,

$s_2' = s_1 = 6.029 = 0.422 + x(8.051),$

$$x = \frac{6.029 - 0.422}{8.051} = 0.696$$

$h_2' = h_f + x.h_{fg} = 121 + (0.696 \times 2433) = 1814.37\,\text{kJ/kg}$

$$\eta_T = \frac{h_1 - h_2}{h_1 - h_2'},\quad 0.85 = \frac{2798 - h_2}{2798 - 1814.37}$$

$h_2 = 2798 - 0.85(2798 - 1814.37) = 1962\,\text{kJ/kg}$

$P = \dot{m}(h_1 - h_2) = 2(2798 - 1962) = \underline{1672\,\text{kW}}$

Example 2.6.13

A steam turbine expands steam from a pressure and temperature of 30 bar and 400°C respectively to 0.05 bar. The feedwater leaves the condenser at a temperature of 25°C and the turbine produces 947.4 kJ of work for every kg of steam. Calculate:

(a) the quality of steam entering the condenser;
(b) the isentropic efficiency of the turbine.

Figure 2.6.21 shows the plant and the cycle on *h*/*s* and *T*/*s* axes.

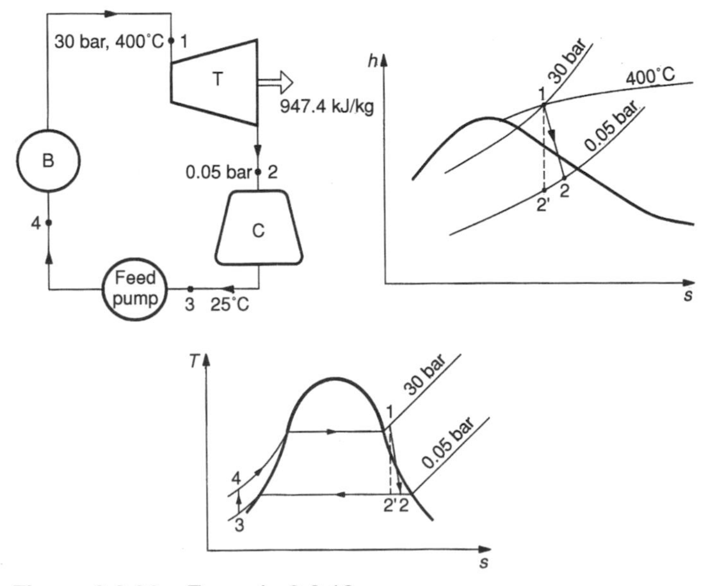

Figure 2.6.21 *Example 2.6.13*

In this case we know the work done for every kg of steam. This is the actual enthalpy drop through the turbine, i.e.,

$h_1 - h_2 = 947.4\,\text{kJ/kg}$

From the superheat tables at 30 bar, 400°C,

$h_1 = 3231\,\text{kJ/kg}$

$3231 - h_2 = 947.4,\ h_2 = 3231 - 947.4 = 2283.6\,\text{kJ/kg}$

$h_2 = 2283.6 = h_f + x.h_{fg}$ at 0.05 bar,

$2283.6 = 138 + x.2423,\quad \underline{x = 0.886}$ (a)

For isentropic expansion,

$s_1 = 6.921\,\text{kJ/kgK} = s_2'$

Note that at 0.05 bar, $s_g = 8.394\,\text{kJ/kgK}$, which is greater, therefore the steam is wet after the expansion.

$s_2' = s_f + x.s_{fg},\ 6.921 = 0.476 + x(7.918),\ x = 0.814$

$h_2' = h_f + x.h_{fg} = 138 + (0.814 \times 2423) = 2110.3\,\text{kJ/kg}$

$$\eta_T = \frac{h_1 - h_2}{h_1 - h_2'} = \frac{3231 - 2283.6}{3231 - 2110.3} = 0.845 = \underline{84.5\%} \qquad \text{(b)}$$

The h/s *chart*

An *h/s* chart can be obtained suitable for reading off values for these calculations, and is a useful alternative to using steam tables, especially where interpolation would be required.

The *h/s* chart is a portion of the *h/s* diagram we have already looked at, but drawn to a larger scale and with values added.

Problems 2.6.4

In each case, show the process on *T/s* and *h/s* axes, and make a line diagram of the plant.

(1) Steam enters a turbine at 30 bar, 500°C and exhausts at 0.5 bar. The isentropic efficiency of the turbine is 0.75. Determine the power for a steam mass flow rate of 2 kg/s. If the condensate leaving the condenser is saturated, determine the heat energy loss to the condenser cooling water.

(2) Steam at a pressure and temperature of 70 bar, 400°C is expanded in a turbine and then condensed to saturated water at 26.7°C. The isentropic efficiency of the turbine is 0.7 and the feedpump work may be neglected. Calculate the thermal efficiency of the cycle.

(3) Steam expands in a turbine from a pressure and temperature of 40 bar, 450°C with an enthalpy drop of 1020 kJ/kg to 0.06 bar. The feedwater leaves the condenser at a temperature of 35°C. Find:

(a) the quality of the steam entering the condenser;
(b) the thermal efficiency of the cycle;
(c) the isentropic efficiency of the cycle.

Steam non-flow processes

The NFEE gave us

$$Q = W + (U_2 - U_1)$$

We have used it so far for gas processes and have calculated heat energy transferred and work done.

The processes of constant volume, constant pressure, adiabatic, polytropic and isothermal are also carried out using steam, and the same changes to the NFEE made (refer to page 30).

However, because in this case the fluid is steam, we cannot use some of the expressions which were valid for gas.

As a rule of thumb, those expressions which include temperature or c_p and c_v are not valid. A re-read of the section on gases will make clear why this is the case.

Since the expressions for work done were derived by finding areas under curves for each of the processes, the same expressions can be used for the equivalent steam processes, but not of course the '$m.R.T$' versions of these.

In the case of steam, we have the complication of latent heat which, for instance, means that in the case of saturated steam we can have only one pressure corresponding to a particular temperature. However, it is worth remembering that superheated steam approximates to the behaviour of a perfect gas.

For a perfect gas, the equation of the curve in an isothermal expansion is $pV = C$, but in the case of steam, a process can follow this equation and yet not be isothermal. Such a process is known as *hyperbolic*. This is an extra process possible in steam expansions and compressions which is not relevant to gas processes.

Most steam processes in industry are flow processes in heaters and turbines, but it is useful to work through some examples of steam non-flow processes.

Example 2.6.14

1 kilogram of steam at 8 bar, dryness 0.9 is expanded until the pressure is 4 bar. If expansion follows the law $pV^{1.25} = C$, find the final dryness fraction of the steam.

$$V_1 = x.V_g = 0.9 \times 0.2403 = 0.2163\,\text{m}^3/\text{kg}$$

$$p_1 V_1^n = p_2 V_2^n$$

$$8 \times 0.2613^{1.25} = 4 \times V_2^{1.25}$$

$$V_2 = \sqrt[1.25]{\frac{8 \times 0.2613^{1.25}}{4}} = 0.455\,\text{m}^3/\text{kg}$$

$V = x.V_g$ at 4 bar

$$0.455 = x \times 0.4623$$

$$x = \frac{0.455}{0.4623} = \underline{0.984}$$

Example 2.6.15

1 kg of steam at 6 bar, 250°C is expanded according to the law $pV^{1.13} = C$ to a pressure of 1.5 bar. Determine:

(a) the dryness fraction of the steam after expansion;
(b) the work done;
(c) the heat energy transferred during the process.

The temperature of the steam before expansion is greater than the saturation temperature corresponding to the pressure. The steam is therefore superheated.

$$V_1 = 0.394\,\text{m}^3/\text{kg}$$

$$p_1 V_1^n = p_2 V_2^n$$

$$6 \times 0.394^{1.13} = 1.5 \times V_2^{1.13},$$

$$V_2 = \sqrt[1.13]{\frac{6 \times 0.394^{1.13}}{1.5}} = \underline{1.344\,\text{m}^3} \qquad \text{(a)}$$

Examination of the steam tables at 1.5 bar shows that this volume lies between 150°C and 200°C.

By interpolation, temperature after expansion = $\underline{168.2°\text{C}}$

This is the condition of the steam after expansion. If it was wet after expansion we would need to quote the dryness fraction.

$$\text{Work done} = \frac{p_1 V_1 - p_2 V_2}{n - 1}$$

$$= \frac{(6 \times 10^5 \times 0.394) - (1.5 \times 10^5 \times 1.344)}{1.13 - 1}$$

$$= 267\,692.3\,\text{J}$$

$$= \underline{267.7\,\text{kJ}} \qquad \text{(b)}$$

Using the non-flow energy equation,

$$Q = W + (U_2 - U_1)$$

where:

$W = 267.7$ kJ/kg

$U_1 = 2722$ kJ/kg

$U_2 = 2607.7$ kJ/kg, by interpolation.

$Q = 267.7 + (2607.7 - 2722) = \underline{153.4 \text{ kJ/kg}}$ (c)

This is +ve, therefore heat energy has been added to the steam.

Maths in action

Steam tables interpolation

If a required value lies midway between two given values, e.g. if a value of enthalpy for superheated steam at 5 bar, 225°C is required, the two values either side can be added together and divided by 2,

$$h \text{ at 5 bar, 225°C} = \frac{2857 + 2962}{2} = \underline{2909.5 \text{ kJ/kg}}$$

If the value is not midway it is necessary to interpolate between values using simple proportions.

For example, say we need the enthalpy of steam at 5 bar, 285°C.

At 5 bar, 250°C, $h = 2962$ kJ/kg

At 5 bar, 300°C, $h = 3065$ kJ/kg

50°C temperature difference corresponds to (3065 – 2962) = 103 kJ/kg

Our temperature difference from 250°C = 285 – 250 = 35°C

$$\text{The corresponding enthalpy difference} = \frac{35}{50} \times 103 = 72.1 \text{ kJ/kg}$$

This must be added to the enthalpy at 250°C to give our required h value,

$$h = 2962 + 72.1 = \underline{3034.1 \text{ kJ/kg}}$$

For approximate calculations, it is possible to estimate reasonably accurate values without interpolation.

Example 2.6.16

$0.03\,m^3$ of steam 0.7 dry, at 16 bar expands during a non-flow hyperbolic process to a volume of $0.16\,m^3$. Find the final pressure, dryness after expansion and the work and heat energy transferred.

$V_1 = x.V_g = 0.7 \times 0.1237 = 0.086\,59\,m^3/kg$.

Since our initial volume is $0.03\,m^3$, the mass of steam we have is,

$$\frac{0.03}{0.08659} = \underline{0.346\,kg}$$

For a hyperbolic process, $p_1V_1 = p_2V_2$

$16 \times 0.03 = p_2 \times 0.16,\ p_2 = 3$ bar

At 3 bar, $V = x.V_g$. Convert V_g from a specific value to suit our mass of 0.346 kg, then,

$0.16 = x \times (0.6057 \times 0.346)$

$x = \underline{0.763}$ Dryness after expansion.

$$\text{Hyperbolic work} = p_1V_1 \ln\left(\frac{V_2}{V_1}\right)$$

$$= 16 \times 10^5 \times 0.03 \times \ln\left(\frac{0.16}{0.03}\right)$$

$$= 80\,351\,J = \underline{80.35\,kJ}$$

$u_1 = u_f + x.u_{fg} = 857 + 0.7(2596 - 857) = 2074.3\,kJ/kg$

$u_2 = u_f + xu_{fg} = 561 + 0.763(2544 - 561) = 2074\,kJ/kg$

$Q = W + (U_2 - U_1) = 80.35 + (2074 - 2074.3)0.346$

$= \underline{80.25\,kJ}$

Example 2.6.17

$0.1\,m^3$ of steam at a pressure of 14 bar is expanded at constant volume to a pressure of 6 bar. Find:

(a) the mass of steam;
(b) the dryness after expansion;
(c) the change in internal energy;
(d) the heat energy transferred.

Specific volume of steam at 14 bar, $x = 0.9$, is given by,

$V = x.V_g = 0.9 \times 0.1408 = 0.127\,m^3/kg$

Our volume is 0.1 m^3,

therefore mass of steam $= \dfrac{0.1}{0.127} = \underline{0.787\,\text{kg}}$ (a)

After expansion at 6 bar, the volume is unchanged = 0.127 m^3/kg

$V = x.V_g$ at 6 bar

$0.127 = x.0.3156, \quad x = \dfrac{0.127}{0.3156} = \underline{0.4}$ (b)

$u_1 = u_f + x.u_{fg} = 828 + 0.9(2593 - 828) = 2416.5\,\text{kJ/kg}$

$u_2 = u_f + x.u_{fg} = 669 + 0.4(2568 - 669) = 1428.6\,\text{kJ/kg}$

$$\begin{aligned}\text{Change in internal energy} &= U_2 - U_1 \\ &= (1428.6 - 2416.5)0.787 \\ &= \underline{-778\,\text{kJ/kg}}\end{aligned}$$ (c)

Heat energy transferred, $Q = W + (U_2 - U_1)$

This is a constant volume process in which $W = 0$, therefore the heat energy transferred is the change in internal energy.

Heat energy transferred $= \underline{-778\,\text{kJ/kg}}$ (d)

Problems 2.6.5

(1) 1 kg of steam at 7 bar, 0.95 dry, is expanded polytropically to a pressure of 3.5 bar. The index of expansion is 1.3. Calculate the final dryness fraction of the steam.

(2) Steam at 40 bar, 400°C is expanded isentropically to a pressure of 0.07 bar. Find the final condition of the steam.

(3) Steam at 80 bar, dryness 0.95, expands to a pressure of 10 bar according to the law $pV^{1.2} = C$. Calculate the dryness fraction after expansion, the work done and the heat energy transferred.

(4) Steam of mass 1 kg at 15 bar with a dryness of 0.85 is expanded in a cylinder to a pressure of 10 bar according to the law $pV^{1.3} = C$. Find:

 (a) the volume after expansion;
 (b) the condition of the steam after expansion;
 (c) the work done by the steam.

(5) In a non-flow hyperbolic process ($pV = C$), 1 kg of steam, initially 0.9 dry and 12 bar, is expanded to a pressure of 2.7 bar. Calculate the final dryness fraction and the work done.

(6) In a non-flow process, 1 kg of steam at 15 bar, $x = 0.77$, expands at constant pressure until its volume is doubled. Calculate:

(a) the final condition of the steam;
(b) the work done;
(c) the change in internal energy.

2.7 Refrigeration

For this section, you will need tables of refrigerant properties.

The vapour-compression refrigeration cycle is the basis of almost all domestic and industrial refrigeration plant. In this chapter, we look at the cycle and its operation, and determine useful parameters of refrigeration performance. The operation of the actual refrigeration plant is examined and tables of properties for various refrigerants are used.

The second law and coefficient of performance

The second law of thermodynamics states that 'it is impossible for heat to be transferred from a lower to a higher temperature without the expenditure of external work'.

Therefore, in a refrigerator, or heat pump, external work must be supplied to effect the heat energy transfer from the space to be cooled, in the case of a refrigerator, to a higher temperature body such as the atmosphere.

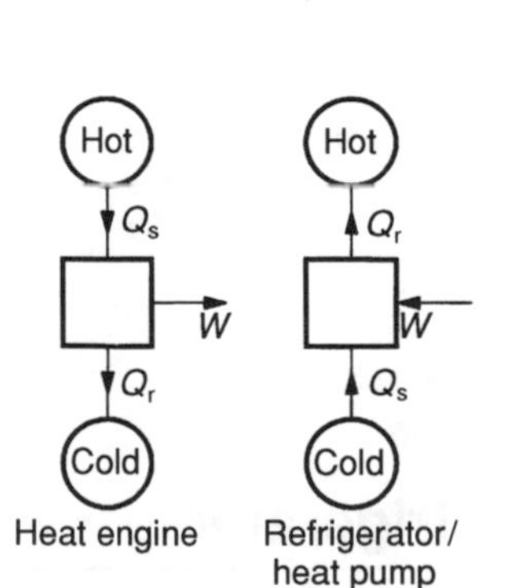

Figure 2.7.1 *Diagrammatic representations*

The diagrammatic representation of the refrigerator/heat pump and the heat engine are shown in Figure 2.7.1. Note that the arrows are in the opposite direction for the refrigerator.

The heat energy rejected in this cycle is to the hot reservoir and the heat energy supplied is from the cold reservoir. The cycle is a reversed heat engine cycle, and the ideal cycle is the reversed Carnot cycle, shown in Figure 2.7.2

For the refrigerator we are interested in the refrigeration effect (heat energy extracted from the cold reservoir) which can be obtained for the work input, i.e.

$$\frac{Q_s}{W}$$

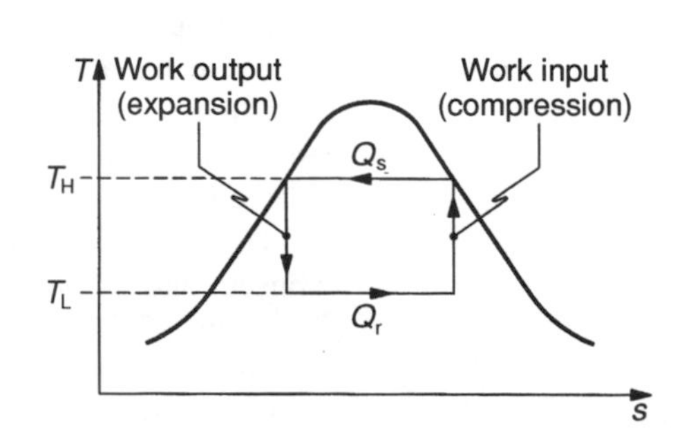

Figure 2.7.2 *Reversed Carnot cycle*

For the heat pump we are interested in the heat energy rejected, because this is used for heating, i.e.

$$\frac{Q_r}{W}$$

This gives rise to two criteria for the effectiveness of the plant, called '*coefficient of performance*', COP_{ref} for the refrigerator and COP_{hp} for the heat pump.

The Carnot coefficients of performance are shown to be,

$$COP_{ref} = \frac{T_L}{T_H - T_L} \quad \text{and} \quad COP_{hp} = \frac{T_H}{T_H - T_L}$$

where:

T_L = lowest cycle temperature

T_H = highest cycle temperature.

By applying the first law we can establish a relationship between the coefficients,

$$Q_s + W = Q_R$$

$$\frac{Q_s}{W} + 1 = \frac{Q_R}{W}$$

giving

$$COP_{hp} = COP_{ref} + 1$$

The vapour-compression refrigerator

Almost all domestic and industrial refrigeration plants are of the vapour-compression type in which work input is required. In the vapour absorption cycle there is, instead, heat energy input. We are looking here only at the vapour-compression cycle.

In this cycle, heat energy is extracted from the low temperature area by evaporation of a fluid (refrigerant), requiring latent heat, and rejection to the higher temperature area by condensation of the vapour.

It is therefore necessary to use refrigerants which will evaporate at low temperatures, i.e. have low boiling points or saturation temperature, because we are making use of the heat demanded by evaporation to produce the refrigeration effect. Choice of refrigerant is also determined by considerations such as corrosivity, inflammability and ease of leak detection.

By increasing the operating pressure, the saturation temperature is raised, and in the case of carbon dioxide plant where the boiling point is about –78°C, a high pressure system is required in order to achieve evaporation at temperatures more usually required in refrigeration plant. This represents a major disadvantage to its use.

The main refrigerants are:

Tetrafluoroethane ($CH_2F{-}CF_3$) Refrigerant 134a
Freon (CF_2Cl_2)
Ammonia (NH_3)
Methyl chloride (CH_3Cl)
Carbon dioxide (CO_2)

Freon (a trade name) is a CFC, and therefore is being phased out because of environmental concerns.

The vapour-compression refrigeration cycle

Figure 2.7.3 shows a diagrammatic arrangement of items in the basic cycle, and Figure 2.7.4 gives an idea of the actual plant.

Starting at point 1, vapour is drawn into the compressor from the low pressure side and compressed to form, usually, a dry or superheated vapour. The vapour passes through the condenser coils where heat energy is extracted by air circulation (e.g. domestic refrigerator) or by circulating water around the coils (e.g. some industrial and marine plant), to produce a saturated or sub-cooled liquid at point 3.

The compressed liquid is then expanded through a regulating valve (throttle), or expansion valve, to form a very wet vapour at 4. Because this is a throttling process, from the SFEE, the enthalpy before and after the expansion is the same.

The wet vapour passes through the evaporator coils where it absorbs heat energy from the warmer surroundings. In so doing, the vapour becomes drier, i.e. its dryness fraction increases as the latent heat energy is absorbed.

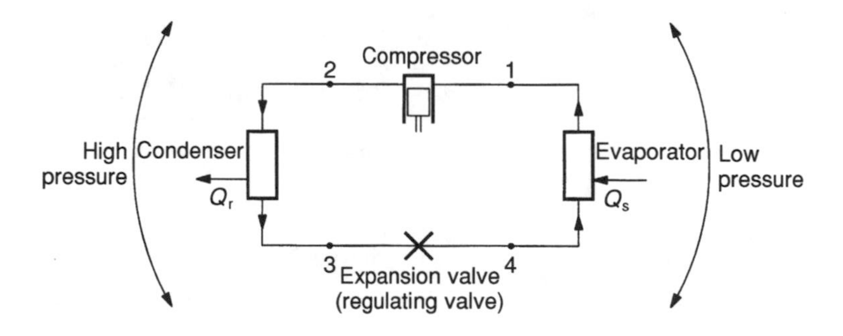

Figure 2.7.3 *Basic vapour compression refrigeration plant*

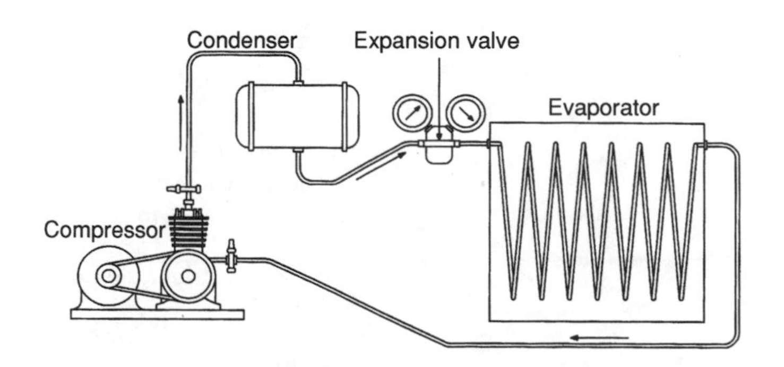

Figure 2.7.4 *Simple vapour-compression refrigeration plant*

The evaporator coils are situated around the freezer cabinet in a domestic refrigerator, and in large industrial and marine plants they are arranged in 'batteries' with a fan to provide chilled air circulation.

Looking at the basic cycle, we can see that there are only two pressures to consider – a high pressure on one side of the compressor and a lower pressure on the other. It is clear that the mass flow of refrigerant around the circuit is constant at all points.

The main refrigerant effect occurs through the evaporator, but because a very wet vapour is produced at the regulating valve (also called the expansion valve), a small refrigeration effect is created, and inspection of the plant would show this pipe ice covered if it was not insulated.

We have quoted the reversed Carnot cycle as the ideal refrigeration cycle. In the practical refrigeration cycle, the major departure from this is that the expansion cannot be isentropic, and in fact occurs by throttling through the expansion valve giving a constant enthalpy process.

Figure 2.7.5 shows a practical refrigeration cycle on *T/s* and *p/h* axes, assuming a dry vapour enters the compressor.

Key points

- There is an increase in entropy during expansion and compression.
- There is *undercooling* in the condenser, i.e. the liquid leaving the condenser is at a temperature lower than the saturation temperature.
- The *p/h* diagram is particularly useful for refrigeration plant to show heat and work energy transfers around the cycle.

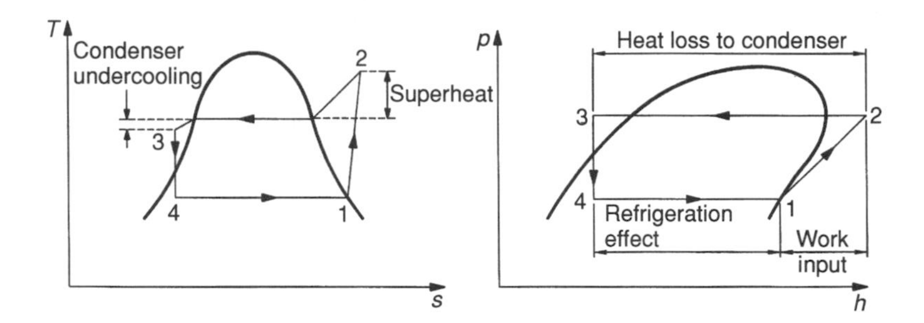

Figure 2.7.5 *Refrigeration cycle on T/s and p/h axes*

Fridge plant performance

From the steady flow energy equation, the following expressions apply,

Refrigeration effect $= (h_1 - h_4)$ (kJ/kg)

Cooling load $= \dot{m}(h_1 - h_4)$ (kW)

Heat energy to condenser cooling $= \dot{m}(h_2 - h_3)$ (kW)

Work input to compressor $= (h_2 - h_1)$ (kJ/kg)

Power input to compressor $= \dot{m}(h_2 - h_1)$ (kW)

where $\dot{m}$ = mass flow of refrigerant (kg/s)

The coefficient of performance (see page 89, the second law and coefficient of performance) for the practical cycle is,

$$\text{COP}_{\text{ref}} = \frac{\text{refrigeration effect}}{\text{work input}} = \frac{h_1 - h_4}{h_2 - h_1}$$

Compressor isentropic efficiency

Figure 2.7.6 shows the h/s diagram for the compression part of the cycle, in which we can see that if the process is not isentropic, the enthalpy rise across the compressor is increased, and therefore the work input is greater.

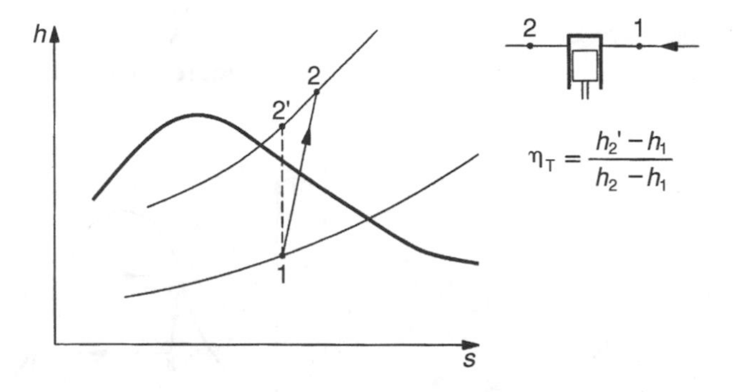

Figure 2.7.6 *Compressor isotropic efficiency*

The compressor isentropic efficiency is given by

$$\eta_T = \frac{h'_2 - h_1}{h_2 - h_1}$$

Key points

- The layout of the tables is simplified, and values of h_{fg} are not given in a separate column and must be calculated from $h_{fg} = h_g - h_f$.
- Temperature is normally in the first column and corresponding saturation pressure in the second.
- Superheat values at 50K and 100K only are given.
- Another significant difference between the tables is that the entropy and enthalpy values are given values of zero at a datum of −40°C in the refrigeration tables for freon and ammonia, and 0°C in the steam tables, for convenience. For refrigerant 134a, the datum is 0°C.

Refrigeration tables.

The theory and nomenclature for the refrigeration tables is the same as for steam tables. If you have not covered steam, it will be necessary for you to go over the section on use of the steam tables.

Example 2.7.1

An ammonia refrigerating plant operates between temperature limits of 10.34 bar and 2.265 bar. The refrigerant leaves the evaporator as a vapour 0.95 dry and leaves the condenser as a saturated liquid. If the refrigerant mass flow rate is 4 kg/min, find:

(a) dryness fraction at evaporator inlet;
(b) the cooling load;
(c) volume flow rate entering the compressor.

Figure 2.7.7 shows the plant and Figure 2.7.8 shows the cycle on T/s and p/h diagrams.

Since the refrigerant leaves the condenser as a saturated liquid, it must be at saturation temperature and its enthalpy will be h_f at 10.34 bar.

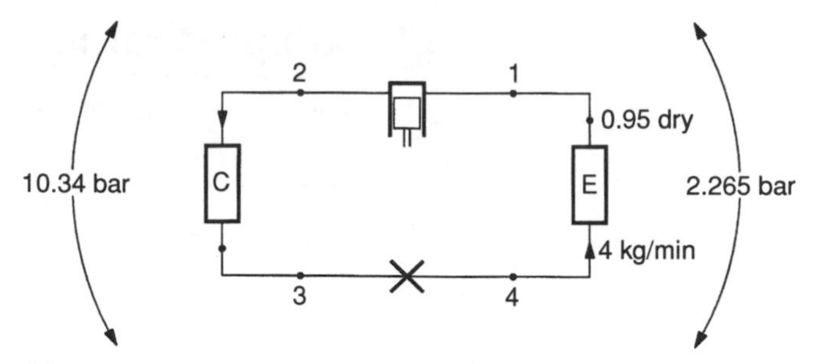

Figure 2.7.7 *Example 2.7.1*

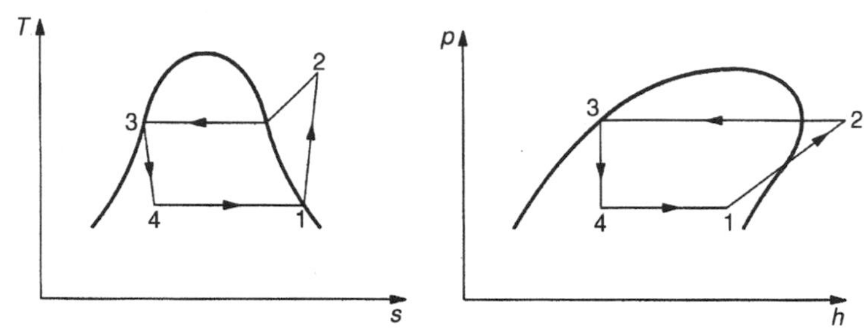

Figure 2.7.8 *Example 2.7.1*

Using the ammonia tables,

$h_3 = 303.7\,\text{kJ/kg}$

The expansion is a constant enthalpy process, i.e. $h_3 = h_4$,

$303.7 = h_f + x.h_{fg}$ at 2.265 bar

$303.7 = 107.9 + x(1425.3 - 107.9)$

$\underline{x = 0.149}$ (a)

The refrigerant leaving the evaporator is wet.

$h_1 = h_f + x.h_{fg} = 107.9 + (0.95 \times 1317.4) = 1359.43\,\text{kJ}$

$$\text{Cooling load} = \dot{m}(h_1 - h_4) = \frac{4}{60}(1359.43 - 303.7)$$

$$= \underline{70.38\,\text{kW}} \quad \text{(b)}$$

The specific volume of the refrigerant at compressor inlet is

$v = x.v_g$ at 2.265 bar $= 0.95 \times 0.5296 = 0.50312\,\text{m}^3/\text{kg}$

Volume flow into compressor = mass flow × specific volume

$= 4 \times 0.503\,12$

$= \underline{2.012\,\text{m}^3/\text{min}}$ (c)

Key point

In this example we do not need to know the condition of the vapour after compression. The diagrams show that we assume it to be superheated.

Example 2.7.2

An ammonia vapour-compression refrigeration cycle operates between saturation temperatures of 20°C and –30°C. The refrigerant is dry saturated at the compressor inlet and the compression is isentropic. Find:

(a) the refrigeration effect;
(b) the coefficient of performance.

Figure 2.7.9 shows the plant and Figure 2.7.10 shows the p/h and T/s axes.

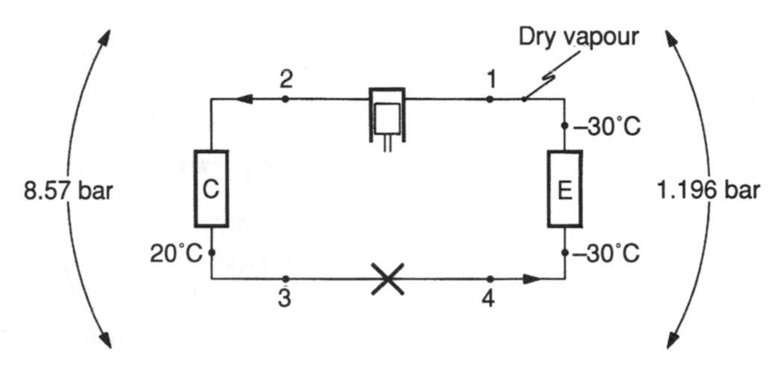

Figure 2.7.9 *Example 2.7.2*

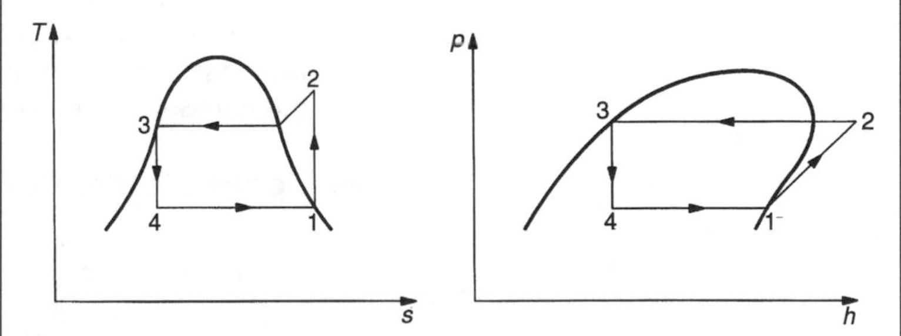

Figure 2.7.10 *Example 2.7.2*

In this case we are given the saturation temperatures. As in the case of steam, this means that we have the pressures also, since there can be only one pressure corresponding to the saturation temperature.

From the ammonia tables.

At – 30°C, p = 1.196 bar

At 20°C, p = 8.57 bar

$h_1 = h_g$ at 1.196 bar = 1405.6 kJ/kg

$s_1 = s_2 = s_g$ at 1.196 bar = 5.785 kJ/kgK

$h_4 = h_3 = h_f$ at 8.57 bar = 275.1 kJ/kg

After compression there is between 50K and 100K of superheat.

Interpolating using entropy values,

$$h_2 = 1597.2 + \frac{5.785 - 5.521}{5.854 - 5.521} \times (1719.3 - 1597.2)$$

$$= 1694 \text{ kJ/kg}$$

$$\text{Refrigeration effect} = h_1 - h_4 = 1405.6 - 275.1$$

$$= \underline{1130.5 \text{ kJ/kg}} \qquad \text{(a)}$$

$$\text{COP} = \frac{h_1 - h_4}{h_2 - h_1} = \frac{1130.5}{1694 - 1405.6} = \underline{3.92} \qquad \text{(b)}$$

Example 2.7.3

An ammonia refrigerating plant produces a cooling load of 13.3 kW. The refrigerant leaves the evaporator dry saturated at 1.902 bar, and leaves the compressor at 7.529 bar, 66°C. At the exit from the condenser, the temperature of the liquid refrigerant is 12°C. Find:

(a) the degree of undercooling in the condenser;
(b) the mass flow rate of the refrigerant;
(c) the heat rejected in the condenser in kW;
(d) the compressor power in kW.

See Figures 2.7.11 and 2.7.12.

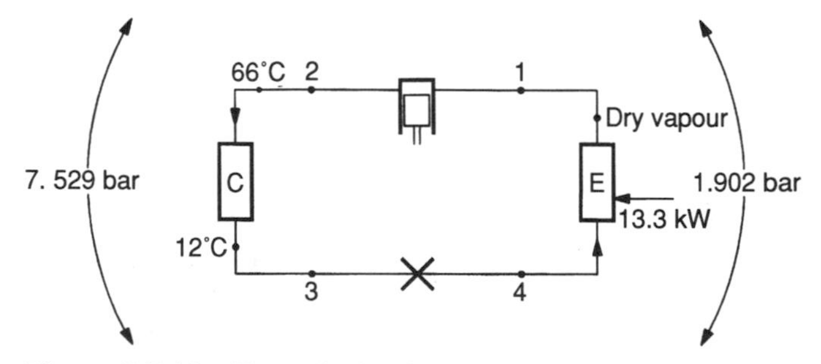

Figure 2.7.11 *Example 2.7.3*

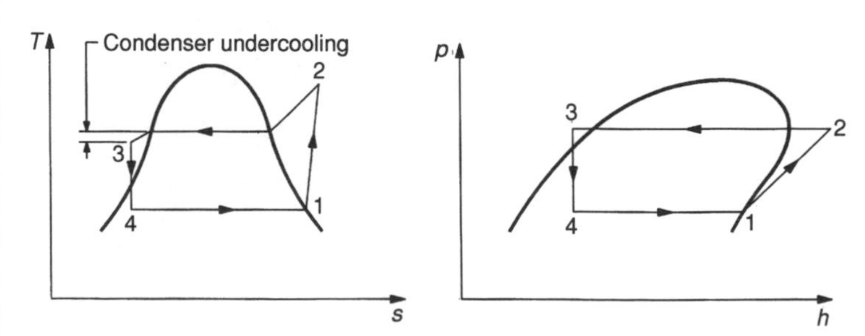

Figure 2.7.12 *Example 2.7.3*

Using the ammonia tables,

$h_1 = h_g$ at 1.902 bar $= 1420\,\text{kJ/kg}$

$h_2 =$ enthalpy of superheated vapour, since $66 - 16 = 50\text{K}$ of superheat

$h_2 = 1591.7\,\text{kJ/kg}$

At 3, for the pressure of 7.529 bar, the saturation temperature is 16°C.

We have a temperature of 12°C.

Degree of condenser undercooling = $\underline{4\text{K}}$ (a)

To find the enthalpy in this case, we can neglect the pressure and read off the enthalpy at 12°C = 237.2 kJ/kg = $h_3 = h_4$.

An alternative is to subtract from the enthalpy at 16°C the value of 4 × specific heat. (From $Q = m.c.\delta T$.)

Cooling load $= \dot{m}(h_1 - h_4) = 13.3$

$= \dot{m}(1420 - 237.2)$, $m = \underline{0.0112\,\text{kg/s}}$ (b)

Condenser heat rejection $= \dot{m}(h_2 - h_3)$

$= 0.0112(1591.7 - 237.2)$

$= \underline{15.2\,\text{kW}}$ (c)

Compressor power $= m(h_2 - h_1)$

$= 0.0112(1597.1 - 1420)$

$= \underline{1.98\,\text{kW}}$ (d)

Example 2.7.4

In a vapour compression refrigeration plant, freon 12 enters the compressor at a pressure and temperature of 1.826 bar and –10°C respectively. It is compressed to a pressure and temperature of 7.449 bar and 45°C respectively. The refrigerant leaves the condenser at 25°C. Calculate:

(a) the refrigeration effect;
(b) the compressor work per kg of refrigerant;
(c) the coefficient of performance;
(d) the Carnot coefficient of performance.

See Figures 2.7.13 and 2.7.14.

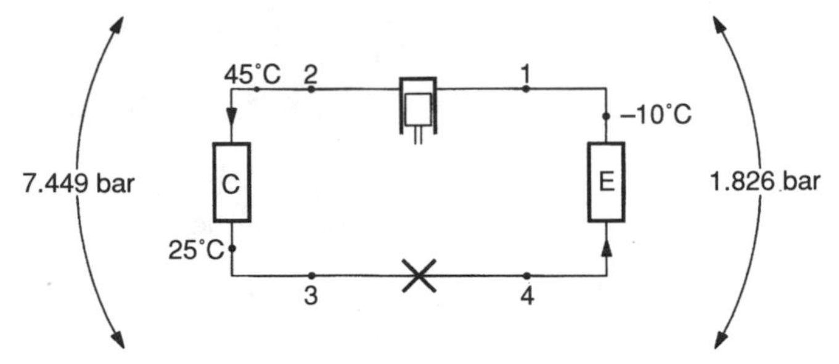

Figure 2.7.13 *Example 2.7.4*

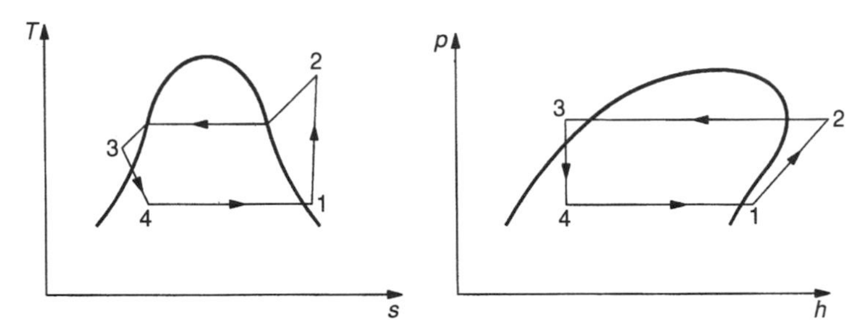

Figure 2.7.14 *Example 2.7.5*

Using the freon tables,

h_1 = enthalpy at 1.826 bar with 5K of superheat

$$h_1 = 180.97 + \frac{5}{15}(190.15 - 180.97) = 184.03\,\text{kJ/kg}$$

$$h_4 = h_3 = h_f \text{ at } 25°\text{C} = 59.7\,\text{kJ/kg}$$

$$\text{Refrigeration effect} = h_1 - h_4 = 184.03 - 59.7$$

$$= \underline{124.33\,\text{kJ/kg}} \qquad \text{(a)}$$

$$\text{Compressor work} = h_2 - h_1$$

h_2 is 15K of superheat = 210.63 kJ/kg

$$\text{Compressor work} = 210.63 - 184.03 = \underline{26.6\,\text{kJ/kg}} \qquad \text{(b)}$$

$$\text{COP} = \frac{h_1 - h_4}{h_2 - h_1} = \frac{124.33}{26.6} = \underline{4.67} \qquad \text{(c)}$$

$$\text{Carnot COP}_{\text{ref}} = \frac{T_L}{T_L - T_H} = \frac{(-10 + 273)}{(45 + 273) - (-10 + 273)}$$

$$= \underline{4.78} \qquad \text{(d)}$$

Example 2.7.5

In a vapour compression plant, 4.5 kg/min of ammonia leaves the evaporator dry saturated at a temperature of 2°C and is compressed to a pressure of 12.37 bar. The isentropic efficiency of the compressor is 0.85. Assuming no undercooling in the condenser, calculate:

(a) the cooling load;
(b) the power input;
(c) the coefficient of performance.

Referring to previous diagrams showing no undercooling in the condenser and superheat after the compressor,

$h_1 = h_f$ at 2°C (4.625 bar) = 1446.5 kJ/kg,

$h_3 = 332.8\,\text{kJ/kg} = h_4$

$s_1 = s'_2 = 5.314\,\text{kJ/kg K}$

$$\text{Cooling load} = \dot{m}(h_1 - h_4) = \frac{4.5}{60}(1446.5 - 332.8)$$

$$= \underline{83.5\,\text{kW}} \qquad \text{(a)}$$

Interpolating to find the enthalpy if the compression was isentropic,

$$h'_2 = 1469.9 + \frac{(5.314 - 4.962)}{(5.397 - 4.962)} \times (1613 - 1469.9)$$

$$= 1585.7\,\text{kJ/kg}$$

$$0.85 = \frac{h'_2 - h_1}{h_2 - h_1} = \frac{1585.7 - 1446.5}{h_2 - 1446.5}, \quad h_2 = 1610.3\,\text{kJ/kg}$$

$$\text{Power} = m(h_2 - h_1) = \frac{4.5}{60}(1610.3 - 1446.5)$$

$$= \underline{12.29\,\text{kW}} \qquad \text{(b)}$$

$$\text{COP} = \frac{\text{cooling load}}{\text{compressor power}} = \frac{83.51}{12.29} = \underline{6.8} \qquad \text{(c)}$$

Problems 2.7.1

In each case, sketch the cycle on *p*/*h* and *T*/*s* axes.

(1) In an ammonia refrigeration plant the refrigerant leaves the condenser as a liquid at 9.722 bar and 24°C. The evaporator pressure is 2.077 bar, and the refrigerant circulates at the rate of 0.028 kg/s. If the cooling load is 25.2 kW, calculate:

(a) the dryness fraction at the evaporator inlet;
(b) the dryness at the evaporator outlet.

(2) A vapour-compression ammonia refrigeration plant operates between pressures of 2.265 bar and 9.722 bar. The vapour at the entry to the compressor is dry saturated and there is no undercooling in the condenser. The coefficient of performance of the plant is 3.5. Calculate the work input required per kilogram flow of refrigerant.

(3) A Refrigerant 134a vapour compression cycle operates between 7.7 bar and 1.064 bar. The temperature of the refrigerant after the compressor is 40°C and there is no undercooling in the condenser. If the mass flow rate of the refrigerant is 0.2 kg/s find the power input to the compressor, the cooling load, and the coefficient of performance.

(4) A freon refrigerator cycle operates between −15°C and 25°C, and the vapour is dry saturated at the end of compression. If there is no undercooling in the condenser and the compression is isentropic, calculate:

(a) the dryness fraction at compressor suction;
(b) the refrigerating effect per kg of refrigerant;
(c) the coefficient of performance;
(d) the Carnot coefficient of performance.

(5) An ammonia refrigeration plant operates between 1.447 bar and 10.34 bar. The refrigerant enters the throttle as a saturated liquid and enters the compressor as a saturated vapour. The compression is isentropic. Calculate the coefficient of performance of the plant.

(6) A refrigeration plant using Refrigerant 12 operates between −5°C and 40°C. There is no undercooling in the condenser and the vapour is dry saturated after isentropic compression. If the cooling load is 3 kW, calculate:

(a) the coefficient of performance;
(b) the refrigerant mass flow rate;
(c) the power input to the compressor.

(7) A vapour compression refrigerating plant using freon 12 operates between 12.19 bar and 2.61 bar, and the temperatures before and after the compressor are 0°C and 75°C respectively. The refrigerant leaves the condenser at 40°C. Calculate:

(a) the isentropic efficiency of the compressor;
(b) the refrigeration effect;
(c) the coefficient of performance.

(8) An ammonia refrigeration plant operates between 2.265 bar and 10.34 bar. The ammonia leaves the evaporator at –10°C and leaves the condenser as a liquid at 26°C. The refrigerant mass flow rate is 0.4 kg/min and the compressor power requirement is 1.95 kW. Calculate:

(a) the dryness fraction at the evaporator inlet;
(b) the cooling load;
(c) the coefficient of performance.

2.8 Heat transfer

Heat transfer by conduction is a major consideration in plant such as boilers and heat exchangers, and through insulation requirements in buildings. Thermal insulation has important benefits in reducing energy costs, and in increasing efficiency of plant where heat transfer is demanded. The purpose of this section is to introduce a practical approach to the estimate of heat transfer in common situations, by imparting an understanding of the factors which influence the rate of heat energy transfer, and applying expressions to calculate heat energy loss through plane walls and in pipework.

Heat transfer through a plane wall

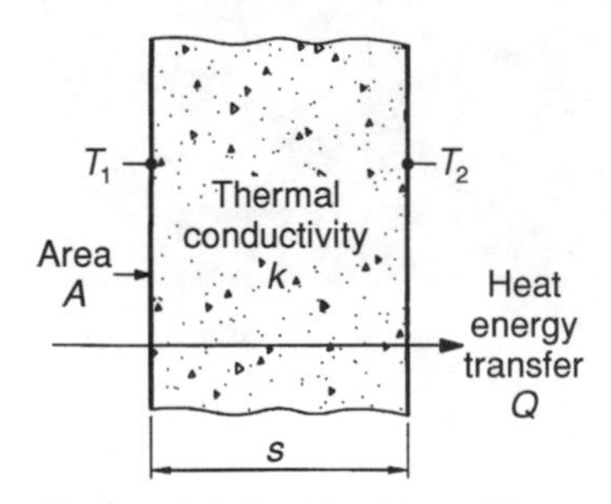

Figure 2.8.1 *Heat transfer through plain wall. Surface temperatures*

Figure 2.8.1 shows a plane wall. The rate of heat transfer across the faces of the wall will depend upon:

- Area of wall, A (m^2).
- Thermal conductivity of the wall, k (W/m^2K). As the unit suggests, this expresses the rate at which heat energy passes through 1 m^2 of the wall area for each degree K of temperature difference across its surfaces.
- Thickness of the wall, s (m).
- The temperature difference across the wall, $(T_1 - T_2)$ (K).

We are dealing here with cases in which the direction of heat energy flow is perpendicular to the plane surfaces of the wall, and there is no temperature variation across the surfaces.

For these cases, Fourier's equation expresses the heat transfer by conduction as

$$\frac{Q}{t} = - k.A. \frac{dt}{ds}$$

the negative sign is because the heat energy flow is towards the direction of temperature fall.

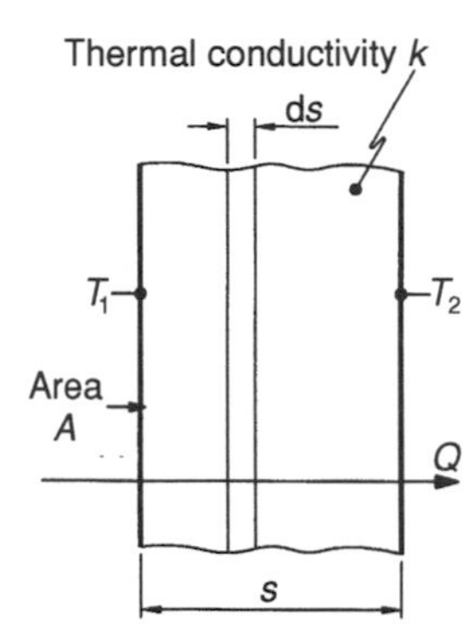

Figure 2.8.2 *Heat transfer through plain wall. Surface temperatures*

Maths in action

It is necessary to integrate Fourier's equation to give an expression for heat transfer into which we can put our values.

Referring to Figure 2.8.2,

$$\frac{Q}{t} = -k.A.\frac{dT}{ds}$$

rearranging,

$$\frac{Q}{t} ds = -k.A.dT$$

Integrating both sides,

$$\int_1^2 \frac{Q}{t}.ds = \int_1^2 -k.A.dT$$

These are definite integrals, i.e. between limits.

$$\frac{Q}{t}(s_2 - s_1) = -k.A\,(T_2 - T_1)$$

$$\frac{Q}{t} s = k.A\,(T_1 - T_2)$$

where the wall thickness, $(s_2 - s_1) = s$.

Heat transfer between fluids

If we consider the heat transfer rate between, say, two fluids, one either side of the wall, we must take into account the rate of heat energy lost from the surface by convection. The type and condition of the surface (and the velocity of flow over the surface, which we are not considering here) will influence the rate of convection, which is quantified by a value of surface heat transfer coefficient, h (W/m^{2}K).

By an analysis similar to that for the plane wall we can show that the heat transfer rate from the surface is given by,

$$\frac{Q}{t} = h.A.\,(T_1 - T_2)$$

where $(T_1 - T_2)$ is the temperature difference between the fluid and the surface.

We can apply the relevant formula in each case for each part of the wall.

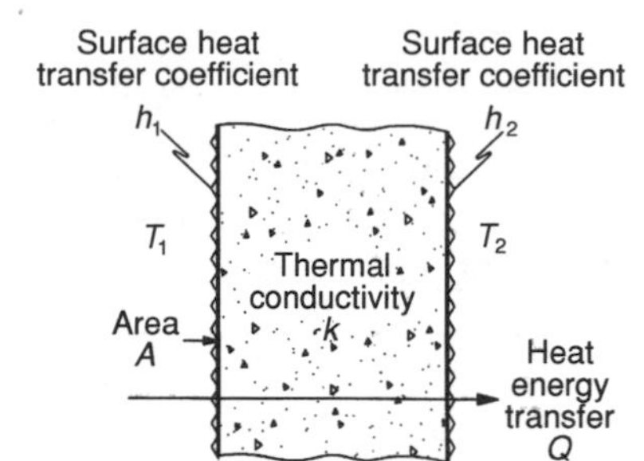

Figure 2.8.3 *Heat transfer through plain wall. Fluid temperatures*

Referring to Figure 2.8.3 and starting from the left,

for the surface,

$$\frac{Q}{t} = h_1 A(T_1 - T_2), \quad T_1 - T_2 = \frac{Q}{t.h_1.A}$$

for the wall,

$$\frac{Q}{t} s = k.A.(T_2 - T_3), \quad T_2 - T_3 = \frac{Q.s}{t.k.A}$$

for the surface,

$$\frac{Q}{t} = h_2.A.(T_3 - T_4), \quad T_3 - T_4 = \frac{Q}{t.h_2.A}$$

Adding,

$$(T_1 - T_2) + (T_2 - T_3) + (T_3 - T_4) = \frac{Q}{t.h_1.A} + \frac{Q.s}{t.k.A} + \frac{Q}{t.h_2.A}$$

giving,

$$(T_1 - T_4) = \frac{Q}{A.t}\left(\frac{1}{h_1} + \frac{s_1}{k} + \frac{1}{h_2}\right)$$

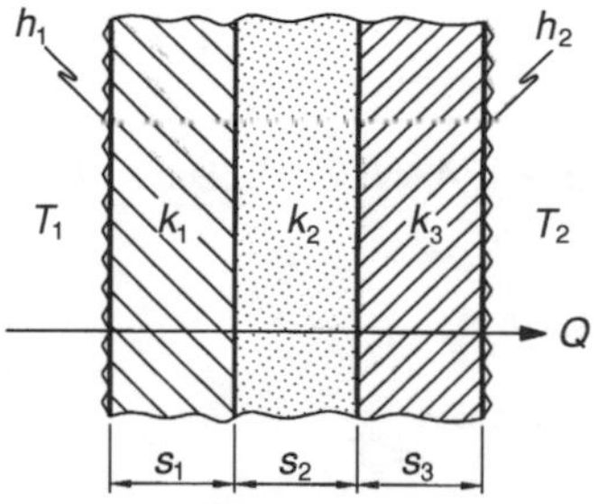

and, rearranging,

$$Q = \frac{1}{\left(\frac{1}{h_1} + \frac{s_1}{k} + \frac{1}{h_2}\right)} A.t(T_1 - T_4)$$

Figure 2.8.4 *Heat transfer through composite wall*

Students may find it more convenient to use the first equation rather than the transposed equation.

In Figure 2.8.4 we have a wall consisting of three parts. Taking into account the surface heat transfer coefficients, and with fluid temperatures T_1 and T_2, the equations become,

$$(T_1 - T_2) = \frac{Q}{A.t}\left(\frac{1}{h_1} + \frac{s_1}{k_1} + \frac{s_2}{k_2} + \frac{s_3}{k_3} + \frac{1}{h_2}\right)$$

and,

$$Q = \frac{1}{\left(\frac{1}{h_1} + \frac{s_1}{k_1} + \frac{s_2}{k_2} + \frac{s_3}{k_3} + \frac{1}{h_2}\right)} A.t.(T_1 - T_2)$$

The value which results from the large bracketed term is the reciprocal of the overall heat transfer coefficient, U. This is the value which is often quoted for insulation materials, since it gives a better idea of the insulation properties than the thermal conductivity alone.

Key point

These equations are easily modified for any configuration by inserting or removing elements from the large brackets.

Exercise 2.8.1

Use reference sources to list values of thermal conductivity, k, for common materials, and look for overall heat transfer coefficients, U, for insulation and lagging used in buildings.

Example 2.8.1

A 9.5 mm thick steel plate in a heat exchanger has a thermal conductivity of 44 W/mK, and the surface temperatures on either side are 504°C and 204°C. Find the rate of heat transfer through 1 m^2 of the plate.

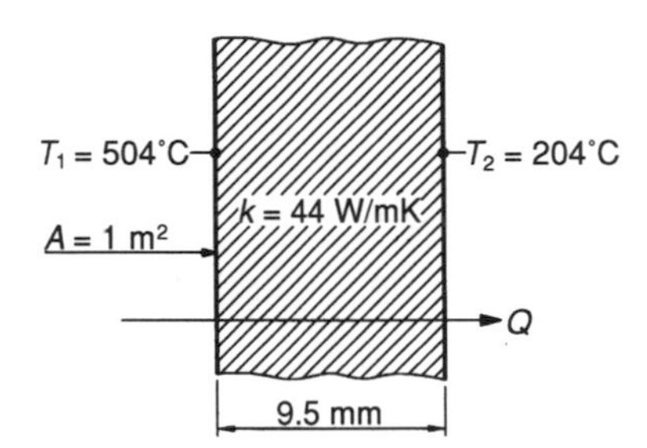

Figure 2.8.5 *Example 2.8.1*

Referring to Figure 2.8.5,

$$T_1 - T_2 = \frac{Q}{At}\left(\frac{s}{k}\right)$$

Note that t is 1 s, because we want the rate of heat energy transfer, i.e. J/s = W.

$$(504 - 204) = \frac{Q}{1 \times 1}\left(\frac{0.0095}{44}\right)$$

$$Q = \frac{(504-204) \times 1 \times 1 \times 44}{0.0095} = 1\,389\,474 \text{ W} = \underline{1389.5 \text{ kW}}$$

Putting the units into the original equation,

$$\text{K} = \frac{\text{J}}{\text{m}^2 \times \text{s}} \times \frac{\text{m}}{\frac{\text{W}}{\text{m.K}}} = \frac{\text{J}}{\text{m}^2 \times \text{s}} \times \frac{\text{m.m.K.s}}{\text{J}} = \text{K}$$

Example 2.8.2

Calculate the heat transfer per hour through a solid brick wall 6 m long, 2.9 m high and 225 mm thick when the outer surface temperature is 5°C and the inner surface temperature is 17°C, the thermal conductivity of the brick being 0.6 W/mK.

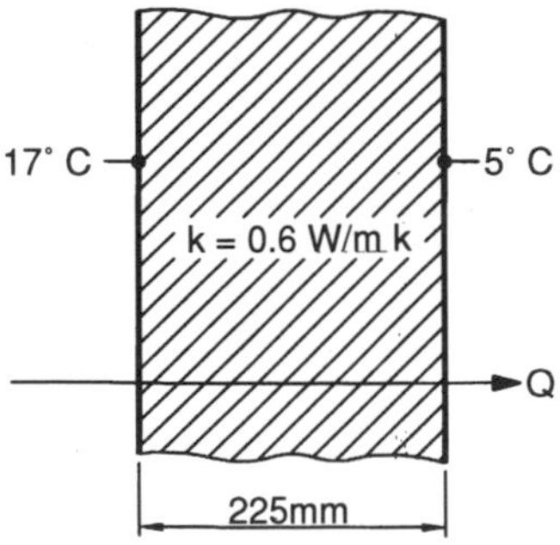

Figure 2.8.6 *Example 2.8.2*

See Figure 2.8.6.

$$T_1 - T_2 = \frac{Q}{At}\left(\frac{s}{k}\right)$$

$$(17-5) = \frac{Q}{(6 \times 2.9) \times (60 \times 60)}\left(\frac{0.225}{0.6}\right)$$

$$Q = \frac{(17 - 5) \times (6 \times 2.9) \times (60 \times 60) \times 0.6}{0.225} = 200\,448\,\text{J}$$

$$= \underline{2004.5\,\text{kJ}}$$

Note that in this case the time is (60 × 60) seconds.

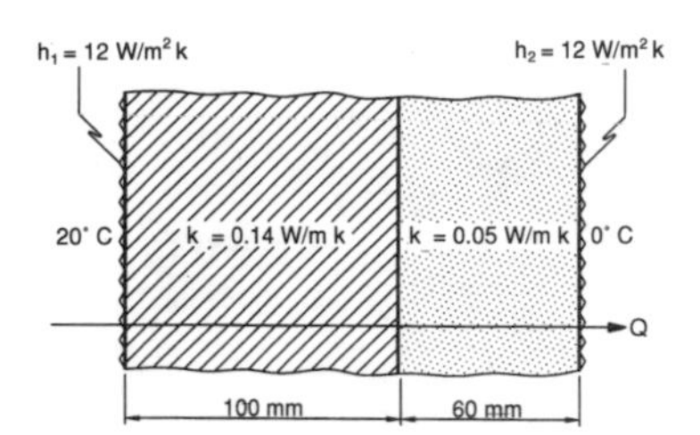

Figure 2.8.7 *Example 2.8.3*

Example 2.8.3

A refrigerated cold room wall has a thickness of 100 mm and a thermal conductivity of 0.14 W/mK. The room wall has a 60 mm thick internal lining of cork having a thermal conductivity of 0.05 W/mK. The thermal conductance between the exposed faces and the respective atmospheres is 12 W/m^{2}K.

If the room is maintained at 0°C and the external atmospheric temperature is 20°C, calculate the heat loss rate through 1 m^2 of the wall.

Figure 2.8.7 shows the wall.

$$T_1 - T_2 = \frac{Q}{A.t}\left(\frac{1}{h_1} + \frac{s_1}{k_1} + \frac{s_2}{k_2} + \frac{1}{h_2}\right)$$

$$(20 - 0) = \frac{Q}{1 \times 1}\left(\frac{1}{12} + \frac{0.1}{0.14} + \frac{0.06}{0.05} + \frac{1}{12}\right)$$

$$20 = Q(2.08)$$

$$Q = \underline{9.6\,\text{W}}$$

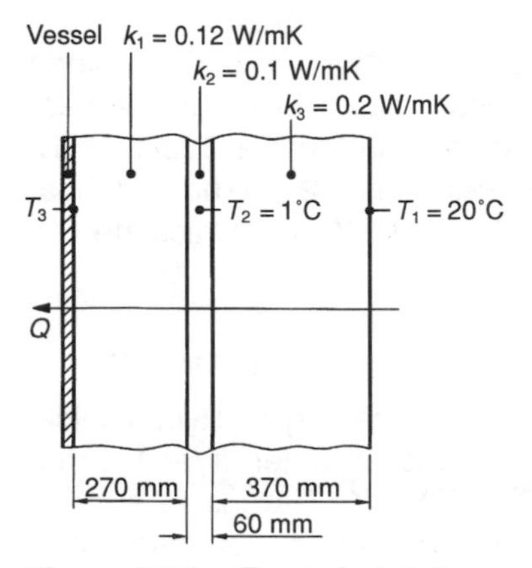

Figure 2.8.8 *Example 2.8.4*

Example 2.8.4

A cold storage vessel has its outer flat surfaces insulated with three layers of lagging. The innermost layer is 270 mm thick, the centre layer is 60 mm thick and the outer layer is 370 mm thick. The thermal conductivities of the lagging materials are 0.12, 0.1 and 0.2 W/mK respectively. A temperature sensor embedded half-way through the centre lagging indicates a temperature of 1°C.

Calculate the temperature of the outer surface of the cold storage vessel when the outer surface temperature of the outer layer of lagging is 20°C.

Figure 2.8.8 shows the wall.

This problem demonstrates working through part of the wall only.

Working from the midpoint of the inner layer, and calling this temperature T_2,

$$T_1 - T_2 = \frac{Q}{A.t}\left(\frac{s_1}{k_1} + \frac{s_2}{k_2}\right)$$

$$20 - 1 = \frac{Q}{1 \times 1}\left(\frac{0.37}{0.2} + \frac{0.03}{0.1}\right)$$

$$Q = 8.84\,\text{W}$$

Remember that Q is the same through all sections of the wall.

Working from the centre to the inner surface,

$$T_2 - T_3 = \frac{Q}{A.t}\left(\frac{s_1}{k_1} + \frac{s_2}{k_2}\right)$$

$$1 - T_3 = 8.84\left(\frac{0.03}{0.1} + \frac{0.27}{0.12}\right)$$

$$1 - T_3 = 8.84(2.55),$$

$$T_3 = \underline{-21.54°\text{C}}$$

Key point

Always make temperature differences positive.

Problems 2.8.1

(1) A brick wall is 3 m high and 5 m wide with a thickness of 150 mm. If the coefficient of thermal conductivity of the brick is 0.6 W/mK, and the temperatures at the surfaces of the wall are 25°C and 5°C, find the heat energy loss through the brickwork in kW.

(2) A cold storage compartment is 4.5 m long by 4 m wide by 2.5 m high. The four walls, ceiling and floor are covered to a thickness of 150 mm with insulating material which has a coefficient of thermal conductivity of 5.8×10^{-2} W/mK. Calculate the quantity of heat energy leaking through the insulation per hour when the outside and inside temperatures of the insulation are 15°C and –5°C.

(3) The walls of a cold room are 89 mm thick and are lined internally with cork of thickness 75 mm. The surface heat transfer coefficient for both exposed surfaces is 11.5 W/m^{2}K. The external ambient temperature is 22°C and the heat transfer rate through the wall is 34.5 W/m^2. Calculate the temperature inside the cold room.

Thermal conductivity of cork = 0.52 W/mK
Thermal conductivity of wall material = 0.138 W/mK

(4) A cold room wall consists of an inner layer 15 mm thick, thermal conductivity 0.18 W/mK and an outer layer 150 mm thick, thermal conductivity 0.045 W/mK. The inside surface temperature is 0°C and the outside surface temperature is 11°C. Calculate:

(a) the heat transfer rate per unit area of wall;
(b) the interface temperature.

(5) A glass window is to be double-glazed by adding a second sheet of glass 3 mm thick with an air gap of 20 mm. The window is 2.3 m high by 1.4 m wide and the original glass is also 3 mm thick. Basing your calculations on a room temperature of 23°C and an outside temperature of 2°C, find:

(a) the percentage heat reduction after double-glazing;
(b) the temperature of the outside surface of the outer glass.

Thermal conductivity of glass = 0.76 W/mK
Thermal conductivity of air = 0.026 W/mK
Inner heat transfer coefficient = 5.7 W/m²K
Outer heat transfer coefficient = 9.1 W/m²K

Heat transfer through pipe lagging

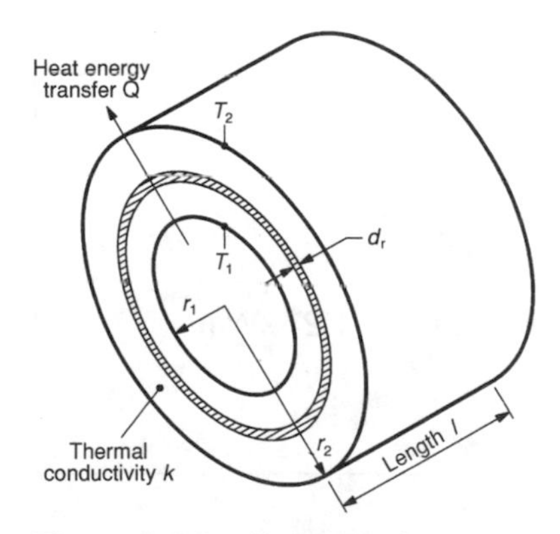

Figure 2.8.9 *Heat transfer through pipe. Surface temperature*

Figure 2.8.9 shows a pipe with a surface temperature T_1 and a layer of insulation with surface temperature T_2. The diameter of the pipe is r_1 and the radius to the outer surface of the lagging is r_2. The thermal conductivity of the lagging is k.

Fourier's equation for this case is

$$\frac{Q}{t} = -k.2\pi.r.l.\frac{dT}{dr}$$

For length, l, the area of the elemental strip is (circumference × length), i.e. $2.\pi.r.l.$

Integrating between 1 and 2 gives,

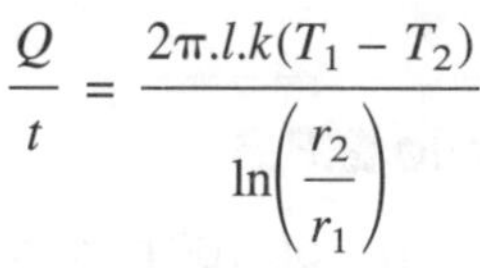

$$\frac{Q}{t} = \frac{2\pi.l.k(T_1 - T_2)}{\ln\left(\frac{r_2}{r_1}\right)}$$

Multi-layer pipe lagging

See Figure 2.8.10.

By the same analysis for a plane wall, and noting that for fluid/lagging interfaces, the areas are $2\pi.r.l$, where r values are for inner pipe radius and radius to outer layer of lagging,

$$\frac{Q}{t} = \frac{2\pi.l(T_1 - T_5)}{\frac{1}{r_1.h_i} + \frac{\ln\frac{r_2}{r_1}}{k_1} + \frac{\ln\frac{r_3}{r_2}}{k_2} + \frac{1}{r_3 h_o}}$$

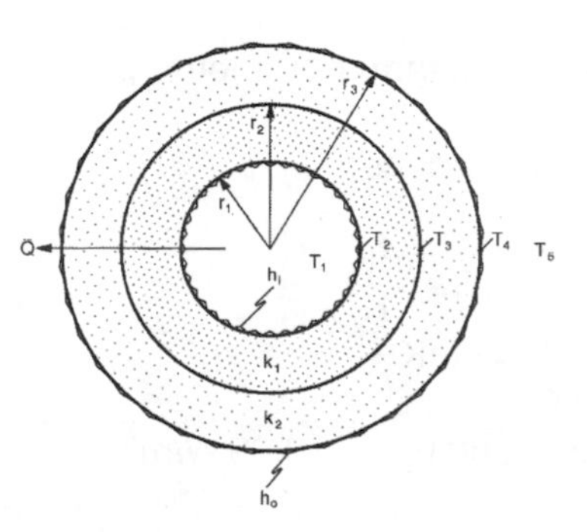

Figure 2.8.10

This equation looks complicated, but examination will show that it has a pattern and can be easily adapted to suit any number of elements.

As for a plane wall, any element of the pipe can be isolated for calculating. Across the inner layer for instance,

$$\frac{Q}{t} = \frac{(T_2 - T_3)2\pi.l}{\dfrac{\ln \dfrac{r_2}{r_1}}{k_1}}$$

Key point

t is usually 1 s to give *Q* in W or kW.

Example 2.8.5

A pipe of inner diameter 0.15 m is lagged with 0.065 m thick material of thermal conductivity 0.6 W/mK. If the inner and outer surface temperatures of the lagging are 260°C and 50°C respectively, calculate the heat loss per metre length of pipe.

Figure 2.8.11 shows the pipe.

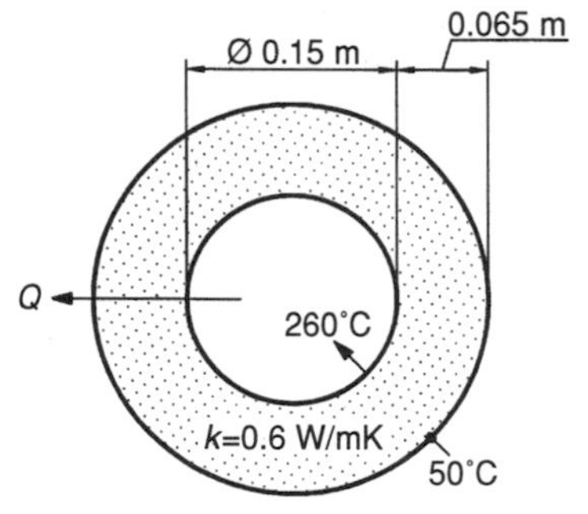

Figure 2.8.11 *Example 2.8.5*

Using

$$Q = \frac{2\pi.k.l(T_1 - T_2)}{\ln \dfrac{r_2}{r_1}}$$

$$Q = \frac{2\pi.0.6.1(260 - 50)}{\ln \dfrac{0.14}{0.075}} = 1268.72\,\text{W} = \underline{1.27\,\text{kW/m}}$$

Example 2.8.6

A steel pipe of 100 mm bore and 10 mm bore thickness, carrying dry saturated steam at 28 bar, is insulated with a 40 mm layer of moulded insulation. This insulation in turn is insulated with a 60 mm layer of felt. The atmospheric temperature is 15°C.
Calculate:

(a) the rate of heat loss by the steam per metre pipe length;
(b) the temperature of the outside surface.

Inner heat transfer coefficient = 550 W/m^{2}K
Outer heat transfer coefficient = 15 W/m^{2}K
Thermal conductivity of steel = 50 W/mK
Thermal conductivity of felt = 0.07 W/mK
Thermal conductivity of moulded insulation = 0.09 W/mK

See Figure 2.8.12.

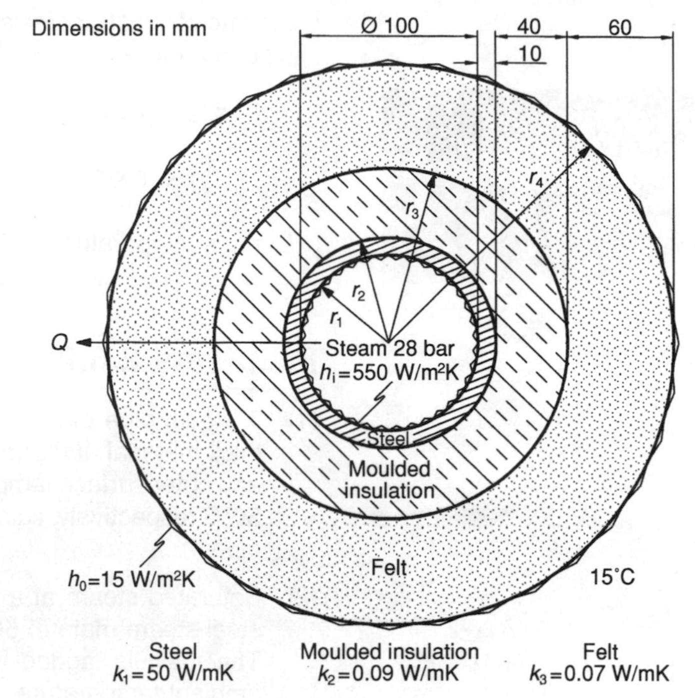

Figure 2.8.12 *Example 2.8.6*

t_s at 28 bar = 230°C

$$Q = \frac{2\pi.l(T_1 - T_6)}{\frac{1}{r_1 h_1} + \frac{\ln\frac{r_2}{r_1}}{k_1} + \frac{\ln\frac{r_3}{r_2}}{k_2} + \frac{\ln\frac{r_4}{r_3}}{k_3} + \frac{1}{r_4\, h_0}} \quad \text{(length, } l = 1\text{)}$$

$$Q = \frac{2\pi.(230 - 15)}{\frac{1}{0.05 \times 550} + \frac{\ln\frac{60}{50}}{50} + \frac{\ln\frac{100}{60}}{0.09} + \frac{\ln\frac{160}{100}}{0.07} + \frac{1}{0.16 \times 15}}$$

$$Q = \frac{1350.89}{0.036\,36 + 0.003\,646 + 5.6758 + 6.7143 + 0.417}$$

$$= \underline{105.15\,\text{W/m length}}$$

To find the outside surface temperature, we work between the surface and the atmosphere,

$$Q = \frac{2\pi.(T_5 - T_6)}{\frac{1}{r_4.h_0}}$$

$$Q = 2\pi.r_4.h_0\,(T_5 - T_6)$$

Key point

When taking $\log_e$, values can be radii or diameters, mm or m.

the rate of heat transfer is the same through all sections of the pipe, therefore,

$$105.15 = 2\pi \times 0.16 \times 15(T_5 - T_6)$$

$$(T_5 - T_6) = 6.97, \qquad (T_5 - 15) = 6.97$$

Surface temperature $= T_5 = \underline{21.97°C}$

Problems 2.8.2

(1) A steam pipe of inner diameter 0.2 m is lagged with 0.08 m thick material of thermal conductivity 0.05 W/mK. If the inner and outer surface temperatures of the lagging are 300°C and 50°C respectively, calculate the heat loss per metre length of pipe.

(2) Saturated steam at a pressure of 10 bar passes through a steel steam main of 50 mm bore and wall thickness 25 mm. The main is lagged with a 13 mm layer of cork, and the ambient temperature is 18°C. Calculate:

(a) the heat transfer rate per metre length of pipe;
(b) the interface temperature between steel and cork;
(c) the surface temperature of the steel;
(d) the surface temperature of the cork.

Thermal conductivity of the steel = 52 W/mK
Thermal conductivity of the cork = 0.04 W/mK
Inside surface heat transfer coefficient = 15 W/m^{2}K
Outside surface heat transfer coefficient = 10 W/m^{2}K

(3) Fresh water at 65°C flows through a 40 m length of steel pipe which has a bore of 50 mm and a thickness of 10 mm. The ambient temperature of the air is 20°C. A 30 mm layer of insulation is added to the pipe. Calculate the reduction in heat transfer rate resulting from the addition of the insulation.

Thermal conductivity of the steel = 52 W/mK
Thermal conductivity of the insulation = 0.17 W/mK
Inside surface heat transfer coefficient = 1136 W/m^{2}K
Outside surface heat transfer coefficient from the pipe = 13 W/m^{2}K
Outside surface heat transfer coefficient from the insulation = 9.7 W/m^{2}K

(4) A steel pipe with an internal diameter of 100 mm and a wall thickness of 10 mm, carries dry saturated steam at 8 bar, and is covered with a 12.5 mm thickness of cork. A layer of moulded polystyrene, with an outer surface temperature of 26°C, covers the cork. The heat loss from the pipe is 100 W/m length. Calculate:

(a) the thickness of the moulded polystyrene;
(b) the outside temperature of the steel pipe.

Thermal conductivity of the polystyrene = 0.01 W/mK
Thermal conductivity of the steel = 450 W/mK
Thermal conductivity of the cork = 0.04 W/mK
Inside surface heat transfer coefficient = 15 W/m^2K

(5) A steam pipe, 80 m long, is lagged with two different materials, each of thickness 50 mm, and carries dry saturated steam which enters the pipe at 12 bar at a rate of 540 kg/h. The pipe is 120 mm diameter with a wall thickness of 10 mm, and the inner heat transfer coefficient is negligible. If the ambient temperature is 16.5°C, calculate:

(a) heat loss from the pipe per hour;
(b) dryness fraction of steam at the pipe exit;
(c) interface temperature between the two layers of insulation.

Coefficient of thermal conductivity of the pipe = 55 W/mK
Coefficient of thermal conductivity of the inner lagging = 0.075 W/mK
Coefficient of thermal conductivity of the outer lagging = 0.15 W/mK
Outer heat transfer coefficient = 12 W/m^2K

3 Fluid mechanics

Summary

Fluid mechanics is the study of the behaviour of liquids and gases, and particularly the forces that they produce. Many scientific disciplines have an interest in fluid mechanics. For example, meteorologists try to predict the motion of the fluid atmosphere swirling around the planet so that they can forecast the weather. Physicists study the flow of extremely high temperature gases through magnetic fields in a search for an acceptable method of harnessing the energy of nuclear fusion reactions. Engineers are interested in fluid mechanics because of the forces that are produced by fluids and which can be used for practical purposes. Some of the well-known examples are jet propulsion, aerofoil design, wind turbines and hydraulic brakes, but there are also applications which receive less attention such as the design of mechanical heart valves.

The purpose of this chapter is to teach you the fundamentals of engineering fluid mechanics in a very general manner so that you can understand the way that forces are produced and transmitted by fluids that are, first, essentially at rest and, second, in motion. This will allow you to apply the physical principles behind some of the most common applications of fluid mechanics in engineering. Most of these principles should be familiar – conservation of energy, Newton's laws of motion – and so the chapter concentrates on their application to liquids.

Objectives

By the end of this chapter, the reader should be able to:

- recognize some fluid properties and types of flow;
- understand the transmission of pressure in liquids and its application to hydraulics;
- use manometry to calculate pressures;
- calculate hydrostatic forces on plane and curved submerged surfaces;
- understand Archimedes' principle and buoyancy;
- employ the concept of continuity of flow;
- define viscosity;
- calculate pressure drops in pipe flow;
- use Bernoulli's equation to measure flow rate and velocity;
- apply the momentum principle to liquids in jets and pipes.

3.1 Hydrostatics – fluids at rest

The first half of this chapter is devoted to hydrostatics, the study of fluids at rest. It is a subject that is most commonly associated with civil engineers because of their interest in dams and reservoirs, but it is necessary for mechanical engineers too as it leads on to the subject of hydrodynamics, fluids in motion.

What are fluids?

Fluids are any substances which can flow. We normally think of fluids as either liquids or gases, but there are also cases where solids such as fine powders can behave as fluids. For example, much of the ground in Japan is made up of fine ash produced by the many volcanoes which were active on the islands until quite recently. Earthquakes are still common there and the buildings run the risk of not only collapsing during the tremors but also sinking into the ground as the powdery ash deposits turn into a sort of fluid due to the vibration. Nevertheless we shall only consider liquids and gases for simplicity, and most of the time we shall narrow the study down even further to liquids because we can look at the basic principles without the complications that apply to gases because of their compressibility. There are only a few major differences between liquids and gases, so let us have a look at them first.

Liquids and gases

It is easiest if we arrange the differences between liquids and gases as a double list (see Figures 3.1.1–3.1.6).

Liquids **Gases**

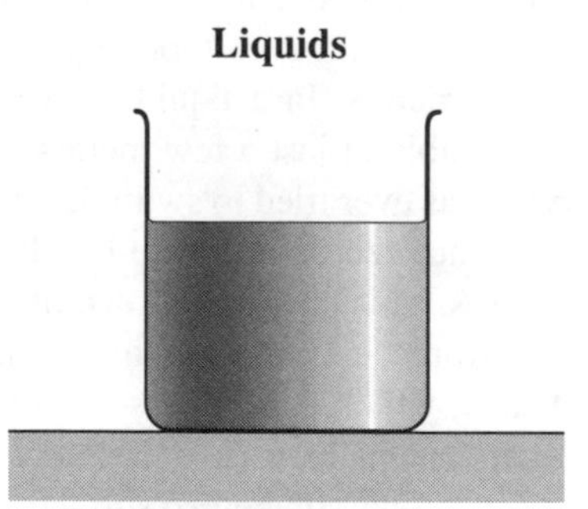

Figure 3.1.1 *Liquids have a fixed volume*

Figure 3.1.2 *Gases expand to fill the container*

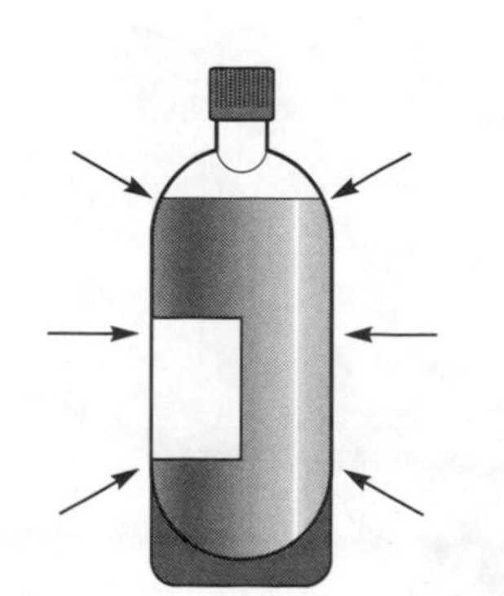

Figure 3.1.3 *Liquids are difficult to compress*

Figure 3.1.4 *Gases are highly compressible*

Figure 3.1.5 *Liquids are not affected by temperature*

Figure 3.1.6 *Gases are very temperature dependent*

From this list of differences it is plain to see that it is much easier to study liquids, and so that is what we shall do most of the time since the basic principles of fluid mechanics apply equally well to liquids and gases. There is only one instance where liquids are more complicated to study than gases and that is to do with the fact that liquids are much more dense.

Pressure in liquids

The drawback to this approach of concentrating on liquids is that liquids are very dense compared with gases and so we do not have to go very far down into a liquid before the pressure builds up enormously due to the weight of all the mass of liquid above us. This variation of pressure with depth is almost insignificant in gases. If you were to climb to the top of any mountain in the UK then you would not notice any difference in air pressure even though your altitude may have increased by about 1000 metres. In a liquid, however, the difference in pressure is very noticeable in just a few metres of height (or depth) difference. Anyone who has ever tried to swim down to the bottom of a swimming pool will have noticed the pressure building up on the ears after just a couple of metres. This phenomenon is, of course, caused by gravity which makes the water at the top of the swimming pool press down on the water below, which in turn presses down even harder on the ears. In order to quantify this increase of pressure with depth we need to look at the force balance on a submerged surface, so let us make that surface an ear drum, as shown in Figure 3.1.7

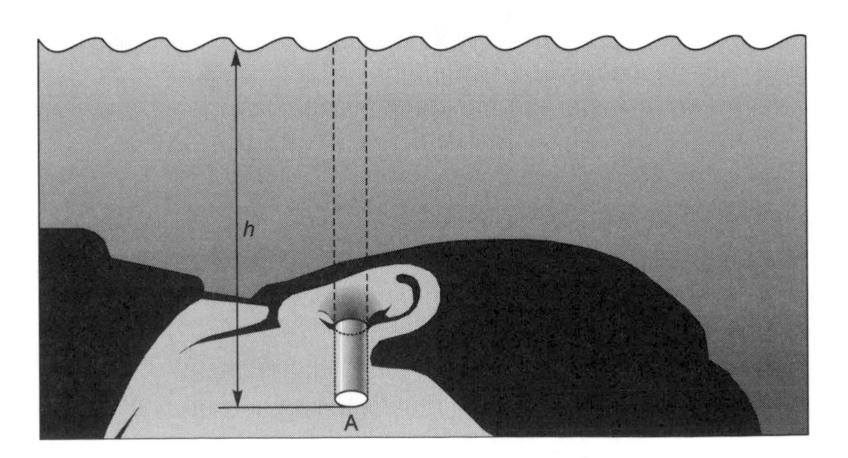

Figure 3.1.7 *Water creating a pressure on an ear drum*

We are dealing with gravitational forces, which always act vertically, and so we only need to consider the effect of any liquid, in this case water, which is vertically above the ear drum. Water which is to either side of the vertical column drawn in the diagram will not have any effect on the pressure on the ear drum, it will only pressurize the cheek or the neck, etc.

The volume of water which is pressing down on the ear drum is the volume of a cylinder of height h, equal to the depth of the ear, and end area A, equal to the area of the ear drum,

$$\text{Volume} = hA$$

Therefore the mass of water involved is

$$\text{Volume} \quad \text{density} = \rho hA \quad \text{where } \rho \text{ is the density in kg/m}^3$$

and the weight of this water is

$$\text{Mass} \quad \text{gravity} = \rho ghA$$

We are interested in the pressure p rather than this force, so that we can apply the result to any shaped surface. This pressure will be uniform across the whole of the area of the ear drum and we can therefore rewrite the force due to the water as *pressure area*. Hence:

$$pA = \rho ghA$$

Cancelling the areas we end up with:

$$p = \rho gh \qquad (3.1.1)$$

Since the area of the eardrum cancelled out, this result is not specific to the situation we looked at; this equation applies to any point in any liquid. We can therefore apply this formula to calculate the pressure at a given depth in any liquid in an engineering situation. There are two further important features that need to be stressed:

- Two points at the same depth in the same liquid must be at the same pressure even if one of them is not directly underneath the full depth (Figure 3.1.8).

Figure 3.1.8 *Pressure at a constant overall depth is constant*

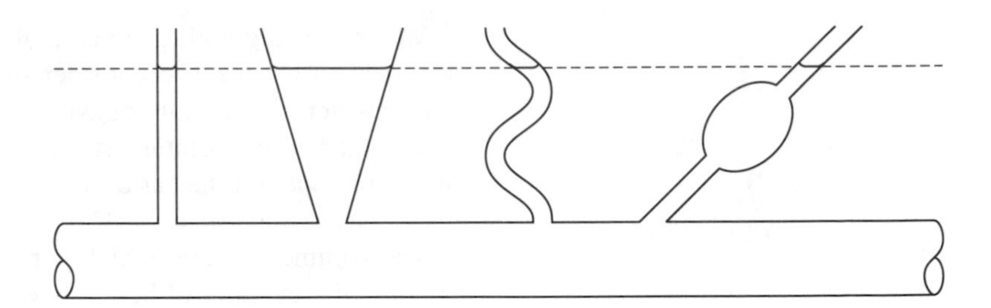

Figure 3.1.9 *Pressure only depends on vertical depth*

- The same pressure due to depth can be achieved with a variety of different shaped columns of a liquid since only the *vertical* depth matters (Figure 3.1.9).

Pressure head

We can relate a liquid's pressure to the height of a column of that liquid whether there really is a column there or not. We could be producing the pressure with a pump, for example, but it can still be useful to talk in terms of a height of liquid since this is a simple measurement which is easier to understand and visualize than the correct units of pressure (pascals or N/m^2). This height can be calculated by rearranging Equation (3.1.1) to give

$$h = p/\rho g$$

and it is known as the *pressure head* or the *static head* (since it refers to liquids at rest). The idea of 'head' comes from the early British engineers who built canals and reservoirs, and realized that the amount of pressure or even power they could get from the water depended on the vertical height difference between the reservoir surface and the place where they were working. The idea was taken up by the steam engineers of the Victorian era who talked about the 'head of steam' they could produce in a boiler, and it is such a useful concept that it is still used today even though it sounds old fashioned.

In order to get a better understanding of the meaning of a head, and to gain practice in converting from head to pressure, we shall now look at ways of using the height of a column of liquid to measure pressure.

Manometry

Manometry is the measurement of pressure using columns of liquid, although more modern electronic pressure measurement devices often also get called manometers. Liquid manometers are still widely used for pressure measurement and so this study is far from being of just historical interest. They are used in a great many configurations, so the four examples we shall look at here are just illustrations of how to go about calculating the conversions from a height of liquid in metres to a pressure in pascals.

Piezometer tube (or simple manometer tube) (Figure 3.1.10)

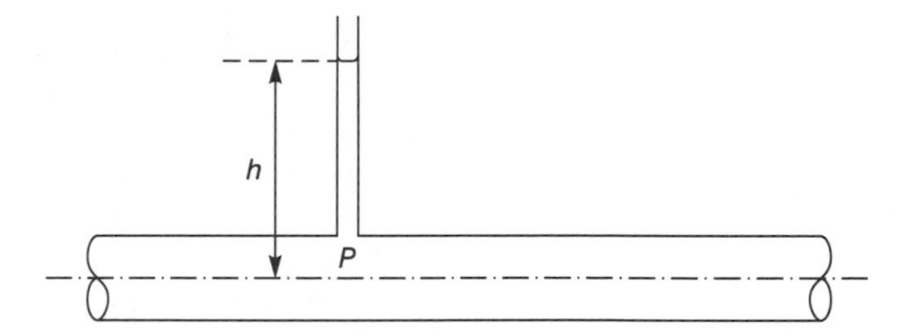

Figure 3.1.10 *A piezometer tube*

The pressurized liquid in the horizontal pipe rises up the vertical glass tube until the pressure from the pipe is balanced by the pressure due to the column of liquid, and the liquid comes to rest. At that point the pressure head is simply the height of the liquid in the tube, measured from the centreline of the pipe. The pressure in the pipe can be found from the formula

$$p = \rho g h$$

where ρ is the density of the liquid filling the pipe and the manometer tube. This is a beautifully simple device that is inherently accurate; as mentioned earlier it is only the vertical height that matters so any inclination of the tube or any variation in diameter does not affect the reading so long as the measuring scale itself is vertical. In practice the piezometer tube is limited to measuring heads of about 1 metre because otherwise the glass tube would be too long and fragile.

U-tube manometer (Figure 3.1.11)

For higher pressures we can use a higher density liquid in the tube. Clearly the choice of liquids must be such that the liquid in the tube does not mix with the liquid in the pipe. Mercury is the most commonly used liquid for the manometer tube because it has a high density (relative

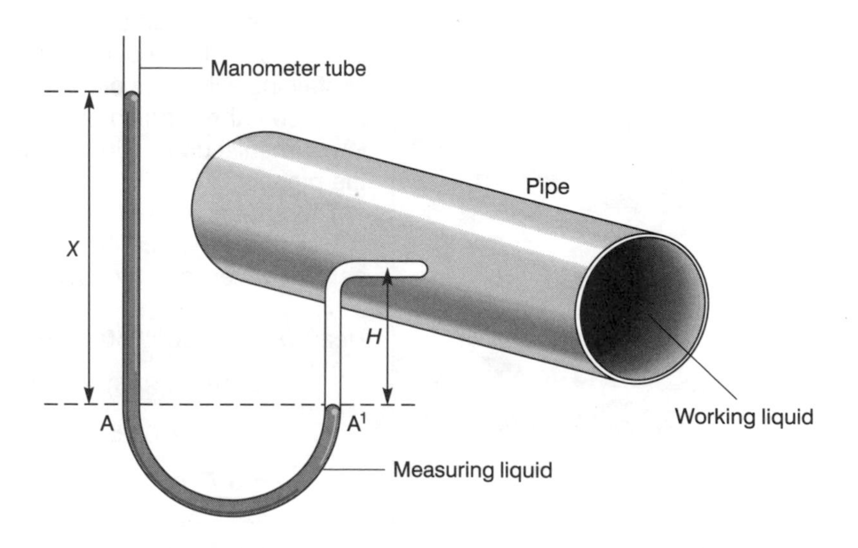

Figure 3.1.11 *A U-tube manometer*

density of 13.6, i.e. mercury is 13.6 times denser than water) and it does not mix with common liquids since it is a metal. To prevent it escaping from the manometer tube a U-bend is used.

Note that in the diagram the height of the mercury column is labelled as x and not as h. This is because the head is always quoted as the height of a column of the *working* liquid (the one in the pipe), rather than the *measuring* liquid (the one in the manometer tube). We therefore need to convert from x to obtain the head of working liquid that would be obtained if we could build a simple manometer tube tall enough. To solve any conversion problem with manometers it is usually best to work from the lowest level where the two liquids meet, in this case along the level AA′.

Pressure at A is due to the left-hand column so

$$p_A = \rho_m g x$$

Pressure at A′ is partly due to the right-hand column and partly due to the pressure in the pipe, so

$$p_{A'} = \rho_{wl} g H + p_{wl}$$

We can interpret the pressure in the pipe as a pressure head using $p_{wl} = \rho_{wl} g h_{wl}$.

Now we know that the pressure in a liquid is constant at a constant depth, so the pressure at A must equal the pressure at A′ just inside the mercury. Therefore

$$\rho_m g x = \rho_{wl} g H + \rho_{wl} g h_{wl}$$

To simplify this and find the pressure head:

$$h_{wl} = (\rho_m/\rho_{wl})x - H \text{ metres of the working liquid} \qquad (3.1.2)$$

Example 3.1.1

A U-tube manometer containing mercury of density 13 600 kg/m^3 is used to measure the pressure head in a pipe containing a liquid of density 850 kg/m^3. Calculate the head in the pipe if the reading on the manometer is 245 mm and the lower meniscus is 150 mm vertically below the centreline of the pipe.

The set-up is identical to the manometer shown in Figure 3.1.11 and so we can use Equation (3.1.2) directly. Remember to convert any measurements in millimetres to metres.

$$\text{Head } h = (13\,600/850) \quad (245/1000) - 150/1000$$
$$= (16 \quad 0.245) - 0.15$$
$$= 3.92 - 0.15$$
$$= \underline{3.77\text{ m}}$$

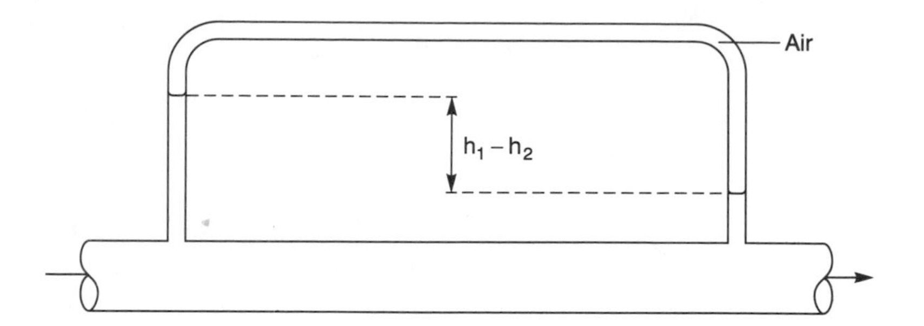

Figure 3.1.12 *Differential inverted U-tube manometer*

Differential inverted U-tube manometer

In many cases it is not just the pressure that needs to be measured, it is the pressure *difference* between two points that is required. This can often be measured directly by connecting a single manometer to the two points and recording the differential head, as shown in Figure 3.1.12.

Since it is the working liquid itself that fills the manometer tubes and there is no separate measuring liquid, the head difference is given directly by the difference in the heights in the tubes $(h_1 - h_2)$. In this case the difference is limited again to about 1 metre because of the need for a glass tube in order to see the liquid levels.

Differential mercury U-tube manometer (Figure 3.1.13)

To make it possible to record higher differential pressures, again we can use mercury for the measuring liquid, giving a device which is by far the most common form of manometer because of its ability to be used on various flow measuring devices.

Again we tackle the problem of converting from the reading x into the head difference of the working liquid $(h_1 - h_2)$ by working from the lowest level where the two liquids meet. So we start by equating the pressures at A and A′ (i.e. at the same depth in the same liquid).

Pressure at A is equal to the pressure at point 1 in the pipe plus the pressure due to the vertical height of the column of working liquid in the left-hand tube: $p_1 + \rho_{wl}gH$.

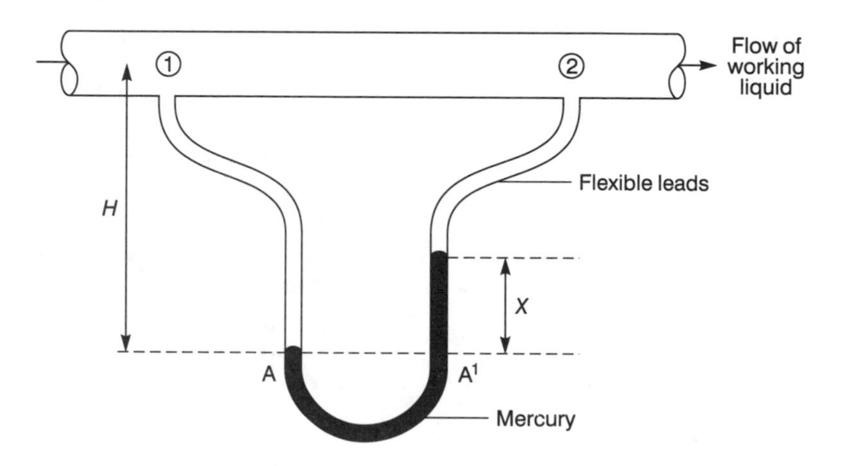

Figure 3.1.13 *Differential mercury U-tube manometer*

Pressure at A′ is equal to the pressure at point 2 in the pipe plus the pressure due to the vertical height of the short column of working liquid sitting on the column of mercury in the right-hand tube:

$$p_2 + \rho_{wl} g(H - x) + \rho_m g x$$

Rewriting the pressure terms to get heads h_1 and h_2, and equating the pressures at A and A' we get:

$$\rho_{wl} g h_1 + \rho_{wl} g H = \rho_{wl} g h_2 + \rho_{wl} g(H - x) + \rho_m g x$$

Cancel g because it appears in all terms and cancel the term $\rho_{wl} g H$ because it appears on both sides.

$$\rho_{wl} h_1 = \rho_{wl} h_2 + \rho_m x - \rho_{wl} x$$

Finally we have

$$h_1 - h_2 = (\rho_m/\rho_{wl} - 1)\ x \qquad (3.1.3)$$

This equation is well worth learning because it is such a common type of manometer, but remember that it only applies to this one type.

Example 3.1.2

A mercury U-tube manometer is to be used to measure the difference in pressure between two points on a horizontal pipe containing liquid with a relative density of 0.79. If the reading on the manometer scale for the difference in height of the two levels is 238 mm, calculate the head difference and the pressure difference between the two points.

We can use Equation (3.1.3) directly to solve this problem and give the head difference, but remember to convert millimetres to metres (divide by 1000) and convert relative density to real density (multiply by 1000 kg/m^3).

Head difference = {(13 600/790) – 1} 238/1000

= (17.215 – 1) 0.238

= 16.215 0.238

= <u>3.86 m of the pipeline liquid</u>

The pressure difference is found by using $p = \rho g h$ and remembering that the density to use is the density of the liquid in the pipeline; once we have converted from x to h in the first part of the calculation then the mercury plays no further part.

Pressure difference = 790 9.81 3.86

= <u>29 909 Pa</u>

Hydrostatic force on plane surfaces

In the previous section we looked at the subject of hydrostatic pressure variation with depth in a liquid, and used the findings to explore the possibilities for measuring pressure with columns of liquids (manometry). In this section we are going to extend this study of hydrostatic pressure to the point where we can calculate the total force due to liquid pressure acting over a specified area. One of the main reasons that we study fluid mechanics in mechanical engineering is so that we can calculate the size of forces acting in a situation where liquids are employed. Knowledge of these forces is essential so that we can safely design a range of devices such as valves, pumps, fuel tanks and submersible housings. Although we shall not look at all the many different situations that can arise, it is vital that you understand the principles involved by studying a few of the most common applications.

Hydraulics

The first thing to understand is the way that pressure is transmitted in fluids. Most of this is common sense but it is worthwhile spelling it out so that the most important features are made clear.

Look at Figures 3.1.14 and 3.1.15. In Figure 3.1.14 a short solid rod is being pushed down onto a solid block with a force F. The cross-sectional area of the rod is A and so a localized pressure of

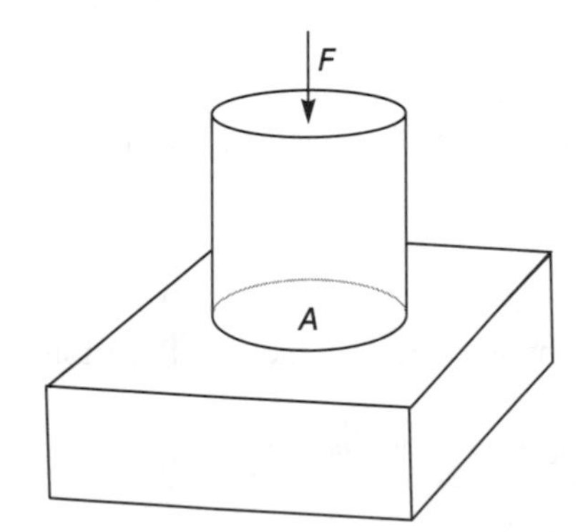

Figure 3.1.14 *Force applied to a solid block*

$$P = F/A$$

is felt underneath the base of the rod. This pressure will be experienced in the block mainly directly under the rod, but also to a much lesser extent in the small surrounding region. It will not be experienced in the rest of the solid block towards the sides.

Now look at Figure 3.1.15 where the same rod is used as a piston to push down with the same force on a sealed container of liquid. The result is very different because the pressure of

$$P = F/A$$

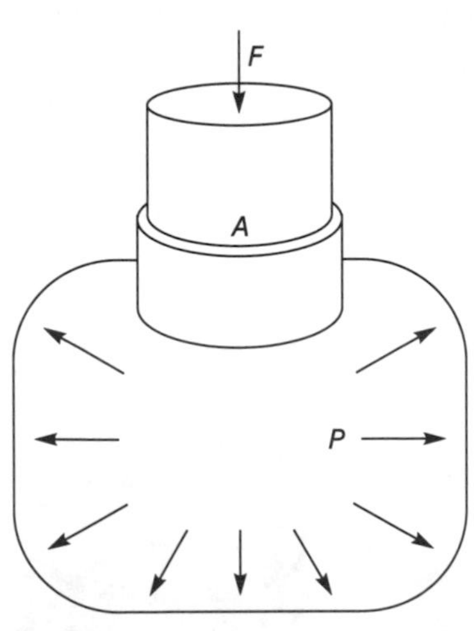

Figure 3.1.15 *Force applied to a liquid*

will now be experienced by all the liquid equally. If the container were to spring a leak we know that the liquid would spurt out normal to the surface that had the hole in it. If the leak were in the top surface of the container then the liquid would spurt upwards, in completely the opposite direction to the force that is being applied to the piston. This means that the liquid was being pushed in that upwards direction and that can only happen if the pressure acts normally to the inside surface.

This simple imaginary experiment therefore has two very important conclusions:

- Pressure is transmitted uniformly in all directions in a fluid.
- Fluid pressure acts normal (i.e. at right angles) to any surface that it touches.

These two properties of fluid pressure are the basis of all hydraulic systems and are the reason why car brakes work so successfully. The force applied by the driver's foot to the small cross-section piston and cylinder produces a large pressure in the hydraulic fluid that completely fills the system. This large pressure is transmitted along the hydraulic pipes under the car to the larger area pistons and cylinders at the wheels where it produces a large force that operates the brakes themselves. The force on the piston faces is at right angles to the face and therefore exactly in the direction where it can have most effect.

The underlying principle behind hydraulics is that by applying a small force to a small area piston we can generate a large pressure in the hydraulic fluid that can then be fed to a large area piston to produce an enormous force. The reason that this principle works is that the hydraulic fluid transmits the applied pressure uniformly in all directions, as we saw above.

It sometimes seems as though we are getting something for nothing in examples of hydraulics because small, applied forces can be used to produce very large output forces and shift massive objects such as cars on jacks. Of course we cannot really get something for nothing and the penalty that we have to pay is that the small piston has to be replaced by a pump which is pushed backwards and forwards many times in order to displace the large volume of fluid required to make the large piston move any appreciable distance. This means that the small force on the driving piston is applied over a very large distance (i.e. many times the stroke length of the pump) in order to make the large force on the big piston move through one stroke length, as illustrated in Figure 3.1.16.

The volume of hydraulic fluid expelled by the small piston in one stroke is given by:

$$V = sa$$

where s and a are the stroke length and area of the small piston respectively.

The work done by the piston during this stroke is given by:

$$\text{Work}_{\text{in}} = \text{force} \quad \text{distance}$$
$$= fs$$

This volume of fluid expelled travels to the large cylinder, which moves up by some distance d such that the extra volume in the cylinder is given by:

$$V = dA$$

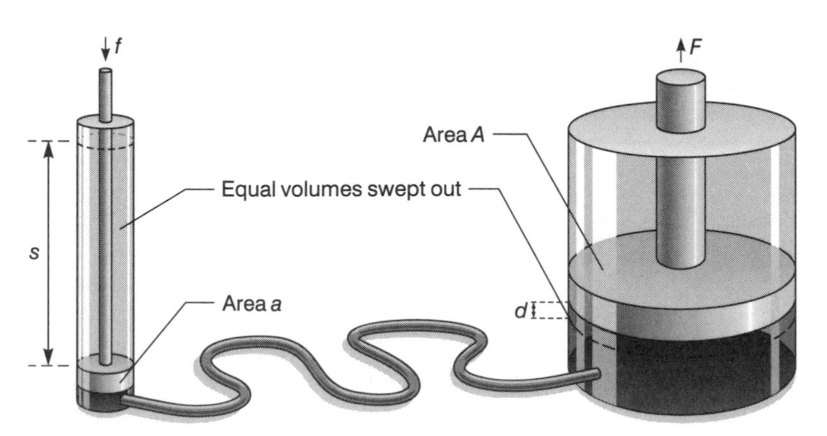

Figure 3.1.16 *A simple hydraulic system*

The work done on the load on the large piston is given by:

$\text{Work}_{out} = Fd$

Now the volume expelled from the small cylinder is the same as the volume received by the large cylinder, because liquids are incompressible for all practical purposes, therefore:

$V = dA = sa$ and hence $d = sa/A$

So $\text{Work}_{out} = Fsa/A = Psa = fs = \text{Work}_{in}$. (Note: pressure is uniform so $P = f/a = F/A$.)

In practice there would be a slight loss of energy due to friction and so it is clear that hydraulic systems actually increase the amount of work to be done rather than give us something for nothing, but their big advantage is that they allow us to carry out the work in a manner which is more convenient to us.

Example 3.1.3

A small hydraulic cylinder and piston, of diameter 50 mm, is used to pump oil into a large vertical cylinder and piston, of diameter 250 mm, on which sits a lathe of mass 350 kg. Calculate the force which must be applied to the small piston in order to raise the lathe.

The pressures in the two cylinders must be equal and so

$P = f/a = F/A$

Where the lower case refers to the small components and the upper case refers to the large components. We are trying to find the small force f and we do not need to calculate P, so this equation can be rearranged:

$f = Fa/A$

The large force F is equal to the mass of the lathe multiplied by the acceleration due to gravity, i.e. it is the lathe's weight. The two areas do not need to be calculated because the ratio of areas is equal to the square of the ratio of the diameters. Note that we do not need to convert the millimetres to metres in the case of a ratio. Therefore

The force on the small piston $= f = (350 \quad 9.81) \quad (50/250)^2$

$= 137.3$ N

Pressure on an immersed surface

We now need to look at fluid forces in a more general way. Suppose that you were required to design a watertight housing for an underwater video camera that is used to traverse up and down the legs of a giant oil rig looking for structural damage, as shown in Figure 3.1.17.

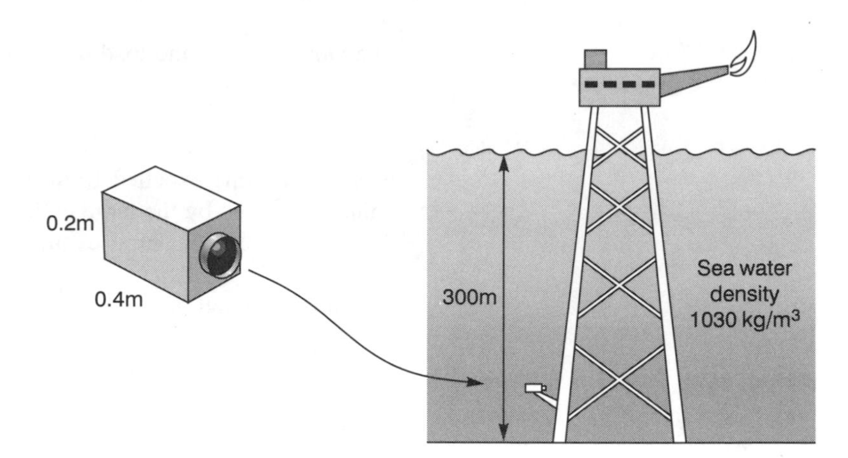

Figure 3.1.17 *An inspection system on an oil rig*

In order to specify the thickness, and hence the strength, of the housing material, you would need to know the maximum force that each face of the housing would be likely to experience. This means that the pressure on the camera at maximum depth would need to be known, using Equation (3.1.1)

$$p = \rho g h$$

The maximum pressure on the camera is therefore:

$$p_{\text{max}} = 1030 \quad 9.81 \quad 300 = 3031\,\text{kPa}$$

The total force on one of the large faces of the camera housing would then be equal to this pressure times the area of the face.

$$F = 3031\,\text{kPa} \quad (0.2 \quad 0.4)\,\text{m}^2$$
$$= 242.5\,\text{kN}$$

This is clearly a very simple calculation and the reason why we do not have to employ anything more complicated is that the dimensions of the camera housing are tiny compared with the depth. This means that we do not need to consider the variation of pressure from top to bottom of the housing, or the fact that the camera might not be in an upright position.

This is therefore a very special solution which would not apply to something like a lock gate in a shipping canal, where the hydrostatic pressure is enormous at the bottom but close to zero at the top of the structure close to the surface of the liquid. We must look a little further to find a general solution that will accurately predict the total force in *any* situation, taking into account the variation of *all* the factors.

Consider the vertical rectangular surface shown in Figure 3.1.18; we need to add up all the pressure forces acting over the entire area in order to find the total force F. This means that we will have to carry out an integration, and so the first step is to identify a suitable area element of integration.

We know that pressure varies with depth, but it is constant at a constant depth. Therefore if we use a horizontal element with an infinitesimal height dy, we can treat it just like the underwater camera above and say that the pressure on it is a constant.

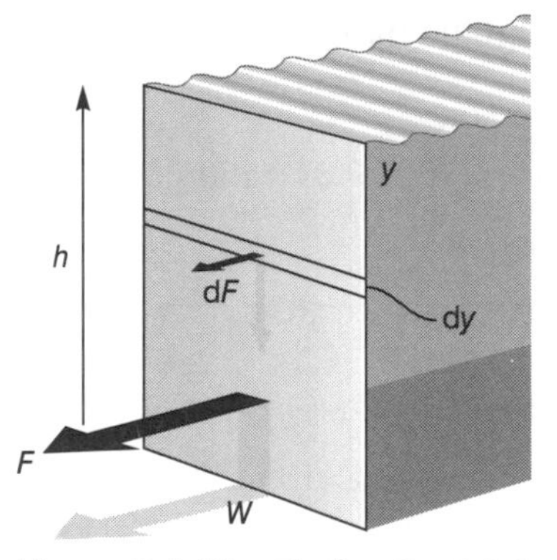

Figure 3.1.18 *Finding the total force on a submerged plane*

The pressure at a depth y is:

$$p = \rho g y$$

Hence the small force dF is given by (pressure area)

$$\mathrm{d}F = (\rho g y)\ (w\ \mathrm{d}y)$$

The total force is now found by integrating from top to bottom.

$$F = \rho g w \int_0^h y\ \mathrm{d}y$$

$$F = \rho g w h^2/2$$

This is a final answer for this particular problem but we can interpret it in a more meaningful way as

$$F = (\rho g h/2)(wh)$$

Any force can be thought of as a pressure times an area, so by taking out the total area of the flat rectangle wh we must be left with a 'mean' pressure as the other term. Therefore $\rho gh/2$ must be an average value of pressure which can be thought to act over the total area. In fact it is the pressure which would be experienced half way down the rectangle, i.e. at the centroid or geometrical centre.

Therefore:

$$F = \bar{p}\ \ A \tag{3.1.4}$$

where $\bar{p}$ is the pressure at the centroid.

It turns out that this expression is completely general for a flat surface and we could apply it to triangles, circles, etc.

Location of the hydrostatic force

Having calculated the size of the total hydrostatic force it would be an advantage now if we could calculate *where* it acts so that we could treat fluid mechanics as we would solid mechanics. The total force would then be considered to act at a single point, rather like the distributed weight of a solid body that is taken to act through the centre of gravity. This corresponding single point for a fluid force is the *centre of pressure* (Figure 3.1.19). For the case of the vertical rectangle with one edge along the surface of the liquid, as used above, this can be found quite easily.

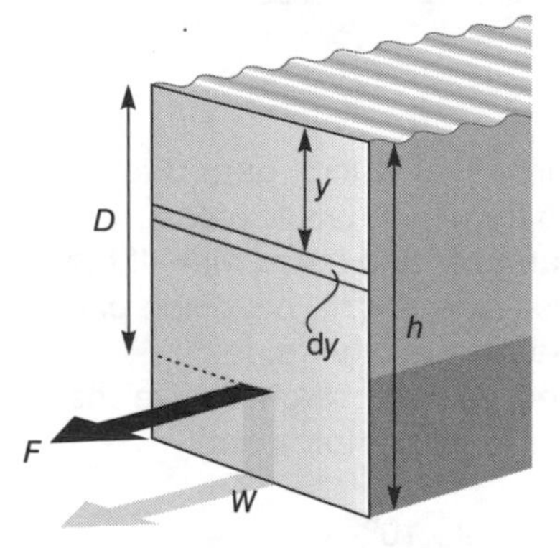

Figure 3.1.19 *Finding the centre of pressure*

Again we need to use integration, but this time we are interested in taking moments because that is the way in which we generally locate the position of a force. We will use the same element of integration as in the first case, and will consider the small turning moment dT caused by the small force dF about the top edge.

$$\mathrm{d}T = y\ \mathrm{d}F$$

$$= y(\rho g y)(w\,\mathrm{d}y)$$

Therefore integrating

$$T = \rho g w \int_0^h y^2 \, dy$$

$$= \rho g w h^3/3$$

This is the total torque about the top edge due to the hydrostatic force. We could also have calculated it by saying that it is equal to the total force F (calculated earlier) multiplied by the unknown depth D to the centre of pressure.

Therefore

$$T = (\rho g w h^2/2)D$$

By comparison with the expression for T found above, it can be seen that

$$D = 2h/3$$

Figure 3.1.20 *Also finding the centre of pressure*

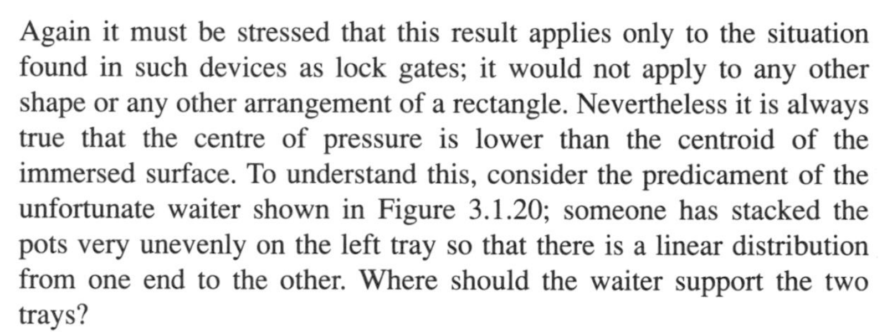

Again it must be stressed that this result applies only to the situation found in such devices as lock gates; it would not apply to any other shape or any other arrangement of a rectangle. Nevertheless it is always true that the centre of pressure is lower than the centroid of the immersed surface. To understand this, consider the predicament of the unfortunate waiter shown in Figure 3.1.20; someone has stacked the pots very unevenly on the left tray so that there is a linear distribution from one end to the other. Where should the waiter support the two trays?

The answer for the evenly stacked tray is easy – he should support it in the middle. The other tray must be supported closer to the heavily stacked end, and in fact the exact position is two-thirds of the way from the lightly stacked end. This is just the same sort of situation as the lock gate, but turned through 90° so that the load due to the water increases linearly from the top to the bottom. If we had to support the gate with a single horizontal force then we would need to locate it two-thirds of the way down, showing that this must be the depth at which all the force distribution due to the water appears to act as a single force.

Example 3.1.4

(a) The maximum resultant force that a lock gate (Figure 3.1.21) can safely withstand from hydrostatic pressure is 2 MN. The lock gate is rectangular and 5.4 m wide. If the height of water on the lower side is 4 m, how high can the water be allowed to rise on the other side?

(b) When this depth is achieved, at what height above the bottom of the gate does the resultant force act?

(a) Resultant force R is $F_2 - F_1 = 2 \quad 10^6$

So $F_2 = F_1 + 2 \quad 10^6$

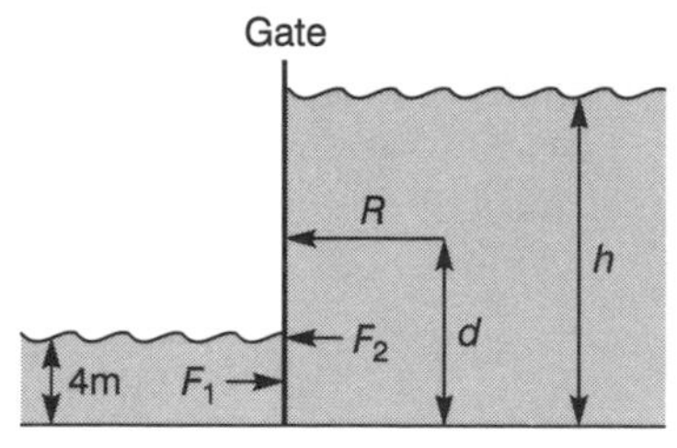

Figure 3.1.21 *A side view of a lock gate*

Now F_1 = mean pressure area

= $(\rho g4/2)$ $(4$ $5.4)$

= 1000 9.81 2 4 5.4

= 423.8 kN

Therefore F_2 = 2 423 800 N

We could also express F_2 as $(\rho gh/2)$ $(h$ $5.4)$

and so 1000 9.81 5.4 $h^2/2$ = 2 423 800

giving h = 9.57 m

(b) Take moments about the base of the gate: the total moment due to the water trying to push the gate over is 2 10^6 d, but this could also be calculated from the individual moments due to the forces F_1 and F_2.

Therefore

2 10^6 $d = (F_2$ $9.57/3) - (F_1$ $4/3)$

= (2 423 800 9.57/3) – (423 800 4/3)

which finally gives

d = 3.59 m

Properties of some common fluids

Fluid	*Density* ρ *kg/m*^{3}	*Relative density RD*	*Dynamic viscosity* μ *Pa s or kg/m/s*
Air	1.3	1.3 10^{-3}	1.85 10^{-5}
Water	1000	1	0.001
Mercury	13 600	13.6	1.55 10^{-3}
Mineral oil	900	0.9	8 10^{-4}

Hydrostatic forces on curved surfaces

In the previous section we found out how to calculate the total hydrostatic force acting on any immersed flat surface, using the equation

$$F = \bar{p}\ \ A$$

We now have to consider the way to calculate the force acting on a curved surface, because many engineering applications such as fuel tanks, pipes and ships involve shapes which cannot be constructed from flat surfaces. Clearly the simple equation above cannot be used without some modification because we need to take into account the fact that pressures will act in different directions at different points on a curved surface. (Hydrostatic pressure always acts perpendicularly to the immersed surface, so as the surface curves around then the direction of the perpendicular must also change.)

This sounds as though we are going to need to employ some fairly complicated integration to add up the effect of the variation of pressure with depth *and* direction, but fortunately there are some short cuts we can take to make this into quite a simple problem.

Resolving the force on a curved surface (Figure 3.1.22)

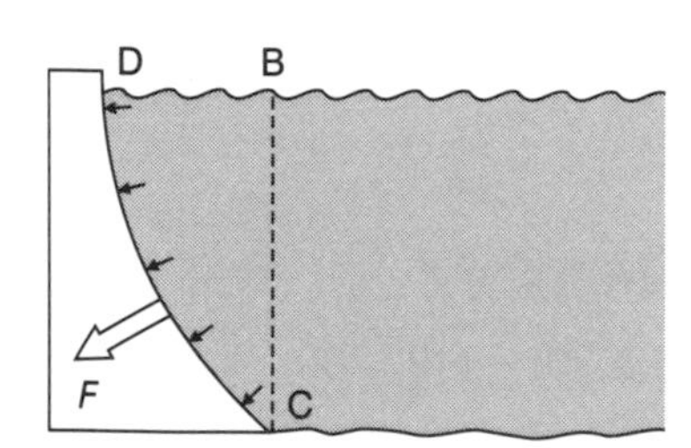

Figure 3.1.22 *Forces on a curved surface*

The key to this simple treatment is the resolution of the total force into *horizontal* and *vertical* components which we can then treat completely separately. Suppose we have the problem of calculating the total hydrostatic force acting on a large dam wall holding the water back in a reservoir.

We are looking for the single force F which represents the sum of all the little forces acting over all the dam wall. We can find this if we can calculate separately the horizontal and vertical forces acting on the wall due to the water. Let us begin with the vertical force.

Vertical force component

The vertical force caused by a static liquid which has a free surface open to the atmosphere results solely from gravity. In other words, the vertical force on this dam wall is simply the weight of the water pressing down on it. Clearly we only need to consider any water which is directly above any part of the wall, i.e. in the volume BCD, since gravity always acts vertically. Any water beyond BC only presses down on the bed of the reservoir, not on the wall.

$$F_{\text{vertical}} = \text{weight of water } W = (\text{volume BCD}) \quad \rho \quad g$$

Calculation of the volume BCD may seem complicated but in engineering the shapes usually have clearly defined profiles such as ellipses, parabolas or circles, which greatly simplify the problem.

Horizontal force component

The horizontal force component is what is left of the total force once we have removed the vertical component. It is therefore exactly the same as the force acting on a flat vertical wall of the same height and width. All we need to do then is calculate the force which would act on an imaginary vertical wall erected along BC, using the method developed in the previous section.

$$F_{\text{horizontal}} = \bar{p} \quad A$$

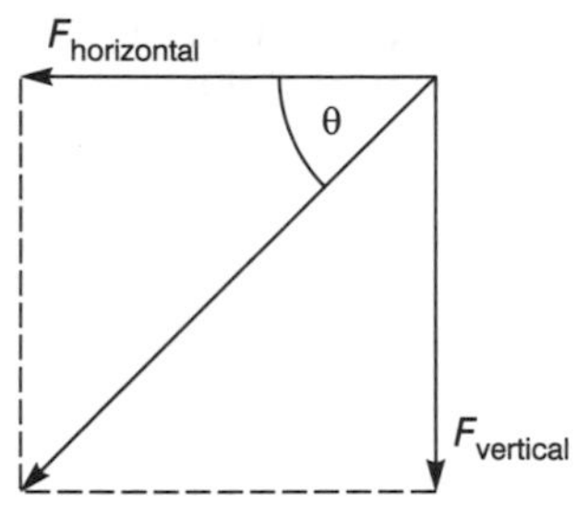

Figure 3.1.23 *Rectangle of forces*

Total force (Figure 3.1.23)

Once the two components have been found it is a simple matter to calculate the single total force acting on the dam wall by recombining them:

$$F_{\text{total}} = \sqrt{(F_{\text{horizontal}}{}^2 + F_{\text{vertical}}{}^2)}$$

It is also possible to calculate the direction of the total force to the horizontal or vertical:

$$\theta = \tan^{-1} (F_{\text{vertical}}/F_{\text{horizontal}})$$

Example 3.1.5

Fuel of relative density 0.8 is stored in a reservoir which is a tank of overall depth 5 m. One end of the tank is composed of two sections: a flat vertical plate of width 1.5 m and height 3 m, and below that is a curved plate that is a quadrant of a horizontal cylinder of radius 2 m. The two plates are welded together so that the curve bulges into the reservoir. If the tank is full to the brim, calculate the total liquid force on the curved plate.

First of all, we shall calculate the vertical force pressing down on the curve. This is the weight of fuel directly above the curve and so the first step is to work out the volume of this liquid. The curve protrudes into the tank by a distance of 2 m horizontally, i.e. its radius. Therefore the volume is the rectangular volume

$5 \quad 1.5 \quad 2 = 15\,\text{m}^3$

minus the volume of the quadrant

$\pi \quad 2^2 \quad 1.5/4 = 4.712\,\text{m}^3$

Hence the total volume over the curve is $10.288\,\text{m}^3$ and the weight pressing down is

$4.712 \quad 0.8 \quad 1000 \quad 9.81 = \underline{36.98\,\text{kN}}$

Now we shall calculate the horizontal force acting on the curve. This is the force which would act on a flat vertical plate of the same width and height as the curved plate. Therefore this force is the product of the flat plate area and the mean pressure acting on it:

$(1.5 \quad 2) \quad (0.8 \quad 1000 \quad 9.81 \quad [3 + 2/2]) = \underline{94.18\,\text{kN}}$

Note that the mean pressure is calculated half way down the flat plate that replaces the curved plate. The top of this is already at a depth of 3 m and so the total depth to the centroid works out to be 4 m.

We have now worked out the vertical and horizontal components of the hydrostatic thrust on the curved plate. It simply remains to combine them using Pythagoras' theorem to find the resultant force F.

$$F = \sqrt{(36.98^2 + 94.18^2)} = \underline{101.2\,\text{kN}}$$

Upthrust

Sometimes it is necessary to calculate the total force on a curved surface which is arranged as in Figure 3.1.24.

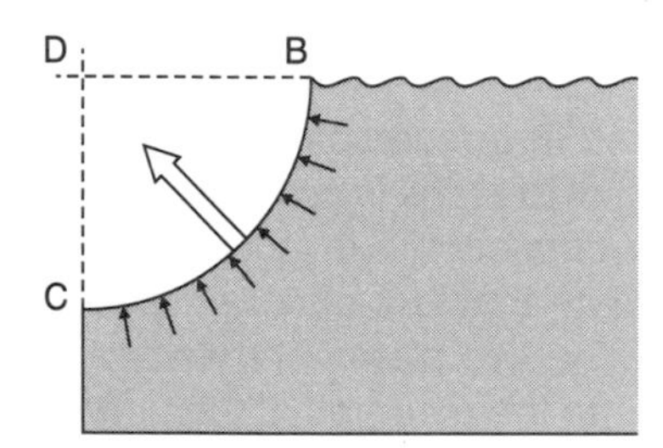

Figure 3.2.24 *Upthrust on a submerged object*

In this case the small pressure forces all have an upwards component, unlike the example on the dam wall. It would seem that we can no longer use the simplification of calculating the vertical force component from the weight of liquid directly above the curved surface because, clearly, there is no liquid directly above to produce any weight.

However, suppose that the curve had not been there and so the end wall was simply a flat surface; the liquid would then occupy the extra volume BCD, and this liquid would be supported by the liquid underneath the curve BC. The liquid beneath BC must therefore be capable of exerting an upwards vertical force to balance the weight of the liquid that we have imagined to be filling the volume BCD. Hence the vertical force on the real curved surface BC is an upthrust whose magnitude is equal to the weight of the liquid which could be put into volume BCD.

This type of problem is therefore simply the inverse of the problem concerning the dam wall, and the calculation method is identical except that the vertical force will be upwards.

Archimedes' principle

Imagine that we now have two surfaces joined back to back to form the hull of a ship (Figure 3.1.25).

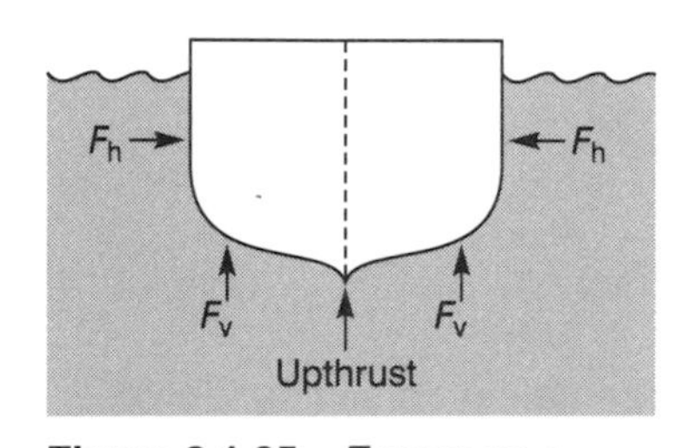

Figure 3.1.25 *Forces on a ship's hull*

The total force on the left side of the hull can be split up into a horizontal component and a vertical component, as can the total force on the right side. Each vertical component is an upthrust equal to the weight of liquid which could be put into half of the hull, up to the outside water level. Each horizontal component is the force which would act on a flat vertical wall with the same area as a side projection of the hull.

If we are interested in the total force on the whole ship, then the two horizontal components can be ignored because they are of equal magnitude and opposite direction, so they cancel each other out. The total force on the ship is therefore simply the sum of the two upthrusts, each one of which is equal to the weight of water which would fit into one half of the hull up to the waterline. In other words:

The upthrust on an immersed object is equal to the weight of liquid displaced.

This is known as Archimedes' principle after the famous Greek philosopher who is said to have discovered it. It shows why large ships made out of very dense materials such as steel can nevertheless float if they are made large enough to displace a sufficient volume of water, contrary to the belief which was popularly held at one time that ships could only be made from materials such as wood which were less dense than water. It is worth noting that large crowds attended the launching of the first steel-hulled ship built by Brunel, fully expecting to witness a disaster caused by building it from a material which would not float.

It is also worth noting that for objects such as submarines which become completely submerged and can go to great depths, there is no increase in the upthrust with depth. Once a submarine is completely under the surface of the water, the upthrust does not get any bigger no matter how deep the vessel may dive since it only depends on the weight of water displaced.

Example 3.1.6

A crane is to lower a vertical cylindrical pillar, of diameter 1.2 m, into position on a plinth submerged in water. The pillar has a mass of 5 tonnes. What will be the tension experienced in the supporting cable when the lower end of the pillar is submerged to a depth of 2.3 m?

We need to calculate the weight of water displaced by the partly submerged pillar in order to find the upthrust and hence calculate the resultant tension in the support cable. The starting point is to calculate the volume of the pillar that is submerged. This is the volume of a cylinder of radius 0.6 m and length 2.3 m.

Volume of pillar submerged $= \pi 0.6^2 \quad 2.3 = 2.60\ \text{m}^3$

Weight of water displaced $= 2.60 \quad 1000 \quad 9.81$

$= 25.506\ \text{kN}$

The tension in the cable
with the pillar in air $= 5 \quad 1000 \quad 9.81$

$= 49.050\ \text{kN}$

Hence the resultant tension in the cable with the pillar partly submerged is:

$49.050 - 25.506 = \underline{23.54\ \text{kN}}$

Buoyancy

This phenomenon of solid objects being able to float is part of a larger subject area known as buoyancy that includes examples of submerged objects which are not allowed to float but nevertheless experience a buoyancy force because of a surrounding liquid. It is one of the most important subjects studied by marine architects because it decides the stability of a ship. It can also become very complex, but it is worth taking a quick look at some of the main features.

The upthrust, or buoyancy force, is a summation of all the small upward forces acting on an immersed object. The usefulness of this summation lies in being able to think of the upthrust as a single force acting at a single point. The location of this single point must be the centre of gravity of the displaced liquid since only then can the upthrust fully cancel out the lost downward force. The point at which the buoyancy force acts is called the *centre of buoyancy.*

Stability of a floating body

Have you ever wondered why the passenger basket on a hot air balloon is suspended underneath and not simply strapped to the top of the balloon? After all, most hot air balloons are used for sightseeing and a basket on the top would give much better views.

Consider the balloon shown in the left-hand diagram of Figure 3.1.26. The balloon floats because the hot air inside it is less dense than the cold air surrounding it, giving rise to a buoyancy force acting upwards through B. When this force equals the total weight of the balloon and basket, acting through the centre of gravity G, the balloon will float at a constant altitude. As the wind changes and the occupants of the basket move around, the balloon will rock through a small angle θ. Since the centre of buoyancy is higher than the centre of gravity, any angular displacement produces a turning moment which acts to *restore* the balloon to an upright position. Such an arrangement is said to be in *stable equilibrium.*

Now look at the bizarre case in the right-hand diagram. The buoyancy force again equals the weight, but here any angular displacement causes a turning moment which makes the basket topple over. The reason for this is that the centre of buoyancy B is below G. The situation is known as *unstable equilibrium.*

Something very similar applies to ships, but there are cases where stable equilibrium can be achieved even where the centre of buoyancy is below the centre of gravity. This occurs because the shape of the

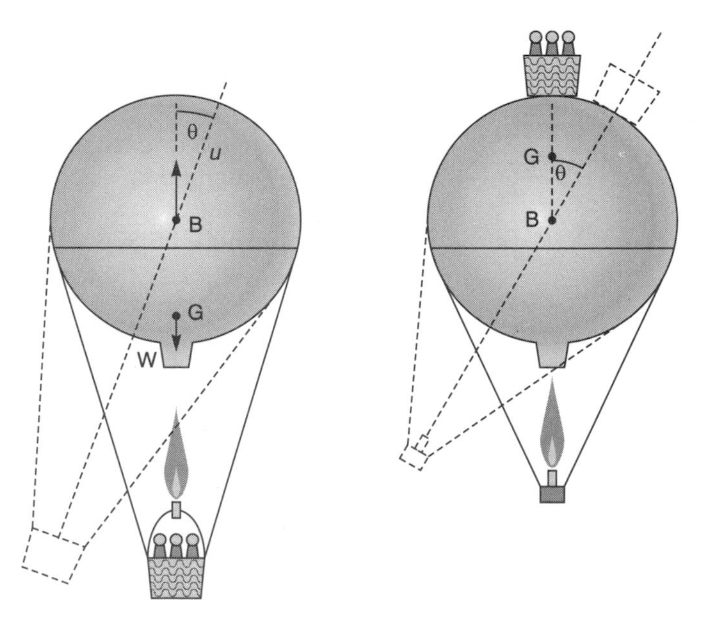

Figure 3.1.26 *Two hot air balloons – which one is safe?*

displaced water alters as the ship rocks and so the centre of buoyancy moves sideways in the same direction as the ship is leaning. Therefore the line of action of the buoyancy force also moves to the side of the ship which is further down in the water, and the buoyancy force tries to lift the ship back to the upright position.

Whether or not the restoring moment is enough to make the ship stable depends on the position of the point where the line of action of the buoyancy force crosses the centreline of the ship, known as the *metacentre, M* (Figure 3.1.27).

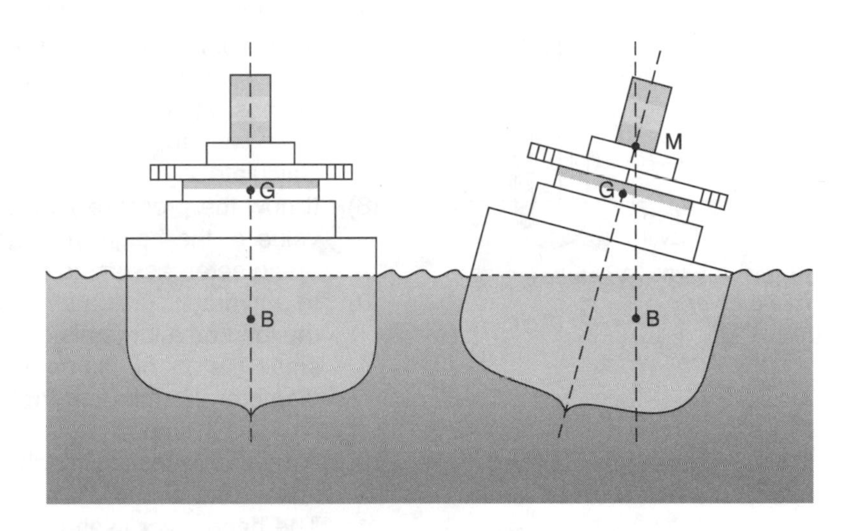

Figure 3.1.27 *Stability of a ship*

The distance between G and M is known as the *metacentric height*.

If M is above G then the metacentric height is positive and the ship is in stable equilibrium.

If G is above M then the metacentric height is negative and the ship is in unstable equilibrium. This is the situation which led to the sinking of King Henry VIII's flagship, the *Mary Rose*, off Portsmouth. This had sailed successfully for a number of years and was just stable as it cast off on its fateful last voyage, even though an unusually large shipment of weapons and soldiers had raised the centre of gravity to danger level. Finally, when the soldiers crowded up onto deck for a last glimpse of land as the ship put out to sea, the centre of gravity rose so high that the first big wave they encountered away from the shelter of the harbour caused the ship to topple completely over.

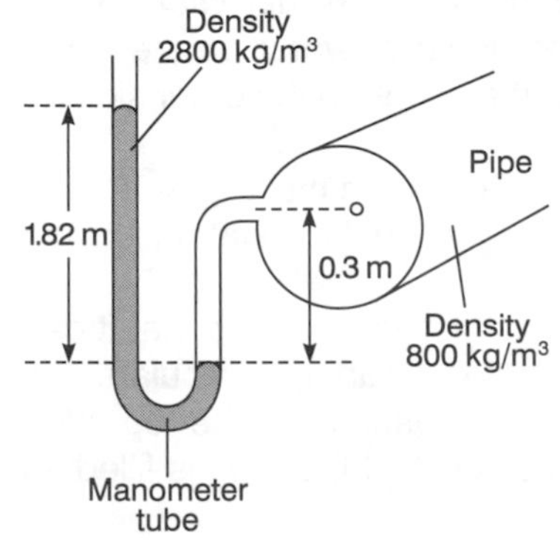

Figure 3.1.28 *A U-tube manometer*

Problems 3.1.1

(1) Find the height of a column of water which will produce a pressure equal to that of the atmosphere.

$p_{\text{atmosphere}} = 101.4\,\text{kPa}$

(2) Find the equivalent height to Problem 1 for a column of mercury.

(3) Referring to Figure 3.1.28, find:

(a) the pressure head in the pipe;
(b) the pressure in the pipe;

(4) Find the difference in pressure head between two points on a horizontal pipe containing water if a mercury U-tube differential manometer gives a reading of 200 mm.

(5) What is the *pressure* difference between the two points in Problem 4?

(6) A mercury U-tube manometer indicates a reading of 135 mm when connected to a pipeline at points A and B, both at the same level but at some distance apart. If the pipe contains oil of density 800 kg/m^3, what is the pressure difference between A and B?

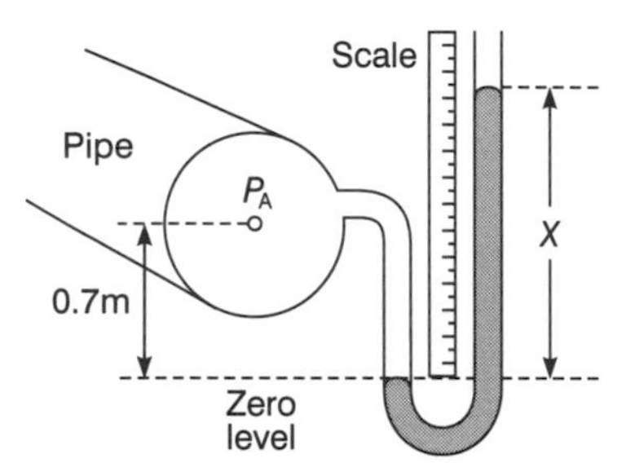

Figure 3.1.29 *A U-tube manometer*

(7) A pipe contains water at a pressure of p_A which is equivalent to a head of 150 mm of water (Figure 3.1.29). The pressure recorded on the manometer filled with liquid of density 2800 kg/m^3 gives a scale reading of x on the right side, with the left-hand level on the zero of the scale. Calculate x.

(8) If now the pressure p_A increases to 2.0 kPa, find the new value of the height reading for the right-hand side of the manometer, assuming the scale is not moved.

(9) In a hydraulic jack a force is applied to a small piston to lift the load on a large piston. For a hydraulic jack which has a small piston of diameter 15 mm and a large piston of diameter 200 mm, determine the force required to lift a car of mass 1 tonne.

(10) A hydraulic testing machine is designed to exert a maximum load of 0.5 MN and utilizes a hand-operated pump. The hand lever itself has a ratio of 8:1 between the force output and the force exerted by the hand. The lever operates a piston of 20 mm diameter. The test head on the machine is operated by a hydraulic piston of 300 mm diameter. The two pistons are connected by a hydraulic pipe and the whole system contains a hydraulic oil. For maximum load, determine the pressure in the hydraulic oil and the force which must be applied to the lever by hand.

(11) A tank 1.5 m high, 1 m wide and 2.5 m long is filled with oil of relative density 0.9. Calculate the total force acting on:

(a) one of the small sides;
(b) one of the large sides;
(c) the base.

(12) A rectangular lock gate 3.6 m wide can just withstand a resultant force of 1 MN. If the depth of water on the lower side is 3 m, to what depth can the water on the other side be allowed to rise?

(13) Calculate the hydrostatic thrust acting on the large side of a fish tank of length 2.8 m and height 1 m if the water is 85 cm deep.

If this side is hinged along its bottom edge so that it can be swung forward and down for cleaning, calculate the horizontal force experienced by a fastener at the top edge when it is holding the side in place and the tank is filled as above.

(14) Find the resultant hydrostatic force on two vertical plates used as gates to two separate pipes leading from a

reservoir. In each case the centre of the plate is 1.2 m below the surface of the water:

(a) a square plate of side 1.8 m;
(b) a circular plate of diameter 1.8 m.

(15) A square section, open irrigation channel of width 0.85 m is normally sealed at the outlet end by a vertical square flap which will swing downwards about an axis along the bottom edge of the channel when the hydrostatic torque is greater than 420 Nm. To what depth can the channel be filled before this flap opens and releases the water to the fields?

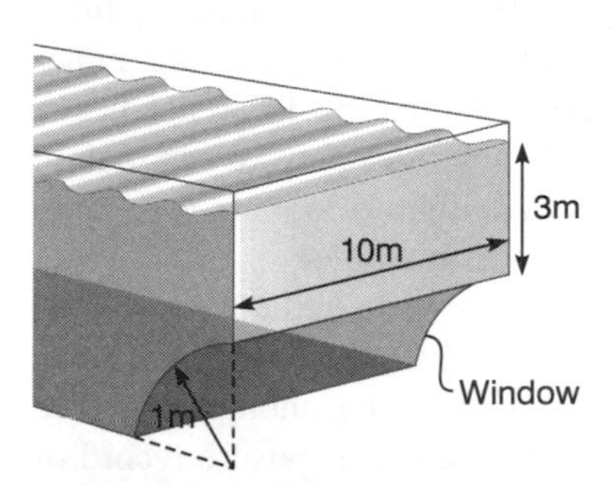

Figure 3.1.30 *Swimming pool with window*

(16) A swimming pool is to have an underwater window consisting of a 10 m length quadrant, of radius 1 m, as shown in Figure 3.1.30. Calculate the magnitude and direction of the resultant hydrostatic thrust on the window.

(17) A sluice gate (Figure 3.1.31) consists of a quadrant of a cylinder of radius 1.5 m, pivoted so that it can be rotated upwards to open. Its centre of gravity is at G, its mass is 6000 kg and its length (into the paper) is 3 m. Calculate:

(a) the magnitude of the resultant force on the gate due to the water;
(b) the turning moment required to start to lift the gate.

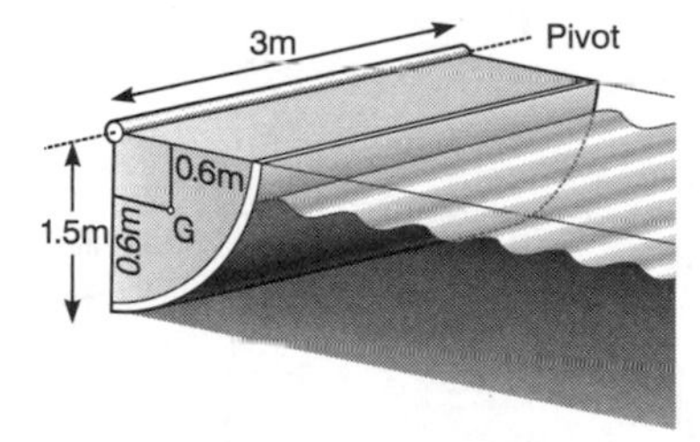

Figure 3.1.31 *Sluice gate*

(18) A 10 m deep dolphin pool is to be fitted with an observation tunnel on the horizontal base of the pool so that visitors can see the dolphins from underneath (Figure 3.1.32). The tunnel is made from plate glass panels of radius 2.4 m and width 0.9 m. If the pool is filled with seawater of relative density 1.06, calculate the magnitude and direction of the hydrostatic thrust on one panel.

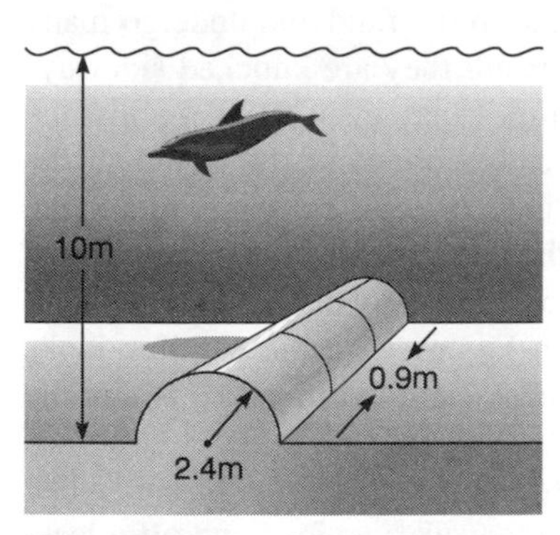

Figure 3.1.32 *A dolphin pool with a viewing tunnel*

(19) A cruise liner is to be fitted with a hemispherical glass observation window, which will be let into the side of the hull below the waterline. The radius of the hemisphere will be 2.6 m and its top edge will be 3 m below the water surface. If the sea water has a density of 1058 kg/m^3, calculate the horizontal and vertical hydrostatic forces acting on the window.

(20) A tank of water of total mass 6 kg stands on a set of scales. An iron block of mass 2.8 kg and relative density 7.5 is lowered into the water on the end of a fine wire hanging from a spring balance. When the block is completely immersed in the water, what will be the reading on the scales and on the balance?

3.2 Hydrodynamics – fluids in motion

This section deals with hydrodynamics – fluids in motion. There are two important concepts that apply here: conservation of energy and Newton's second law. Both of these will be familiar to the reader from earlier studies and so the treatment here concentrates on their application to fluids. The section starts with a detailed description of the types of fluid flow in order to build an understanding of the way in which energy can be lost by a fluid to its surroundings.

Fluid flow

Much of fluid mechanics for engineers is concerned with flowing liquids: pumped flow of fuel and lubricants, flow of water through generators, etc. To understand how we can analyse these flows we must first be able to recognize different types of flow, and fortunately there are only two important types. Flow of a fluid is generally either *streamlined* or *turbulent*. The key to understanding the difference lies in remembering that the fluid is made up of molecules and we can look in detail at the way in which they move in addition to looking on a large scale at the way in which the fluid moves as a whole.

Streamlined flow (Figure 3.2.1) is when a particle in the liquid can be traced and shown to move along a smooth line in the direction of flow. Suppose we could take a series of very high-speed photographs of a single particle in the fluid and build up a multiple exposure record of what the particle does. In streamlined flow we would see that the particle followed exactly the line of flow of the liquid, as in this illustration of a particle flowing along in a tube. In this case the line of flow of the fluid is known as a streamline because the individual particles stream along it.

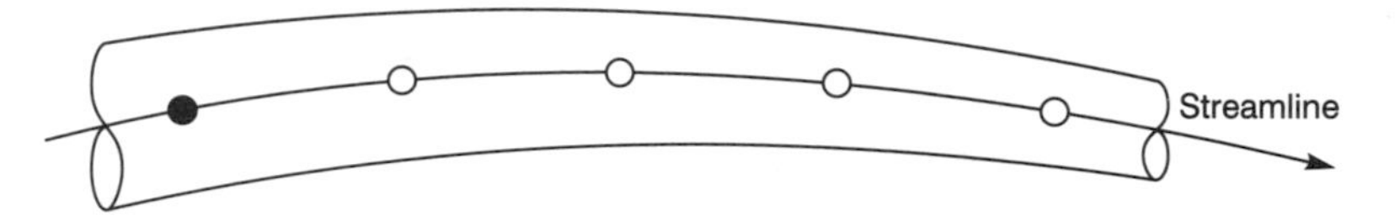

Figure 3.2.1 *Streamlined flow*

Turbulent flow is when the individual particles do not follow any regular path but their overall motion gives rise to the liquid flow. The particles are clearly being stirred around in the fluid and undergo many collisions with their colleagues. As a result they are knocked sideways from the direction of flow, and sometimes may even travel against the direction of flow for a short distance. Nevertheless the liquid in the example shown below is still flowing in the same direction as that in the example above and so we know that the particles must ultimately travel along the tube from left to right. The route taken by the fluid at any point is called a pathline. In the example in Figure 3.2.2 the pathline looks identical to the streamline in Figure 3.2.1, but the difference is that the individual particles do not follow it.

The difference between the two extreme types of flow is rather like the difference between taking a dog for a walk on a lead and off a lead. When the dog is on the lead it follows exactly the same route as its owner but displaced slightly to one side, just like two adjacent molecules in laminar or streamlined flow. When the dog is let off the lead on a country path it will race around exploring its surroundings, sometimes going back down the path for a short distance, but overall it follows the same route as its owner. This is like turbulent flow.

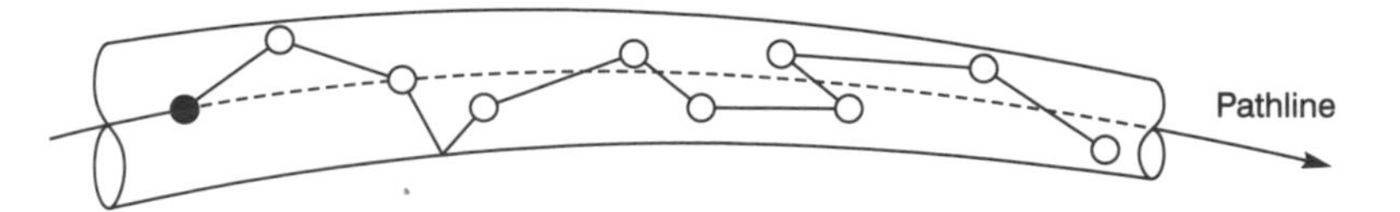

Figure 3.2.2 *Turbulent flow*

Since these two diagrams look so similar and we can only tell the detailed molecular pictures apart with the aid of a high-speed camera connected to a high power microscope, how can we predict when we will get one kind of flow and not the other?

Osborne Reynolds

This question was the big thing on the mind of a Manchester scientist called Osborne Reynolds who became involved in the development work for one of the greatest Victorian engineering projects of the late nineteenth century, Tower Bridge in London. At the time this was conceived as an enormous advance in engineering terms since it involved the bridge's roadway being divided into two halves which could be lifted like drawbridges at the entrance to a castle. The bearings required for supporting the roadways as they were raised when a ship needed to pass into the Pool of London were many times larger than any that had been constructed up to that point for applications such as railway locomotives. Suddenly engineers realized that they had no real idea how fluid bearings worked in detail, largely because they did not understand in detail how fluids flowed, and so nobody could design the new giant bearings. This is where Reynolds came to the rescue, eventually producing a long equation which described how fluid bearings can generate enough pressure to keep the metal surfaces apart. He began, however, by carrying out some simple experiments on fluid flow.

The apparatus he used was something like that shown in Figure 3.2.3.

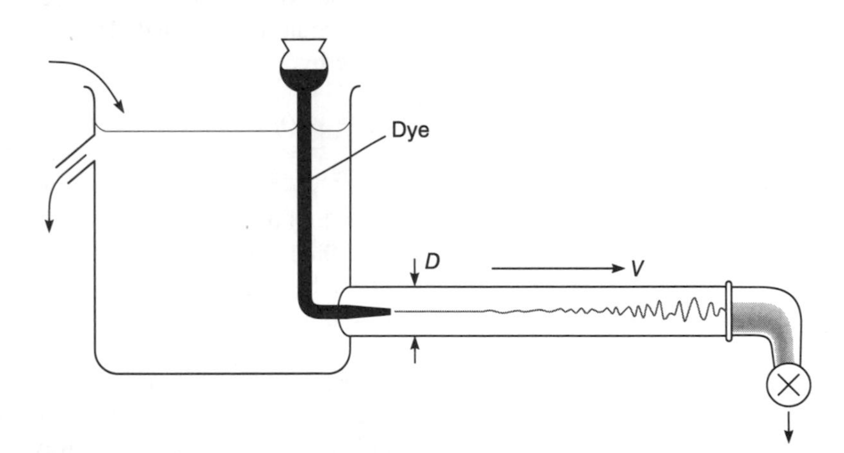

Figure 3.2.3 *Reynolds' experiments on flow*

Reynolds filled the large tank with a range of liquids of different densities and viscosities and allowed them to flow at different rates by adjusting the valve at the end of the glass tube. He also could change the glass tube for others of different diameters. The clever part was to mark the flow by injecting a dye into the entrance of the glass tube via a funnel connected to a fine jet.

Reynolds found that if the liquid velocity was low then the fine stream of dye remained undisturbed all the way along the glass tube. He had discovered streamlined flow (Figure 3.2.4), because the fact that the dye did not get mixed up showed that the fluid itself was flowing very

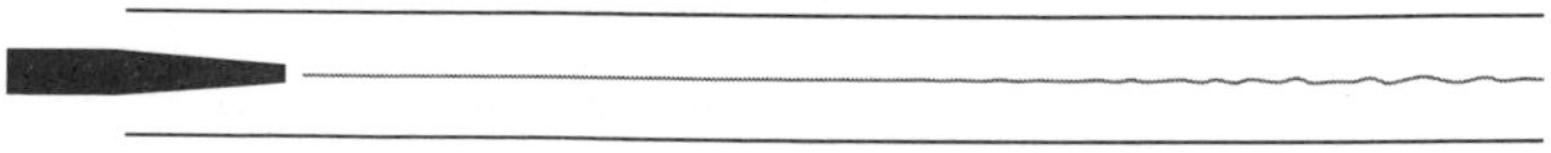

Figure 3.2.4 *Streamlined flow*

smoothly. This type of flow is also known as laminar flow, from a Latin origin which means that it is in distinct layers. Reynolds obtained particularly good results when the tube was of small diameter and the liquid was very viscous.

If the liquid velocity was high, because the valve was opened wide, the dye was stirred up almost as soon as it left the jet on the end of the funnel. This effect was strongest when the tube diameter was large and the liquid had a low viscosity. This was turbulent flow (Figure 3.2.5) and the mixing was caused by the fact that the molecules themselves were flowing along in a very agitated manner.

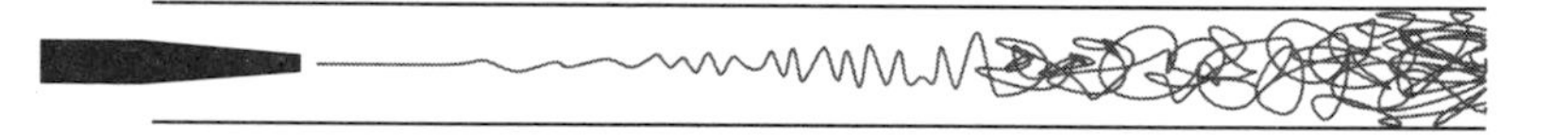

Figure 3.2.5 *Turbulent flow*

In between these two extremes the transition from smooth dye streams to stirred-up dye took place at different lengths along the glass tube as one type of fluid flow gave way to another. Reynolds managed to make sense of the different combinations of velocities, tube diameters, viscosities and even liquid densities (although density does not generally vary much from one liquid to another) by coming up with the idea of a dimensionless quantity which determined what kind of flow could be expected. This is now known as the Reynolds number, given by:

$$Re = \rho v D/\mu \qquad (3.2.1)$$

where:

ρ = fluid density
v = fluid velocity
D = tube diameter
μ = fluid viscosity.

Streamlined flow occurs at low Reynolds numbers, while turbulent flow occurs at high Reynolds numbers. The transition from streamlined to turbulent flow occurs at a Reynolds number of about 2000. This transition can be illustrated by looking at the results of measuring the differential pressure applied to the two ends of a pipe to produce an increasing flow rate (like plotting voltage against current for an electrical circuit) (Figure 3.2.6).

Within the linear, streamlined region at low velocities the flow rate can be doubled by doubling the pressure applied. However, within the turbulent region the flow rate can only be doubled by *quadrupling* the pressure. This gives us one of the big conclusions to be drawn from Reynolds' work:

Turbulent flow requires large applied pressures and that can lead to high energy losses.

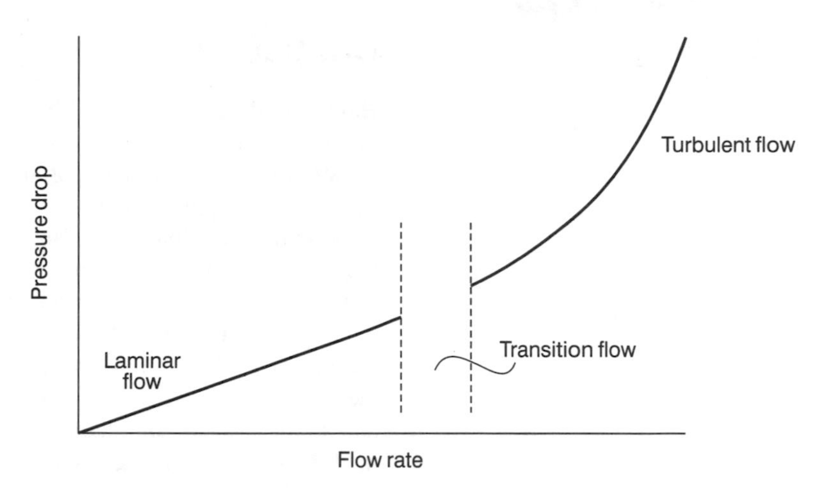

Figure 3.2.6 *Pressure/flow characteristics for flow along a pipe*

This is a big generalization and sometimes turbulent flow is an advantage (e.g. dishwashers use high velocity jets of water to create turbulent flow because the jets of swirling liquid achieve a better cleaning action than a smooth jet of streamlined flow). Nevertheless it is usually advantageous to try to achieve as smooth a flow pattern as possible. Hence there is great emphasis on 'streamlining' of cars, for example. Different definitions of Reynolds number are required to overcome the fact that airflow around a car is not the same as flow in a tube, but the principle is the same.

Figure 3.2.7 shows the idea behind the spoilers which are attached to the tailgate or boot of some cars.

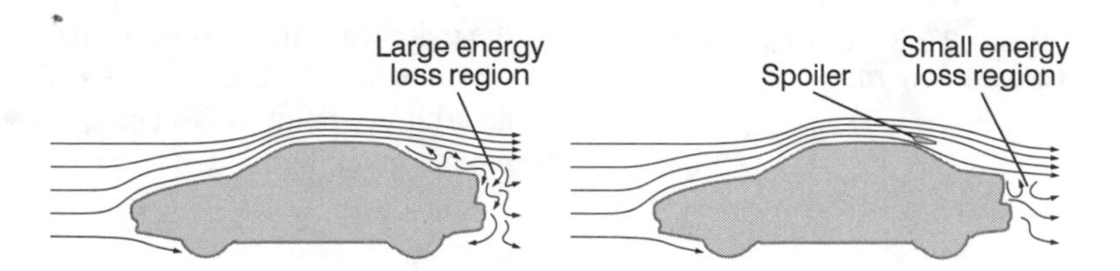

Figure 3.2.7 *The advantages of 'streamlining'*

Example 3.2.1

Liquid of density 780 kg/m^3 and viscosity 0.125 Pa s is flowing along a pipe which has a diameter of 300 mm. Find the approximate flow velocity at which the flow will change from streamlined to turbulent.

The transition from streamlined to turbulent flow takes place approximately at a value of Reynolds number of 2000. Using Equation (3.2.1).

$$\rho vD/\mu = 780\ v\ 0.3/0.125 = 2000$$

$$v = 2000\ \ 0.125/(780\ \ 0.3)$$

$$= \underline{1.07\ \text{m/s}}$$

Continuity

Having studied the work of Reynolds we are now ready to apply some of our knowledge to the flow of real liquids along pipes since this is the situation which is most important to mechanical engineers. For most practical applications the flow of liquids along pipes is *turbulent* because the combination of values which have to be put into Reynolds number leads to it being well over 2000. For example, consider water flowing along a pipe in your home. The density of water is 1000 kg/m^3 and its dynamic viscosity is approximately 0.001 pascal seconds, as shown in the table of fluid properties in Section 3.1 (page 127). Together these two terms account for a total of one million in Reynolds number and so the velocity and diameter would need to be very small to bring the final value down to less than 2000. Typically the water pipe diameter in your home would be 12 mm (i.e. 0.012 m for substitution into any calculation) and the water would leave a fully opened tap at a speed of about 1 m/s. The final value for Reynolds number in this situation is therefore about 12 000 and so the flow is very turbulent. This means that the liquid is stirred around as it flows along a pipe and so any differences in velocity from one side of the pipe to the other are eliminated. Hence all the fluid moves along the pipe with the same mean velocity, except for a tiny fraction around the edge which is slowed down by contact with the pipe wall.

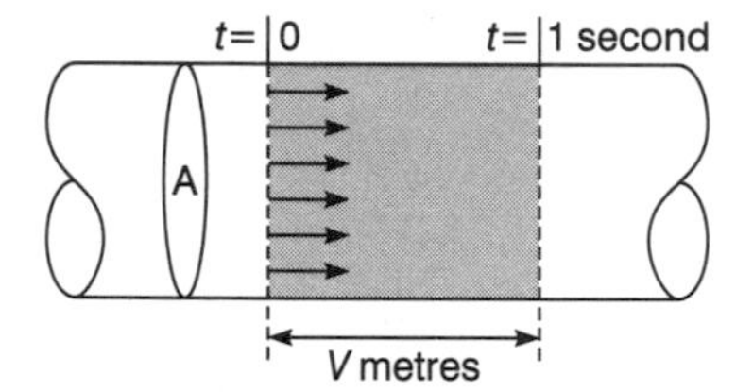

Figure 3.2.8 *Calculation of volume flow rate*

Figure 3.2.8 shows what happens to the liquid in a pipe during a time span of 1 second.

If all the liquid is travelling down the pipe at a velocity of v metres/second then in the time span of 1 second the chosen cross-section in the liquid will move by a distance of v metres along the pipe. Therefore the volume of liquid which has passed the starting position is the volume of the cylinder which has been shaded. This volume is *end area length*, which equals $A \quad v$. Since this is the volume of liquid which has passed an arbitrary point in 1 second, the volume flow rate of liquid along the pipe is given by:

$$Q = Av \qquad (3.2.2)$$

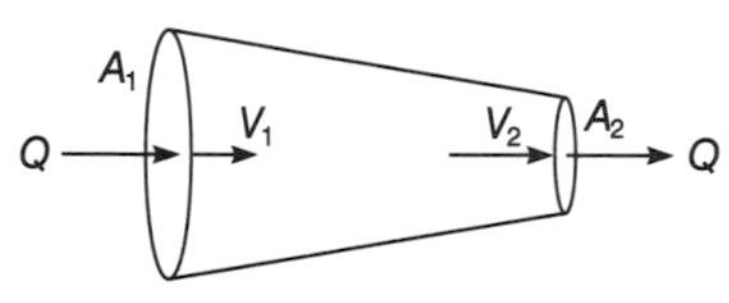

Figure 3.2.9 *The continuity law*

Suppose now that the pipe considered above is connected to a smaller diameter pipe by a tapered section, as shown in Figure 3.2.9.

The volume flow rate Q must stay the same all the way along the pipes, even though the velocities v_1 and v_2 will be different, because whatever enters the pipe at one end must leave it at the other end *at the same rate* otherwise the pipe would inflate and eventually explode. Therefore the flow rate at the entry, $Q = A_1 v_1$, must equal the flow rate at the other end, $Q = A_2 v_2$.

Therefore:

$$Q = A_1 v_1 = A_2 v_2 \qquad (3.2.3)$$

This is known as the *continuity equation* and is very important in calculations involving pipe flow. It is part of a much more universal principle known as the *continuity law*, which is to do with the fact that fluids cannot be torn apart like a solid.

Sometimes it is necessary to work out the velocities and flow rates in a network of pipes, such as the case where a number of pumps are forcing oil into a large undersea pipeline which subsequently splits into several branches at a tanker loading terminal. This situation can be analysed using the continuity law provided it is remembered that what goes in at one end, from all the different inlets, must come out at the same rate at the other end, through all its branches.

Although we deliberately stated at the start of this section on continuity that we were looking at turbulent flow, this equation applies equally well to laminar flow so long as we remember to use the mean or average velocity. This is because the velocity variation across a pipe containing laminar or streamlined flow is very wide. The relationship between the mean velocity and the maximum velocity for laminar flow will be developed in the next section.

Example 3.2.2

A liquid is flowing along a tapering pipe at a volume flow rate of 15 m^3/min. The cross-sectional area of the pipe at the start is 0.05 m^2 and this reduces to 0.02 m^2 at the outlet. Calculate the flow velocity at the inlet and outlet to the pipe.

This is an example of the use of the continuity law represented in Equation (3.2.3). We must remember to convert the volume flow rate to cubic metres per *second* from cubic metres per *minute*.

$$Q = A_1 v_1 = A_2 v_2$$

$$15/60 = 0.05\ v_1 = 0.02\ v_2$$

$$v_1 = 15/(60 \quad 0.05) = 5\,\text{m/s}$$

$$v_2 = 15/(60 \quad 0.02) = 12.5\,\text{m/s}$$

Laminar flow

Viscosity

Before we can consider flow of fluids in much more detail, we must take a closer look at the property of fluids known as viscosity. We have already met viscosity as one of the quantities in Reynolds equation where we have used it to represent the 'oiliness' of a fluid. Fluids such as oils and syrup that flow sluggishly have a high value of viscosity, while thin, watery fluids have a low viscosity. The concept of viscosity is generally credited to Newton who studied it experimentally and came up with an equation which essentially defines viscosity. In fact Newton was not directly interested in the flow of liquids; his life's work was devoted to the study of the stars and planets, eventually coming up with the idea of gravity. He became involved with the study of viscosity almost by chance as a means of explaining the small variations in speed of the planets when they are closest to one of their neighbouring planets. For example, he had studied the fact that the earth temporarily moves faster along its orbit when its quicker neighbour, Venus, overtakes it from time to time. As a result of Newton's later work we now know that this effect is caused by the gravitational attraction between the two planets but initially Newton thought that it was due to viscous drag in the celestial fluid or ether that was held to fill the universe. When Venus passed the earth it would shear this fluid in the relatively small gap between the two planets and there would be a resistance or drag force which would act to slow Venus down while speeding up the earth. Newton pursued this theory to the point of tabletop experiments with liquids and plates and produced an equation which basically describes and defines viscosity, before discarding the idea in favour of gravity.

The equation that Newton developed to define the viscosity of a fluid is:

Viscous shear stress = viscosity velocity gradient

In its simplest form this can be applied to two flat plates, one moving and one stationary, in the following equation:

$$\frac{F}{A} = \mu \frac{u}{h} \tag{3.2.4}$$

Here the force F applied to one of the plates, each of area A and separated by a gap of width h filled with fluid of viscosity μ, produces a difference in velocity of u. The viscosity is correctly known as the *dynamic* viscosity and has the units of Pascal seconds (Pa s). These are identical to N s/m^2 and kg/m/s.

For the case where these two plates are adjacent lamina or layers:

$F = \mu A$ velocity gradient

We have replaced the term u/h with something called the *velocity gradient* which will allow us later to apply Newton's equation to situations where the velocity does not change evenly across a gap.

We can apply this to pipe flow if we now wrap the lamina into cylinders.

Laminar flow in pipes (Figure 3.2.10)

We said earlier that most fluid flow applications of interest to mechanical engineers involve turbulent flow, but there are some increasingly important examples of laminar flow where the pipe diameter is small and the liquid is very viscous. The field of medical engineering has many such examples, such as the flow of viscous blood along the small diameter tubes of a kidney dialysis machine. It is therefore necessary for us to study this kind of pipe flow so that we may be able to calculate the pumping pressure required to operate this type of device.

Consider flow at a volume flow rate Q m^3/s along a pipe of radius R and length L. The liquid viscosity is μ and a pressure drop of P is required across the ends of the pipe to produce the flow. Let us look at the forces on the cylindrical core of the liquid in the pipe, up to a radius of r.

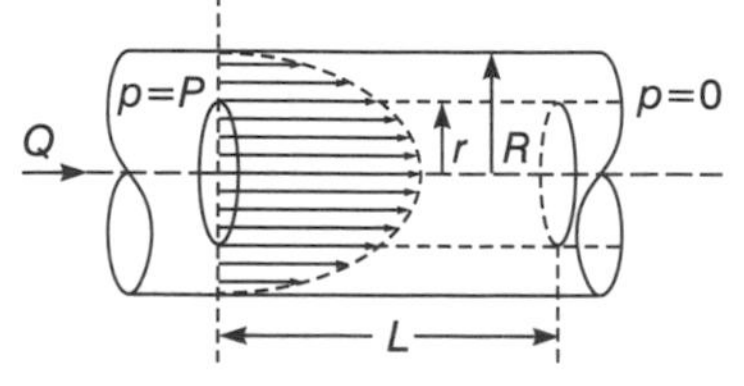

Figure 3.2.10 *Laminar flow along a pipe*

If we take the outlet pressure as 0 and the inlet pressure as P then the force pushing the core to the right, the pressure force, is given by

$$F_p = P \quad A$$
$$= \pi r^2 P$$

The force resisting this movement is the viscous drag around the cylindrical surface of the core. Using Newton's defining equation for viscosity, Equation (3.2.4),

$$F_{\text{drag}} = \mu A_{\text{core surface}} \text{ velocity gradient}$$

$$= \mu 2\pi rL \frac{\mathrm{d}v}{\mathrm{d}r}$$

For steady flow these forces must be equal and opposite,

$$F_{\text{p}} = -F_{\text{drag}}$$

$$\pi r^2 P = -\mu 2\pi r \text{ L} \frac{\mathrm{d}v}{\mathrm{d}r}$$

$$rP = -2L\mu \frac{\mathrm{d}v}{\mathrm{d}r}$$

So the velocity gradient is

$$\frac{\mathrm{d}v}{\mathrm{d}r} = -\frac{rP}{2L\mu}$$

This is negative because v is a maximum at the centre and decreases with radius.

We need to relate the pressure to the flow rate Q and the first step is to find the velocity v at any radius r by integration.

$$v = \int -\frac{rP}{2L\mu} \mathrm{d}r$$

$$= -\frac{r^2P}{4L\mu} + C$$

To evaluate the constant C we note that the liquid is at rest ($v = 0$) at the pipe wall ($r = R$). Even for liquids which are not sticky or highly viscous, all the experimental evidence points to the fact that the last layer of molecules close to the walls fastens on so tightly that it does not slip (Figure 3.2.11).

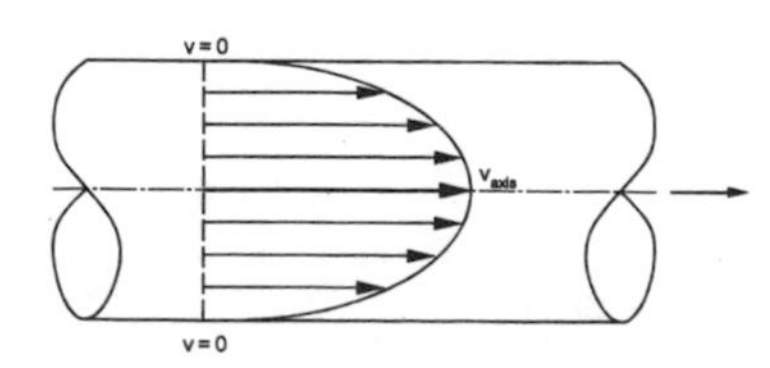

Figure 3.2.11 *Velocity distribution inside the pipe*

Therefore

$$0 = -\frac{R^2P}{4L\mu} + C, \text{ so } C = \frac{R^2P}{4L\mu}$$

Therefore

$$v = \frac{P}{4L\mu}(R^2 - r^2)$$

This is the equation of a parabola so the average or mean velocity equals half the maximum velocity, which is on the central axis.

$$v_{mean} = \frac{1}{2} V_{axis}$$

Now volume flow rate is $Q = V_{mean}$ area.

This is the continuity equation and up to now we have only applied it to turbulent flow where all the liquid flows at the same velocity and we do not have to think of a mean. So

$$Q = \frac{V_{axis}}{2} \pi R^2$$

Finally

$$Q = \frac{\pi P R^4}{8 L \mu} \tag{3.2.5}$$

This is called *Poiseuille's law* after the French scientist and engineer who first described it and this type of flow is known as Poiseuille flow.

Note that this is analogous to Ohm's law with flow rate Q equivalent to current I, pressure drop P equivalent to voltage E and the term

$$\frac{8L\mu}{\pi R^4}$$

representing fluid resistance Ω, equivalent to resistance R.

Hence Ohm's law $E = IR$ becomes $P = Q\Omega$ when applied to viscous flow along pipes. This can be very useful when analysing laminar flow through networks of pipes. For example, the combined fluid resistance of two different diameter pipes in parallel and both fed with the same liquid at the same pressure can be found just like finding the resistance of two electrical resistors in parallel.

Example 3.2.3

A pipe of length 10 m and diameter 5 mm is connected in series to a pipe of length 8 m and diameter 3 mm. A pressure drop of 120 kPa is recorded across the pipe combination when an oil of viscosity 0.15 Pa s flows through it. Calculate the flow rate.

First we must calculate the two fluid resistances, one for each section of the pipe.

$$\Omega_1 = 8 \quad 10 \quad 0.15/(\pi 0.0025^4) = 9.778 \quad 10^{10} \text{ Pa s/m}^3$$

$$\Omega_1 = 8 \quad 8 \quad 0.15/(\pi 0.0015^4) = 60.361 \quad 10^{10} \text{ Pa s/m}^3$$

Total resistance $= 70.139 \quad 10^{10}$ Pa s/m^3

The flow rate is then given by:

$$Q = P/\Omega = 120\,000/70.139 \quad 10^{10} \text{ Pa s/m}^3$$

$$= 1.71 \quad 10^{-7} \text{ m}^3\text{/s}$$

Examples of laminar flow in engineering

We have already touched on one example of laminar flow in pipes which is highly relevant to engineering but it is worthwhile looking at some others just to emphasize that, although turbulent flow is the more important, there are many instances where knowledge of laminar flow is necessary.

One example from mainstream mechanical engineering is the dashpot. This is a device which is used to damp out any mechanical vibration or to cushion an impact. A piston is pushed into a close-fitting cylinder containing oil, causing the oil to flow back along the gap between the piston and the cylinder wall. As the gap is small and the oil has a high viscosity, the flow is laminar and the pressure drop can be predicted using an adaptation of Poiseuille's law.

A second example which is more forward looking is from the field of micro-fluidics. Silicon chip technology has advanced to the point where scientists can build a small patch which could be stuck on a diabetic's arm to provide just the right amount of insulin throughout the day. It works by drawing a tiny amount of blood from the arm with a miniature pump, analysing it to determine what dose of insulin is required and then pumping the dosed blood back into the arm. The size of the flow channels is so small, a few tens of microns across, that the flow is very laminar and therefore so smooth that engineers have had to go to great lengths to ensure effective mixing of the insulin with the blood.

Conservation of energy

Probably the most important aspect of engineering is the energy associated with any application. We are all painfully aware of the cost of energy, in environmental terms as well as in simple economic terms. We therefore now need to consider how to keep account of the energy associated with a flowing liquid. The principle that applies here is the law of conservation of energy which states that energy can neither be created nor destroyed, only transferred from one form to another. You have probably met this before, and so we have a fairly straightforward task in applying it to the flow of liquids along pipes. If we can calculate the energy of a flowing liquid at the start of a pipe system, then we know that the same total of energy must apply at the end of the system even though the values for each form of energy may have altered. The only problem is that we do not know at the moment how to calculate the energy associated with a flowing liquid or even how many types of energy we need to consider. We must begin this calculation therefore by examining the different forms of energy that a flowing liquid can have (Figure 3.2.12).

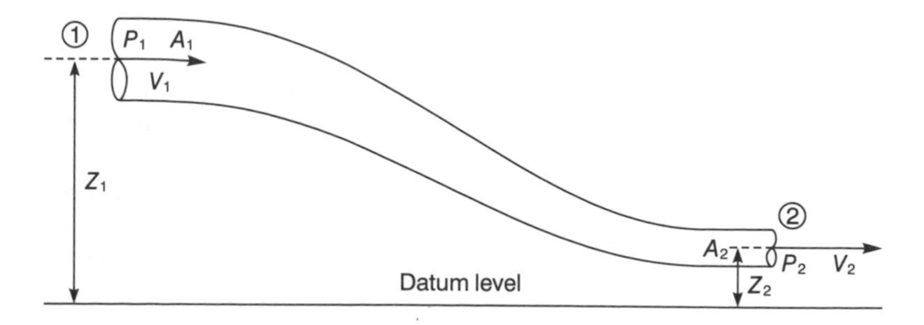

Figure 3.2.12 *Energy of a flowing liquid*

If we ignore chemical energy and thermal energy for the purposes of flow calculations, then we are left with potential energy due to height, potential energy due to pressure and kinetic energy due to the motion.

In Figure 3.2.12 adding up all three forms of energy at point 1 for a small volume of liquid of mass m:

Potential energy due to height is calculated with reference to some datum level, such as the ground, in the same way as for a solid.

$$PE_{height} = mgz_1$$

Potential energy due to pressure is a calculation of the fact that the mass m could rise even higher if the pipe were to spring a leak at point 1. It would rise by a height of h_1 equal to the height of the liquid in a manometer tube placed at point 1, where h_1 is given by $h_1 = p_1/\rho g$. Therefore the energy due to the pressure is again calculated like the height energy of a solid.

$$PE_{pressure} = mgh_1$$

Note that the height h_1 is best referred to as a pressure head in order to distinguish it from the physical height z_1 of the pipe at this point.

Kinetic energy is calculated in the same way as for a solid.

$$KE = \tfrac{1}{2}mv_1{}^2$$

Therefore the total energy of the mass m at point 1 is given by

$$E_1 = mgz_1 + mgh_1 + \tfrac{1}{2}mv_1^2$$

Similarly the energy of the same mass at point 2 is given by

$$E_2 = mgz_2 + mgh_2 + \tfrac{1}{2}mv_2^2$$

From the principle of conservation of energy we know that these two values of the total energy must be the same, provided that we can ignore any losses due to friction against the pipe wall or within the liquid. Therefore if we cancel the m and divide through by g, we produce the following equation:

$$z_1 + h_1 + v_1^2/2g = z_2 + h_2 + v_2^2/2g \tag{3.2.6}$$

This is known as *Bernoulli's equation* after the French scientist who developed it and is the fundamental equation of hydrodynamics. The dimensions of each of the three terms are *length* and therefore they all have units of *metres*. For this reason the third term, representing kinetic energy, is often referred to as the *velocity head*, in order to use the familiar concept of head which already appears as the second term on both sides of the equation. The three terms on each side of the equation, added together, are sometimes known as the *total head*. A second advantage to dividing by the mass m and eliminating it from the equation is that we no longer have to face the problem that it would be very difficult to keep track of that fixed mass of liquid as it flowed along the pipe. Turbulent flow and laminar flow would both make the mass spread out very rapidly after the starting point.

Bernoulli's equation describes the fact that the total energy in an ideal flowing liquid stays constant between two points. It is very much a practical engineering equation and for this reason it is commonly reduced to the form given here where all the terms are measured in metres. A pipeline designer, for example, could use it to keep track of how the pressure head would change along a pipe system as it followed the local terrain over hill and valley, without any need to ever work in joules, the true units of energy.

When carrying out calculations on Bernoulli's equation it is sometimes useful to use the substitution $h = p/\rho g$ to change from head to pressure, and it is often useful to use the substitution $v = Q/A$ because the volume flow rate is the most common way of describing the liquid's speed.

An example of the use of Bernoulli's equation is given later in Example 3.2.5.

Venturi principle

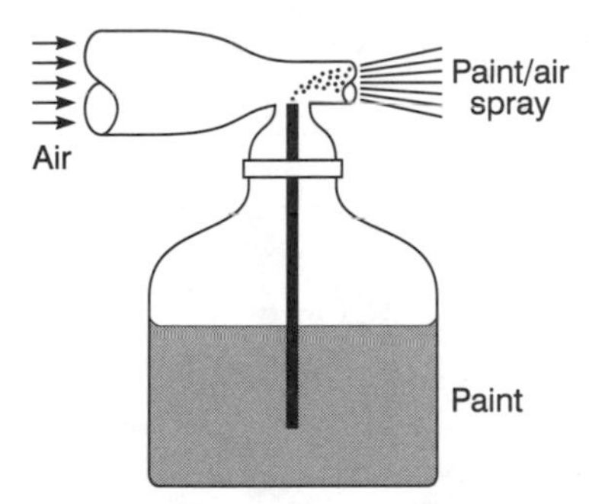

Figure 3.2.13 *Paint sprayer*

Bernoulli's equation can seem very daunting at first sight, but it is worthwhile remembering that it is simply the familiar conservation of energy principle. Therefore it is not always necessary to put numbers into the equation in order to predict what will happen in a given flow situation. Onc of the most useful applications in this respect is the behaviour of the fluid pressure when the fluid, either liquid or gas, is made to go through a constriction.

Consider what happens in Figures 3.2.13 and 3.2.14

In both cases the fluid, air, is pushed through a narrower diameter pipe by the high pressure in the large inlet pipe. The velocity in the narrow pipe is increased according to the relationship $v = Q/A$ since the volume flow rate Q must stay constant. Hence the kinetic energy term $v^2/2g$ in Bernoulli's equation is greatly increased, and so the pressure head term h or $p/\rho g$ must be much reduced if we can ignore the change in physical height over such a small device. The result is that a very low pressure is observed at the narrow pipe, which can be used to suck paint in through a side pipe in the case of the spray gun in Figure 3.2.13, or petrol in the case of the carburettor in Figure 3.2.14. This effect is known as the Venturi principle after the Italian scientist and engineer who discovered it.

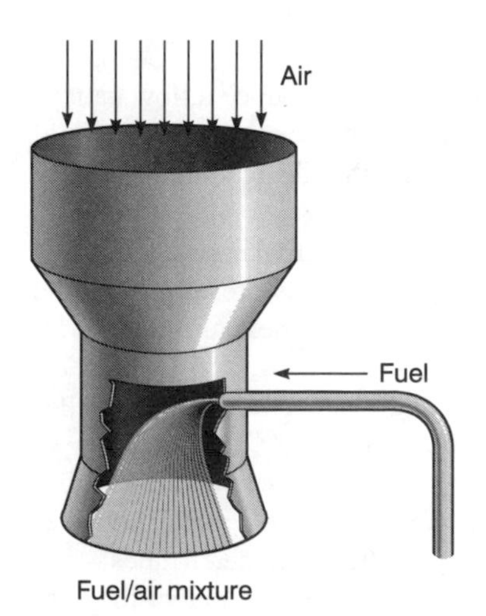

Figure 3.2.14 *Carburettor*

Measurement of fluid flow

Fluid mechanics for the mechanical engineer is largely concerned with transporting liquid from one place to another and therefore it is important that we have an understanding of some of the ways of measuring flow. There are many flow measurement methods, some of which can be used for measuring volume flow rates, others which can be used for measuring flow velocity, and yet others which can be used to measure both. We shall limit ourselves to analysis of one example of a simple flow rate device and one example of a velocity device.

Measurement of volume flow rate – the Venturi meter

In the treatment of Bernoulli's equation we found that changing the velocity of a fluid through a change of cross-section leads to a change in the pressure as the *total head* remains constant. In the Venturi effect the large increase in velocity through a constriction causes a marked reduction in pressure, with the size of this reduction depending on the size of the velocity increase and therefore on the degree of constriction. In other words, if we made a device which forced liquid through a constriction and we measured the pressure head reduction at the constriction, then we could use this measurement to calculate the velocity or the flow rate from Bernoulli's equation.

Such a device is called a *Venturi meter* since it relies on the Venturi effect. In principle the constriction could be an abrupt change of cross-section, but it is better to use a more gradual constriction and an even more gradual return to the full flow area following the constriction. This leads to the formation of fewer eddies and smaller areas of recirculating flow. As we shall see later, this leads to less loss of energy in the form of frictional heat and so the device creates less of a load to the pump producing the flow. A typical Venturi meter is shown in Figure 3.2.15.

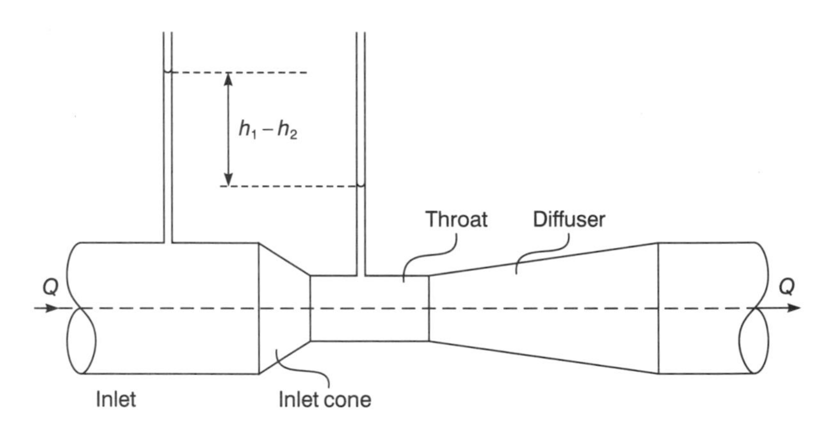

Figure 3.2.15 *A typical Venturi meter*

The *inlet cone* has a half angle of about 45° to produce a flow pattern which is almost free of recirculation. Making the cone shallower would produce little extra benefit while making the device unnecessarily long.

The *diffuser* has a half angle of about 8° since any larger angle leads to separation of the flow pattern from the walls, resulting in the formation of a jet of liquid along the centreline, surrounded by recirculation zones.

The *throat* is a carefully machined cylindrical section with a smaller diameter giving an area reduction of about 60%.

Measurement of the drop in pressure at the throat can be made using any type of pressure sensing device but, for simplicity, we shall consider the manometer tubes shown. The holes for the manometer tubes – the pressure taps – must be drilled into the meter carefully so that they are accurately perpendicular to the flow and free of any burrs.

Analysis of the Venturi meter

The starting point for the analysis is Bernoulli's equation:

$$z_1 + h_1 + v_1^2/2g = z_2 + h_2 + v_2^2/2g$$

Since the meter is being used in a horizontal position, which is the usual case, the two values of height z are identical and we can cancel them from the equation.

$$h_1 + v_1^2/2g = h_2 + v_2^2/2g$$

The constriction will cause some non-uniformity in the velocity of the liquid even though we have gone to the trouble of making the change in cross-section gradual. Therefore we cannot really hope to calculate the velocity accurately. Nevertheless we can think in terms of an average or mean velocity defined by the familiar expression:

$$v = Q/a$$

Therefore, remembering that the volume flow rate Q does not alter along the pipe:

$$h_1 + Q^2/2a_1^2g = h_2 + Q^2/2a_2^2g$$

We are trying to get an expression for the flow rate Q since that is what the instrument is used to measure, so we need to gather all the terms with it onto the left-hand side:

$$(Q^2/2g) \quad (1/a_1^2 - 1/a_2^2) = h_2 - h_1$$

$$Q^2 = 2g(h_2 - h_1)/(1/a_1^2 - 1/a_2^2)$$

Since $h_1 > h_2$ it makes sense to reverse the order in the first set of brackets, compensating by reversing the order in the second set as well:

$$Q^2 = 2g(h_1 - h_2)/(1/a_2^2 - 1/a_1^2)$$

Finally we have:

$$Q = \sqrt{\{2g(h_1 - h_2)/(1/a_2^2 - 1/a_1^2)\}} \qquad (3.2.7)$$

This expression tells us the volume flow rate if we can assume that Bernoulli's equation can be applied without any consideration of head loss, or energy loss, in the device. In practice there will always be losses of energy, and head, no matter how well we have guided the flow through the constriction. We could experimentally measure a head loss and use this in a modified form of Bernoulli's equation, but it is customary to stick with the analysis carried out above and make a final correction at the end.

Any loss of head will lead to the drop in heights of manometer levels across the meter being *bigger* than it should be ideally. Therefore the calculated flow rate will be too large and so it must be corrected by applying some factor which makes it *smaller*. This factor is known as the *discharge coefficient* C_D and is defined by:

$$Q_{real} = C_D \quad Q_{ideal}$$

In a well-manufactured Venturi meter the energy losses are very small and so C_D is very close to 1 (usually about 0.97).

In some situations it would not matter if the energy losses caused by the flow measuring device were considerably higher. For example, if you wanted to measure the flow rate of water entering a factory then even a considerable energy loss caused by the measurement would only result in the pumps at the local water pumping station having to work a little harder; there would be no disadvantage as far as the factory was concerned. In these circumstances it is not necessary to go to the trouble of having a carefully machined, highly polished Venturi meter, particularly since they are complicated to install. Instead it is sufficient just to insert an orifice plate at any convenient joint in the pipe, producing a device known as an *Orifice meter* (Figure 3.2.16). The orifice plate is rather like a large washer with the central hole, or orifice, having the same sort of area as the throat in the Venturi meter.

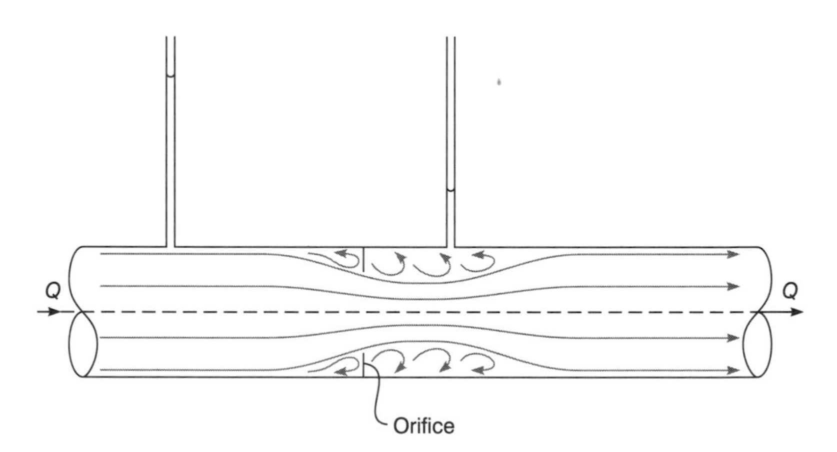

Figure 3.2.16 *An orifice meter*

The liquid flow takes up a very similar pattern to that in the Venturi meter, but with the addition of large areas of recirculation. In particular the flow emerging from the orifice continues to occupy a small cross-section for quite some distance downstream, leading to a kind of throat. The analysis is therefore identical to that for the Venturi meter, but the value of the discharge coefficient C_D is much smaller (typically about 0.6). Since there is not really a throat, it is difficult to specify exactly where the downstream pressure tapping should be located to get the most reliable reading, but guidelines for this are given in a British Standard.

Measurement of velocity – the Pitot-static tube

Generally it is the volume flow rate which is the most important quantity to be measured and from this it is possible to calculate a mean flow velocity across the full flow area, but in some cases it is also important to know the velocity at a point. A good example of this is in a river where it is essential for the captain of a boat to know what

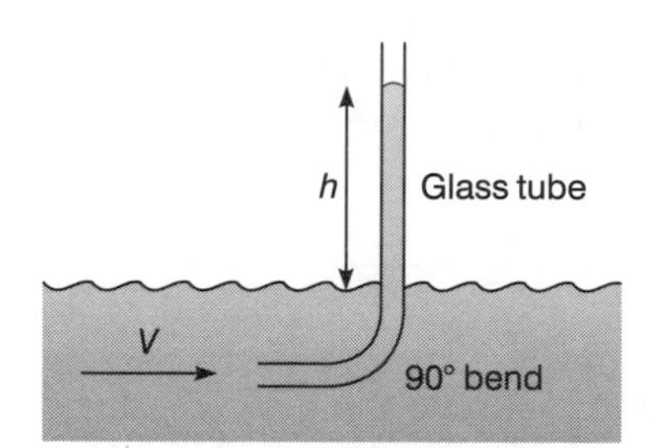

Figure 3.2.17 *The Pitot tube*

strength of current to expect at any given distance from the bank; calculating a mean velocity from the volume flow rate would not be much help even if it were possible to measure the exact flow area over an uneven river bed.

It was exactly this problem which led to one of the most common velocity measurement devices. A French engineer called Pitot was given the task of measuring the flow of the River Seine around Paris and found that a quick and reliable method could be developed from some of the principles we have already met in the treatment of Bernoulli's equation. Figure 3.2.17 shows the early form of Pitot's device.

The horizontal part of the glass tube is pointed upstream to face the oncoming liquid. The liquid is therefore forced into the tube by the current so that the level rises above the river level (if the glass tube was simply a straight, vertical tube then the water would enter and rise until it reached the same level as the surrounding river). Once the water has reached this higher level it comes to rest.

What is happening here is that the velocity head (kinetic energy) of the flowing water is being converted to height (potential energy) inside the tube as the water comes to rest. The excess height of the column of water above the river level is therefore equal to the velocity head of the flowing water:

$$h = v^2/2g$$

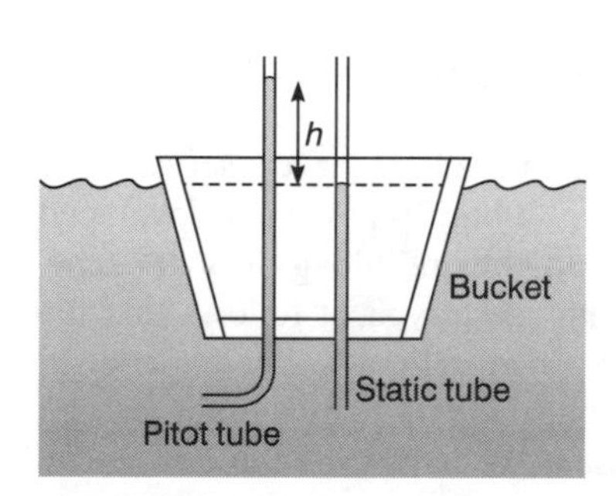

Figure 3.2.18 *An early Pitot-static tube*

Therefore the velocity measured by a Pitot tube is given by:

$$v = \sqrt{(2gh)} \tag{3.2.7}$$

In practice Pitot found it difficult to note the level of the water in the glass tube compared to the level of the surrounding water because of the disturbances on the surface of the river. He quickly came up with the practical improvement of using a straight second tube (known as a static tube) to measure the river level because the capillary action in the narrow tube damped down the fluctuations (Figure 3.2.18).

The Pitot-static tube is still widely used today, most notably as the speed measurement device on aircraft (Figure 3.2.19).

The two tubes are now combined to make them co-axial for the purposes of 'streamlining', and the pressure difference would be

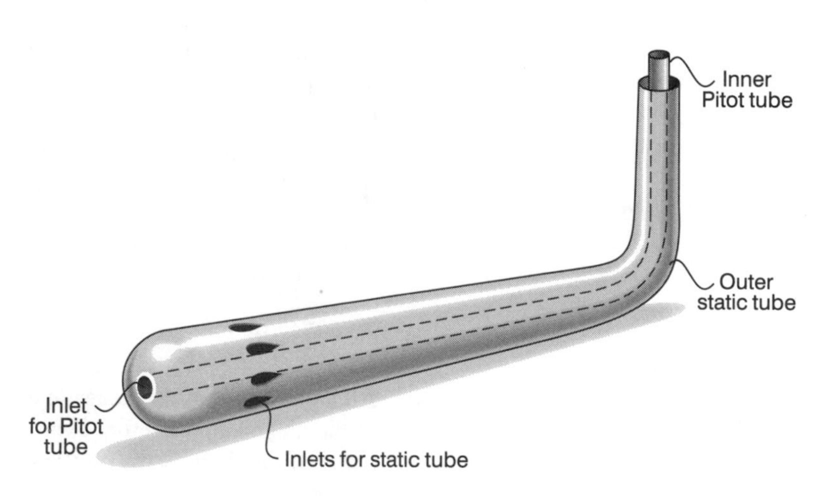

Figure 3.2.19 *A modern Pitot-static tube*

measured by an electronic transducer, but essentially the device is the same as Pitot's original invention. Because the Pitot tube and the static tube are united, the device is called a *Pitot-static tube*.

Example 3.2.4

A Pitot-static tube is being used to measure the flow velocity of liquid along a pipe. Calculate this velocity when the heights of the liquid in the Pitot tube and the static tube are 450 mm and 321 mm respectively.

The first thing to do is calculate the manometric head difference, i.e. the difference in reading between the two tubes.

$$\text{Head difference} = 450\,\text{mm} - 321\,\text{mm} = 129\,\text{mm} = 0.129\,\text{m}$$

Then use Equation (3.2.7),

$$\text{Velocity} = \sqrt{(2 \quad 9.81 \quad 0.129)} = 1.59\,\text{m/s}$$

Losses of energy in real fluids

So far we have looked at the application of the familiar 'conservation of energy' principle to liquids flowing along pipes and developed Bernoulli's equation for an ideal liquid flowing along an ideal pipe. Since energy can neither be created nor destroyed, it follows that the three forms of energy associated with flowing liquids – height energy, pressure energy and kinetic energy – must add up to a constant amount even though individually they may vary. We have used this concept to understand the working of a Venturi meter and recognized that a practical device must somehow take into account the loss of energy from the fluid in the form of heat due to friction. This was quite simple for the Venturi meter as the loss of energy is small but we must now consider how we can take into account any losses in energy, in the form of heat, caused by friction in a more general way. These losses can arise in many ways but they are all caused by friction within the liquid or friction between the liquid and the components of the piping system. The big problem is how to include what is essentially a thermal effect into a picture of liquid energy which deliberately sets out to exclude any mention of thermal energy.

Modified Bernoulli's equation

Bernoulli's equation, as developed previously, may be stated in the following form:

$$z_1 + h_1 + v_1^2/2g = z_2 + h_2 + v_2^2/2g$$

All the three terms on each side of Bernoulli's equation have dimensions of length and are therefore expressed in *metres*. For this reason the total value of the three terms on the left-hand side of the equation is known as the *initial total head* in just the same way as we used the word *head* to describe the height h associated with any pressure p through the expression

$$p = \rho g h$$

Similarly the right-hand side of the equation is known as the *final total head*. Bernoulli's equation for an ideal situation may also be expressed in words as:

Initial total head = final total head

What happens when there is a loss of energy due to friction with a real fluid flowing along a real pipe is that the final total head is smaller than the initial total head. The loss of energy, as heat generated by the friction and dissipated through the liquid and the pipe wall to the surroundings, can therefore be expressed as a loss of head. Note that we are not destroying this energy, it is just being transformed into thermal energy that cannot be recovered into a useful form again. As far as the engineer in charge of the installation is concerned this represents a definite loss which needs to be calculated even if it cannot be reduced any further.

What happens in practice is that manufacturers of pipe system components, such as valves or couplings, will measure this loss of head for all their products over a wide range of sizes and flow rates. They will then publish this data and make it available to the major users of the components. Provided that the sum of the head losses of all the components in a proposed pipe system remains small compared to the total initial head (say about 10%) then it can be incorporated into a modified Bernoulli's equation as follows:

Initial total head – head losses = final total head

With this equation it is now possible to calculate the outlet velocity or pressure in a pipe, based on the entry conditions and knowledge of the energy losses expressed as a head loss in metres. Once again we see the usefulness of working in metres since engineers can quickly develop a feel for what head loss might be expected for any type of fitting and how it could be compensated. This would be extremely difficult to do if working in conventional energy units.

Example 3.2.5

Water is flowing downwards along a pipe at a rate of $0.8\,m^3/s$ from point A, where the pipe has a diameter of 1.2 m, to point B, where the diameter is 0.6 m. Point B is lower than point A by 3.3 m. The pipe and fittings give rise to a head loss of 0.8 m. Calculate the pressure at point B if the pressure at point A is 75 kPa (Figure 3.2.20).

Since the information in the question gives the flow rate Q rather than the velocities, we shall use the substitution

$$Q = a_1 v_1 = a_2 v_2$$

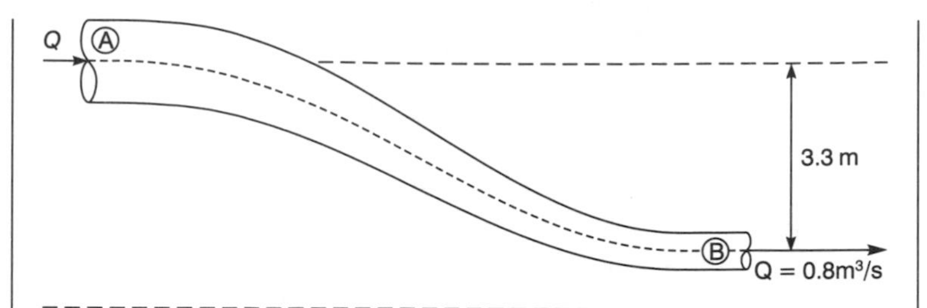

Figure 3.2.20 *A pipe system with energy losses*

Therefore the *modified* Bernoulli's equation becomes:

$$z_1 + p_1/\rho g + Q^2/2a_1^2 g - \text{losses} = z_2 + p_2/\rho g + Q^2/2a_2^2 g$$

The absolute heights of point A and point B do not matter, it is only the relative difference in heights which is important. Therefore we can put $z_1 = 3.3\,\text{m}$ and $z_2 = 0$

$$3.3 + 75\,000/(1000\ \ 9.81) + 0.8^2/(2\ \ (\pi\ \ 0.6^2)^2\ \ 9.81) - 0.8$$
$$= 0 + p_2/(1000\ \ 9.81) + 0.8^2/(2\ \ (\pi\ \ 0.3^2)^2\ \ 9.81)$$

$$3.3 + 7.645 + 0.0255 - 0.8 = 0 + p_2/9810 + 0.4080$$

$$10.71 = p_2/9810 + 0.4080$$

Therefore:

$$p_2 = 9810\ \ (10.71 - 0.4080)$$
$$= 101.1\,\text{kPa}$$

The cause of energy losses

Earlier we looked at the way that liquids flow along pipes and we showed that almost all practical cases involved turbulent flow where the liquid molecules continually collide with each other and with the walls. It is the collisions with the walls which transfer energy from the liquid to the surroundings; the molecules hit a roughness point on the wall and lose some of their kinetic energy as a tiny amount of localized heating of the material in the pipe wall.

The molecules therefore bounce off the walls with slightly lower velocity, but this is rapidly restored to its original value in collisions with other molecules. If this velocity remained lower following a collision then the liquid would not flow along the pipe at the proper rate. This cannot happen since it would violate the continuity law, which states that the flow rate must remain constant. In fact the energy to keep the molecules moving at their original speed following a collision with the wall comes from the pressure energy, which is why the effect of friction appears as a loss of head.

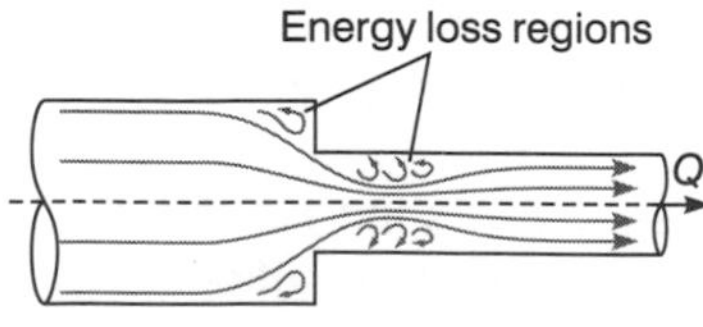

Figure 3.2.21 *Flow patterns through a typical pipe fitting*

Losses in pipe fittings

Let us look at a typical pipe fitting to see where the energy loss arises (see Figure 3.2.21).

The sudden contraction of the flow caused by joining two pipes of different diameters gives rise to regions of recirculating flow or eddies. The liquid which enters these regions is trapped and becomes separated from the rest of the flow. It goes round and round, repeatedly hitting the pipe walls and losing kinetic energy, only to be restored to its original speed by robbing the bulk flow of some of its pressure energy. The energy is dissipated as heat through the pipe walls. If the overall pressure drop was critical and the head loss needed to be kept to a minimum, then a purpose-built pipe fitting could be designed to connect the two pipes with much less recirculation. Essentially this would round off the sharp corners (Figure 3.2.22).

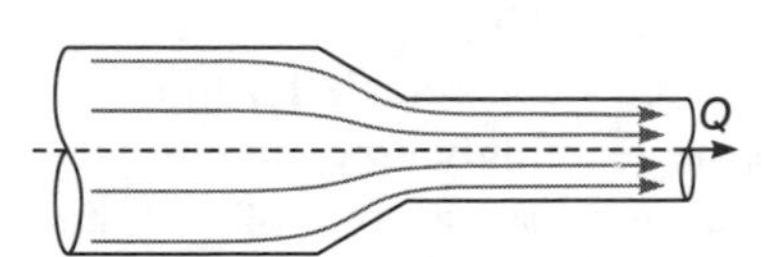

Figure 3.2.22 *Flow patterns through a streamlined pipe fitting*

Since it is kinetic energy which is lost in the collisions which are a feature of recirculating eddies, it follows that faster liquids will lose more energy than slower liquids in the same situation. In extensive experiments it has been found that the energy loss in fact depends on the overall kinetic energy of the liquid as it meets the obstruction. The proportion of the kinetic energy that is lost is approximately a constant for any given shape of obstruction, such as a valve or a pipe fitting, irrespective of the size.

For the purposes of calculations involving Bernoulli's equation it is convenient to work in terms of the *velocity head* (i.e. the third term $v^2/2g$ in Bernoulli's equation) when considering kinetic energy. Therefore a head loss for a particular type of pipe fitting is usually expressed as:

Head loss = loss coefficient velocity head

$$h_{\text{loss}} = k \quad (v^2/2g) \qquad (3.2.8)$$

Some typical values of k are shown below, but it must be remembered that they are only approximate.

Approximate loss coefficient k *for some typical pipe fittings*

90° threaded elbow	0.9
90° mitred elbow	1.1
45° threaded elbow	0.4
Globe valve, fully open	10
Gate valve,	
fully open	0.2
3/4 open	1.15
1/2 open	5.6
1/4 open	24

Turbulent flow in pipes – frictional losses

One of the basic things that engineers need to know when designing and building anything involving flow of fluids along pipes is the amount of energy lost due to friction for a given pipe system at a given flow rate.

We have just looked at the losses in individual fittings as these are of most importance in a short pipe system, such as would be found in the fuel or hydraulic system in a car. Increasingly, however, mechanical engineers are becoming involved in the design of much larger pipe systems such as would be found in a chemical plant or an oil pipeline. For these pipe lengths the head loss due to friction becomes appreciable for the pipes themselves, even though modern pipes are seemingly very smooth.

The energy loss due to friction appears as a loss of pressure (remember pressure and kinetic energy are the two important forms of energy for flow along a horizontal pipe – loss of kinetic energy becomes transformed into a loss of pressure). Therefore if we use simple manometers to record the pressure head along a pipeline then we observe a gradual loss of head. The slope of the manometer levels is known as the hydraulic gradient (Figure 3.2.23).

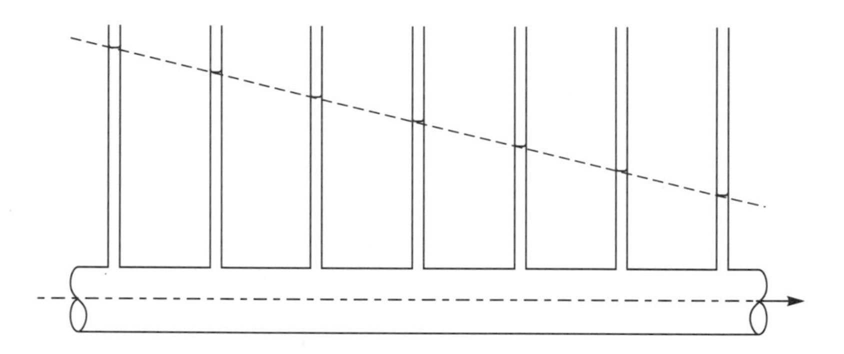

Figure 3.2.23 *Hydraulic gradient*

The head reading on the manometers can be held constant if the pipe itself goes downhill with a slope equal to the hydraulic gradient. Of course we do not really need to install manometer tubes every few metres; we can simply calculate the total head at any point and keep track of its gradual reduction mathematically.

There is no way of predicting the loss of head completely analytically and so we rely on an empirical law based on the results of a large number of experiments carried out by a French engineer Henri Darcy (or d'Arcy) in the nineteenth century. His results for turbulent flow were summed up as follows:

$$h_f = \frac{4fL}{d}\left(\frac{v^2}{2g}\right) \qquad (3.2.9)$$

where:

h_f = head loss due to friction (m)
v = flow velocity (m/s)
L = pipe length (m)
d = pipe diameter (m)
f = friction factor (no units).

Intuitively we can see where each of these terms comes from in the overall equation, as follows:

- The frictional head loss clearly depends on the length of the pipe, L, as we would expect a pipe that is twice as long to have a head loss that is also twice as great.

- Head loss decreases with increasing pipe diameter because a smaller proportion of the liquid comes into contact with the pipe wall.
- Friction arises from loss of kinetic energy and so the expression must have velocity head ($V^2/2g$) in it (which is why we do not cancel the 2 and the 4).
- Head loss also depends on the resistance offered by the roughness of the pipe wall, as represented by the friction factor f.

The friction factor is generally quoted by a pipe manufacturer. It depends on the material and the type of production process (both of which affect the roughness), on the diameter and the flow velocity, and on the amount of turbulence (Reynolds number).

D'Arcy's equation was used successfully for almost 100 years, relying on values of the friction coefficient f that were found experimentally for the very few types of pipes that were available and mostly for gravity feed systems. Once pumping stations and standardized pipes came into common use, however, a more accurate estimate of head loss at the much higher flow rates was required. It became apparent that the friction coefficient f varied quite considerably with type of pipe, diameter, flow rate, type of liquid, etc. The problem was solved by an American engineer called Moody who carried out a vast number of experiments on as many combinations of pipes and liquids as he could find. He assembled all the experimental data into a special chart, now called a Moody chart, as shown in Figure 3.2.24.

This has a series of lines on a double logarithmic scale. Each line corresponds to a pipe of a given 'relative roughness' and is drawn on axes which represent friction factor and Reynolds number. To

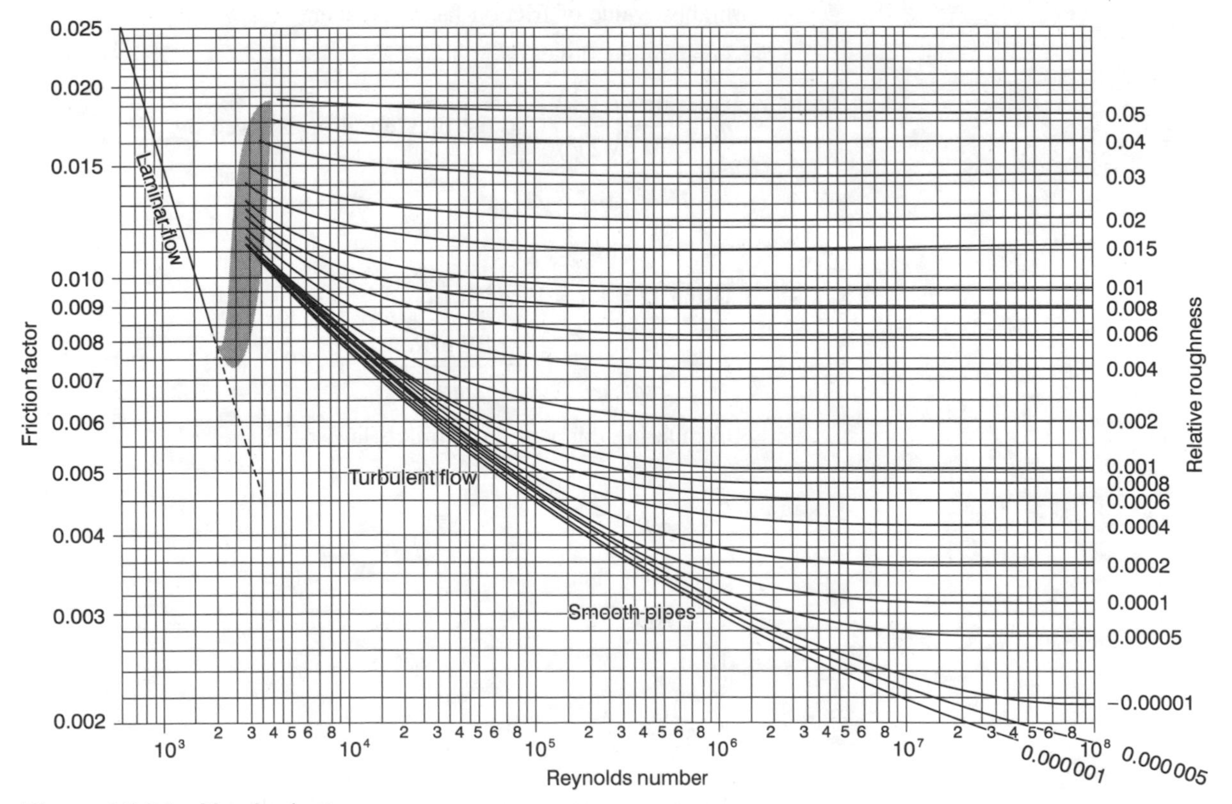

Figure 3.2.24 *Moody chart*

understand the chart we must first look at the pipe roughness as this is the parameter which gives rise to most of the friction.

The roughness on the inside of the pipe is firstly expressed as an 'equivalent height' of roughness. This is an averaging process which imagines the actual randomly scattered roughness points being replaced by a series of identical rough points, all of one height, which produce the same effect. The pipe manufacturers will produce this information using a surface measurement probe linked to a computer. This equivalent roughness height K is then divided by the pipe diameter d to give the relative roughness, K/d, plotted down the right-hand edge of the chart. Note that the relative roughness has no units, it is simply a ratio. Therefore the whole chart is dimensionless since the axes themselves are friction factor and Reynolds number, neither of which has units.

To use the chart, calculate the relative roughness if it is not given, and then locate the nearest line (or lines) at the right-hand edge. Be prepared to estimate values between the curves. Then move to the left along the curve until the specified value of Reynolds number Re is reached along the bottom edge. If the flow is turbulent ($Re > 2$–4000) then it is possible to read off the value of friction factor f corresponding to where the experimental line cuts the Re line. This value is then used in d'Arcy's equation. For example,

For relative roughness $= 0.003$ and $Re = 1.3 \quad 10^5, f = 0.0068$

Notes on the Moody chart

(1) The shaded region is for transitional flows, neither fully turbulent nor fully laminar, so generally the worst case is assumed, i.e. the highest value of friction factor is taken.

(2) For laminar flow:

$$h_f = \frac{4fL}{d}\left(\frac{V^2}{2g}\right) \quad \text{so the pressure drop is given by}$$

$$P = \rho g \cdot \frac{4fL}{d} \cdot \frac{V^2}{2g}$$

Using $Q = VA = V\pi d^2/4$,

$$P = \frac{8fL\rho v}{\pi d^3} \qquad Q = \frac{8fL\mu \cdot Re}{\pi d^4} Q$$

Comparing this with Poiseuille's law,

$$P = \frac{8\,\mu L}{\pi R^4} Q$$

and noting that

$d^4 = 16R^4$

Then

$$f = \frac{16}{Re}$$

This is the straight line on the left of the chart.

This is an artificial use of friction factor because the pressure drop in laminar flow is really directly proportional to V, whereas in turbulent flow it is proportional to V^2 (i.e. kinetic energy).

Nevertheless, the laminar flow line is included on the Moody chart so that the design engineer can quickly move from the common turbulent situations to the less frequent applications where the flow rate is small and the product in the pipeline is very viscous. If the pressure drop along a pipe containing laminar flow is required then normally Poiseuille's equation would be used.

The momentum principle

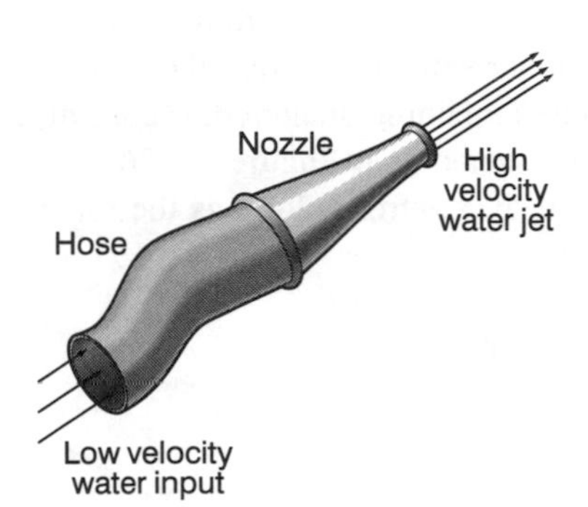

Figure 3.2.25 *Fire hoses use the momentum principle*

This section deals with the forces associated with jets of fluid and therefore has applications to jet engines, turbines and compressors. It is an application of Newton's second law which is concerned with acceleration. Since acceleration covers changes of direction as well as changes of speed, the forces on pipe bends due to the fluid turning corners is also covered. As usual we shall concentrate on liquids for the sake of simplicity.

When a jet of liquid is produced, for example from a fire hose as shown in Figure 3.2.25, it is given a large velocity by the action of the nozzle; the continuity law means that the liquid has to go faster to pass through the small cross-section at the same flow rate. The idea behind giving it this large velocity in this case is to allow the jet of water to reach up into tall buildings.

In the short time it takes to flow along the nozzle the liquid therefore receives a great deal of momentum. According to Newton's second law of motion, the rate of change of this momentum is equal to the force which must be acting on the liquid to make it accelerate. Similarly from Newton's third law there will be a reaction force on the hose itself. That is why it usually takes two burly fire fighters to hold a hose and point it accurately when it is operating at full blast.

At the other end of the jet there is a similar change of momentum as the jet is slowed down when it hits a solid object. Again this can be a considerable force; in some countries the riot police are equipped with water cannon which project a high velocity water jet to knock people off their feet!

Calculation of momentum forces

To calculate the forces associated with jets we must go back to Newton's original definition of forces in order to see how liquids differ from solids.

What Newton stated was that the force on an accelerating body was equal to the rate of change of the body's momentum.

Force = rate of change of momentum

$$F = \mathrm{d}/\mathrm{d}t\ (mv)$$

For a solid this differentiation is straightforward because the mass m is usually constant. Therefore we get:

$$F = m\ \mathrm{d}v/\mathrm{d}t$$

Since dv/dt is the definition of acceleration, this becomes:

$F = ma$ Newton's second law for a solid.

For a liquid, things are very different because we have the problem that we have met before – it is impossible to keep track of a fixed mass of liquid in any process which involves flow, especially where the flow is turbulent, as in this case. We need some method of working with the mass flow rate rather than the mass itself, and this is provided by the *control volume* method.

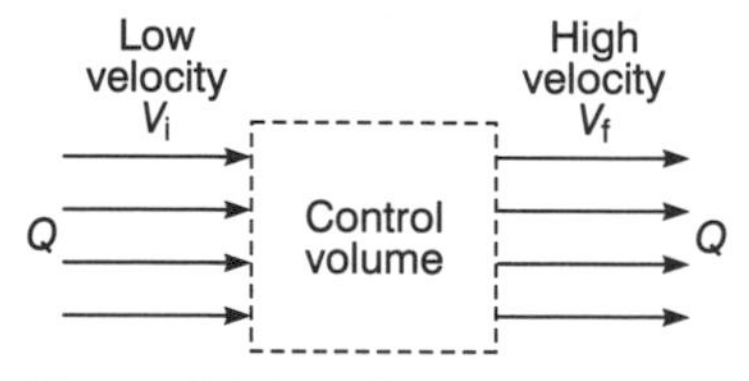

Figure 3.2.26 *Control volume for the nozzle*

We can think of the liquid entering a control volume with a constant uniform velocity and leaving with a different constant velocity but at the same volume flow rate. What happens inside the control volume to alter the liquid's momentum is of no concern, it is only the effect that matters. In the case of the fire hose considered above, the control volume would be the nozzle; water enters at a constant low velocity from the large diameter hose and leaves at a much higher velocity through the narrow outlet, with the overall volume flow rate remaining unaltered, according to the continuity law. This can be represented as in Figure 3.2.26.

The rate of change of momentum for the control volume is the force *on the control volume* and is given by:

$$\begin{aligned} F_{\text{control volume}} &= (\text{rate of supply of momentum}) \\ &\quad - (\text{rate of removal of momentum}) \\ &= (\text{mass flow rate} \quad \text{initial velocity}) \\ &\quad - (\text{mass flow rate} \quad \text{final velocity}) \\ &= \dot{m}v_i - \dot{m}v_f \\ &= \dot{m}(v_i - v_f) \end{aligned}$$

For the fire hose this would be the force on the nozzle, but normally we calculate in terms of the force on the liquid. This is equal in size but opposite in direction, so finally:

$$F_{\text{liquid}} = \dot{m}(v_f - v_i) \qquad (3.2.10)$$

We shall now go on to consider three applications of liquid jets, but it is important to note that this equation is the only one to remember.

Normal impact on a flat plate

A jet of liquid is directed to hit a flat polished plate with a normal impact, i.e. perpendicular to the plate in all directions so that the picture is symmetrical, as shown in Figure 3.2.27.

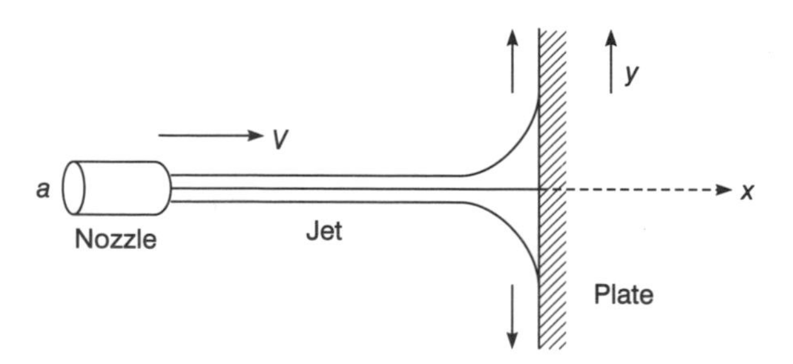

Figure 3.2.27 *Normal impact of a liquid jet on a flat plate*

If this is carried out with the nozzle and the plate fixed firmly, then the water runs exceptionally smoothly along the plate without any rebound. Therefore in the x-direction the final velocity of the jet is zero since the water ends up travelling radially along the plate. Due to the symmetry, the forces along the plate must cancel out and leave only the force in the x-direction.

$$\begin{aligned} F_{\text{liquid}_x} &= \dot{m}(v_f - v_i) \\ &= \rho Q(0 - v) \\ &= \rho av(-v) \\ &= -\rho av^2 \end{aligned}$$

The negative sign indicates that the force on the liquid is from right to left, so the force on the plate is from left to right:

$$F_{\text{plate}} = \rho av^2$$

In this case the direction is obvious but in later examples it will be less clear so it is best to learn the sign convention now.

Example 3.2.6

A jet of water emerges from a 20 mm diameter nozzle at a flow rate of 0.04 m^3/s and impinges normally onto a flat plate. Calculate the force experienced by the plate.

The equation to use here is Equation (3.2.10), but first we need to calculate the mass flow rate and the initial velocity.

Mass flow rate $= Q\rho = 0.04 \quad 1000 = 40$ kg/s

Flow velocity $= Q/A = 0.04/\pi 0.010^2 = 127.3$ m/s

This is the initial velocity and the final velocity is zero in the direction of the jet. Hence:

Force on the liquid $= 40 \quad (0 - 127.3) = -5.09$kN

The force on the plate is +5.09 kN, in the direction of the jet.

Impact on an inclined flat plate (Figure 3.2.28)

In this case the symmetry is lost and it looks as though we will have to resolve forces in the x- and y-directions. However, we can avoid this by using some commonsense and a little knowledge of fluid mechanics. In the y-direction there could only be a force on the plate if the liquid had a very large viscosity to produce an appreciable drag force. Most of the applications for this type of calculation are to do with water turbines

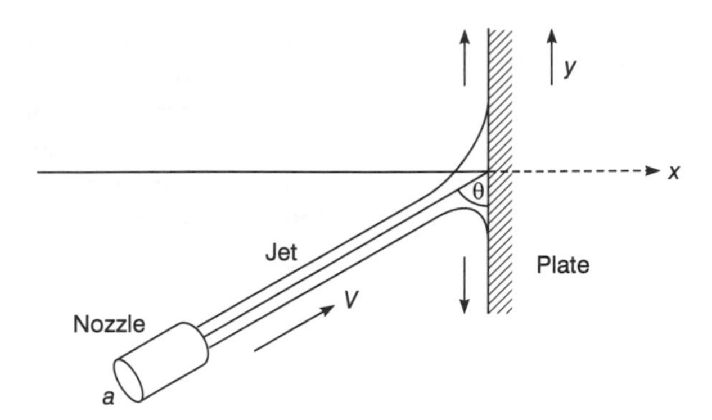

Figure 3.2.28 *Impact of a liquid jet on an inclined flat plate*

and, as we have seen in earlier sections, water has a very low viscosity. Therefore we can neglect any forces in the y-direction, leaving only the x-direction.

$$F_{\text{liquid}_x} = \dot{m}(v_f - v_i)$$

The final velocity in the x-direction is zero, as in the previous example, but the initial velocity is now $v \sin \theta$ because of the inclination of the jet to the x-axis. The velocity to be used in the part of the calculation which relates to the mass flow rate is still the full velocity; however, since the mass flow rate is the same whether the plate is there or not, let alone whether it is inclined.

$$F_{\text{liquid}_x} = \rho av(0 - v \sin \theta)$$
$$= -\rho av^2 \sin \theta$$

Therefore the force on the plate is

$$F_{\text{plate}} = \rho av^2 \sin \theta \quad \text{left to right}$$

Stationary curved vane

This is a simplified introduction to the subject of turbines and involves changes of momentum in the x- and y-directions. Since the viscosity is low and the vanes are always highly polished, we can assume initially that the liquid jet runs along the surface of the vane without being slowed down by any frictional drag (Figure 3.2.29).

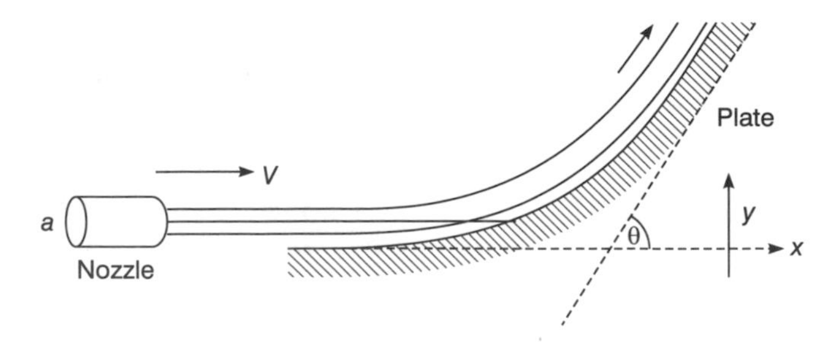

Figure 3.2.29 *Impact of a liquid jet on a curved vane*

F_x and F_y are calculated separately from the formula $F_L = \dot{m}(v_f - v_i)$

$$F_{Lx} = \rho a v(v \cos\theta - v)$$
$$= -\rho a v^2(1 - \cos\theta)$$
$$F_{Ly} = \rho a v(v \sin\theta - 0)$$
$$= \rho a v^2 \sin\theta$$

These two components can then be combined using a rectangle of forces and Pythagoras' theorem to give a single resultant force. The force on the turbine blade is equal and opposite to this resultant force on the liquid.

In practice the liquid does not leave the vane at quite the same speed as it entered because of friction. Even so the amount of slowing is still quite small, and we can take it into account by simply multiplying the *final* velocities by some correction factor. For example, if the liquid was slowed by 20% then we would use a correction factor of 0.8 on the final velocity.

Forces on pipe bends

Newton's second law relates to the forces caused by changes in velocity. Velocity is a vector quantity and so it has direction as well as magnitude. This means that a force will be required to change the direction of a flowing liquid, just as it is required to change the speed of the liquid. Therefore if a liquid flows along a pipeline which has a bend in it then a considerable force can be generated on the pipe by the liquid even though the liquid may keep the same speed throughout. In practical terms this is important because the supports for the pipe must be designed to be strong enough to withstand this force.

The calculation for such a situation is similar to that for the curved vane above except that pipe bends are usually right angles and so the working is easier. The liquid is brought to rest in the original direction and accelerated from rest up to full speed in the final direction. Hence the two components of force are of the same size, as long as the pipe diameter is constant round the bend, and the resultant force on the bend will be outwards and at 45° to each of the two arms of the pipe.

Problems 3.2.1

(1) Look back in this textbook and find out the SI unit for viscosity.

(2) Using the correct SI units to substitute into the expression for Reynolds number, show that R_e is dimensionless.

(3) What is the Reynolds number for flow of an oil of density 920 kg/m^3 and viscosity 0.045 Pa s along a tube of radius 20 mm with a velocity of 2.4 m/s?

(4) Liquid of density 850 kg/m^3 is flowing at a velocity of 3 m/s along a tube of diameter 50 mm. What is the approximate value of the liquid's viscosity if it is known that the flow is in the transition region?

(5) For the same conditions as in Problem 4, what would the diameter need to be to produce a Reynolds number of 12 000?

(6) Imagine filling a kettle from a domestic tap. By making assumptions about diameter and flow rate, and looking up values for density and viscosity, calculate whether the flow is laminar or turbulent.

(7) The cross-sectional area of the nozzle on the end of a hosepipe is 1000 mm^2 and a pump forces water through it at a velocity of 25 m/s. Find (a) the volume flow rate, (b) the mass flow rate.

(8) Oil of relative density 0.9 is flowing along a pipe of internal diameter 40 cm at a mass flow rate of 45 tonne/hour. Find the mean flow velocity.

(9) A pipe tapers from an external diameter of 82 mm to an external diameter of 32 mm. The pipe wall is 2 mm thick and the volume flow rate of liquid in the pipe is 45 litres/min. Find the flow velocity at each end of the pipe.

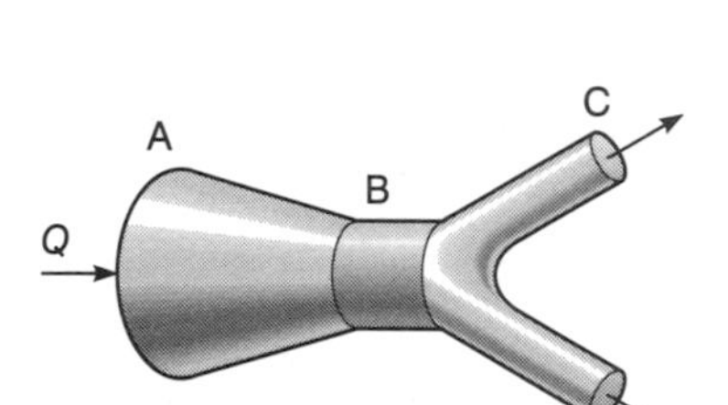

Figure 3.2.30 *A pipe system*

(10) In the pipe system in Figure 3.2.30 Find v_A, v_B, v_C.

Volume flow rate = 0.5 m^3/s
diameter A = 1 m
diameter B = 0.3 m
diameter C = 0.1 m
diameter D = 0.2 m
$v_C = 4v_D$

(11) Calculate the pressure differential which must be applied to a pipe of length 10.4 m and diameter 6.4 mm to make a liquid of viscosity 0.12 Pa s flow through it at a rate of 16 litres/hour (1000 litres = 1 m^3).

(12) Calculate the flow rate which will be produced when a pressure differential of 6.8 MPa is applied across a 7.6 m long pipe, of diameter 5 mm, containing liquid of viscosity 1.8 Pa s.

(13) Referring to Figure 3.2.31, calculate the distance of the join in the pipes away from the right-hand end.

Q = 24 l/hr, μ = 0.09 Pa s, pressure drop = 156 kPa

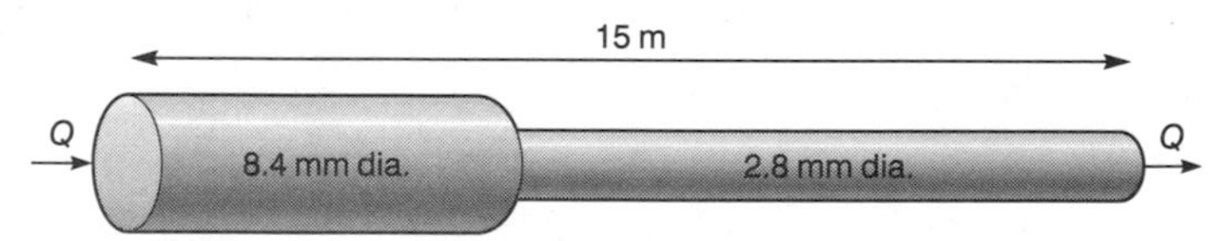

Figure 3.2.31 *A pipe system with laminar flow*

(14) Referring to Figure 3.2.32, calculate the flow rate from the outlet pipe.

(15) Water is flowing along a pipe of diameter 0.2 m at a rate of 0.226 m^3/s. What is the velocity head of the water, and what is the corresponding pressure?

(16) Water is flowing along a horizontal pipe of internal diameter 30 cm, at a rate of 60 m^3/min. If the pressure in the pipe is 50 kN/m^2 and the pipe centre is 6 m above ground level, find the total head of the water relative to the ground.

(17) Oil of relative density 0.9 flows along a horizontal pipe of internal diameter 40 cm at a rate of 50 m^3/min. If the pressure in the pipe is 45 kPa and the centre of the pipe is

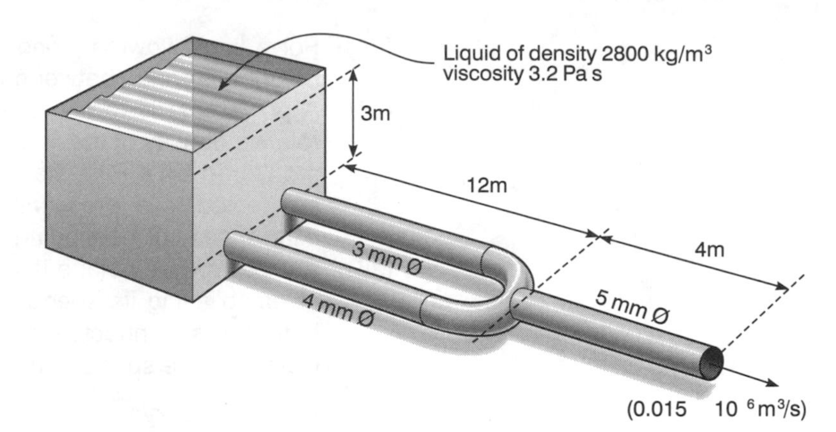

Figure 3.2.32 *A pipe system with laminar flow*

5 m above ground level, find the total head of the oil relative to the ground.

(18) The pipe in Problem 17 joins a smaller pipe of internal diameter 35 cm which is at a height of 0.3 m above the ground. What will be the pressure of the oil in this second pipe?

(19) In a horizontal Venturi meter the pipe diameter is 450 mm, the throat diameter is 150 mm, and the discharge coefficient is 0.97. Determine the volume flow rate of water in the meter if the difference in levels in a mercury differential U-tube manometer connected between the inlet pipe and the throat is 225 mm.

(20) A horizontal Venturi meter with a main diameter of 30 mm and a throat diameter of 16 mm is sited on a new lunar field station. It is being used to measure the flow rate of methylated spirits, of relative density 0.8. The difference in levels between simple manometer tubes connected to the inlet and the throat is 220 mm. Calculate the *mass* rate of flow.

$(C_D = 0.97,\ g_{moon} = g_{earth}/6)$

(21) An orifice meter consists of a 100 mm diameter orifice in a 250 mm diameter pipe and has a discharge coefficient of 0.65. The pipe conveys oil of relative density 0.9 and the pressure difference between the sides of the orifice plate is measured by a mercury U-tube manometer which shows a reading of 760 mm. Calculate the volume flow rate in the pipeline.

(22) An orifice meter is installed in a *vertical* pipeline to measure the upwards flow of a liquid polymer, of relative density 0.9. The pipe diameter is 100 mm, the orifice diameter is 40 mm and the discharge coefficient is 0.6. The pressure difference across the orifice plate, measured from a point 100 mm upstream to a point 50 mm downstream, is 8.82 kPa.

Starting with Bernoulli's equation and remembering to take into account the height difference between the pressure tappings, find the volume rate of flow along the pipe.

(23) A Pitot tube is pointed directly upstream in a fast flowing river, showing a reading of 0.459 m. Calculate the velocity of the river.

(24) For a liquid flowing along a pipe at 10.8 m/s calculate the height difference between levels in a Pitot-static tube.

(25) A powerboat is to be raced on the Dead Sea, where the relative density of the salt water is 1.026. The speed will be monitored by a Pitot-static tube mounted underneath and connected to a pressure gauge. Calculate the pressure differences corresponding to a speed of 70 km/hour.

(26) A prototype aeroplane is being tested and the only means of establishing its speed is a simple Pitot-static tube with both leads connected to a mercury U-tube manometer. Calculate the speed if the manometer reading is 356 mm.

(ρ_{air} = 1.23 kg/m^3)

(27) Water is flowing along a pipe at a rate of 24 m^3/min from point A, area 0.3 m^2, to point B, area 0.03 m^2, which is 15 m lower. If the pressure at A is 180 kN/m^2 and the loss of total head between A and B is 0.6 m, find the pressure at B.

(28) Water flows downwards through a pipe at a rate of 0.9 m^3/min from point A, diameter 100 mm, to point B, diameter 50 mm, which is 1.5 m lower. P_A = 70 kPa, P_B = 50 kPa. Find the head loss between A and B.

(29) Water is flowing along a horizontal pipe of diameter 0.2 m, at a rate of 0.226 m^3/s. The pipe has a 90° threaded elbow, a 90° mitred elbow and a fully open globe valve, all fitted into a short section. Find:

(a) the velocity head;
(b) the loss of head caused by the fittings;
(c) the loss of pressure along the pipe section.

(30) A short piping system is as shown in Figure 3.2.33. Calculate the loss of total head caused by the pipe fittings and hence find the pressure at point B, using the table on p. 155.

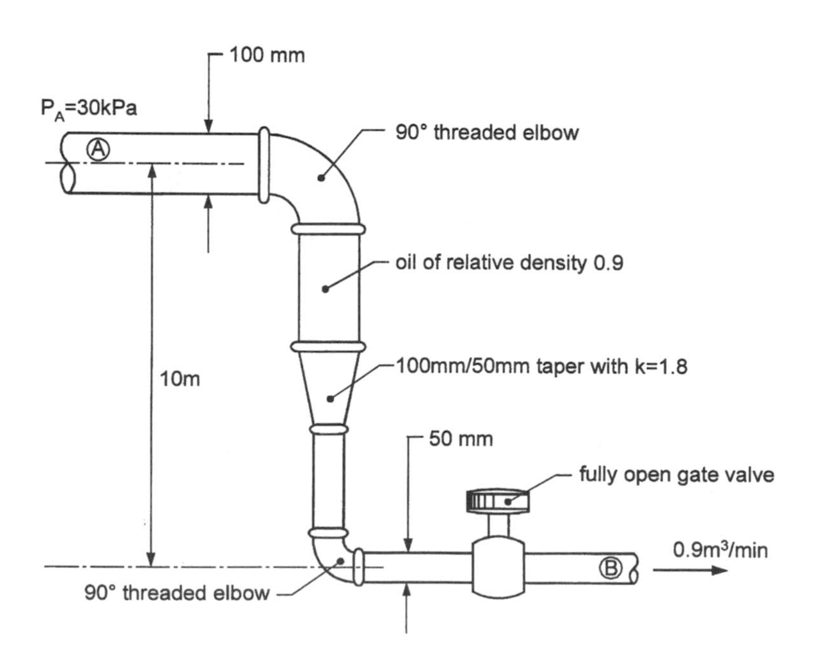

Figure 3.2.33 *A pipe system with pipe fittings*

(31) Calculate the head loss due to friction in a 750 m long pipe of diameter 620 mm when the flow rate is 920 m^3/hour and the friction factor $f = 0.025$.

(32) A pressure drop of 760 kPa is caused when oil of relative density 0.87 flows along a pipe of length 0.54 km and diameter 0.18 m at a rate of 3000 m^3/hour. Calculate the friction factor.

(33) A head loss of 5.6 m is produced when water flows along an 850 m long pipe of radius 0.12 m which has a friction factor of 0.009. Calculate the flow rate of the water in tonnes/hour.

(34) Liquid of relative density 1.18 flows at a rate of 3240 tonnes/hour along a 0.8 km long pipe, of diameter 720 mm and friction factor 0.000 95. Calculate the gradient at which the pipe must fall in order to maintain a constant internal pressure.

(35) The volume rate of flow of a liquid in a 300 mm diameter pipe is 0.06 m^3/s. The liquid has viscosity of $1.783 \quad 10^{-5}$ Pa s and relative density 0.7. The equivalent surface roughness of the pipe is 0.6 mm.

(a) Determine R_e and state whether the flow is turbulent.
(b) Determine f from a Moody chart.
(c) Determine the pressure drop over a 100 m horizontal length of pipe.

(36) Determine the head loss due to friction when water flows through 300 m of 150 mm dia galvanized steel pipe at 50 l/s.

($\mu_{water} = 1.14 \quad 10^{-3}$ Pa s in this case, $K = 0.15$ mm)

(37) Determine the size of galvanized steel pipe needed to carry water a distance of 180 m at 85 l/s with a head loss of 9 m. (Conditions as above in Problem 36.)

This is an example of designing a pipe system where the Moody chart has to be used in reverse. Assume turbulent flow and be prepared to iterate. Finally you must calculate R_e to show that your assumption of the turbulent flow was correct.

(38) A flat rectangular plate is suspended by a hinge along its top horizontal edge. The centre of gravity of the plate is 100 mm below the hinge. A horizontal jet of water of diameter 25 mm impinges normally on to the plate 150 mm below the hinge with a velocity of 5.65 m/s. Find the horizontal force which needs to be applied to the centre of gravity to keep the plate vertical.

(39) The velocity of the jet in Problem 38 is increased so that a force of 40 N is now needed to maintain the vertical position. Find the new value of the jet velocity.

(40) The nozzle of a fire hose produces a 50 mm diameter jet of water when the discharge rate is 0.085 m^3/s. Calculate the force of the water on the nozzle if the jet velocity is 10 times the velocity of the water in the main part of the hose.

(41) A jet of water, of cross-section area 0.01 m^2 and discharge rate 100 kg/s, is directed horizontally into a hemispherical

cup so that its velocity is exactly reversed. If the cup is mounted at one end of a 4.5 m bar which is supported at the other end, find the torque produced about the support by the jet.

(42) A helicopter with a fully laden weight of 19.6 kN has a rotor of 12 m diameter. Find the mean vertical velocity of the air passing through the rotor disc when the helicopter is hovering. (Density of air = 1.28 kg/m^3)

(43) A horizontal jet of water of diameter 25 mm and velocity 8 m/s strikes a rectangular plate of mass 5.45 kg which is hinged along its top horizontal edge. As a result the plate swings to an angle of 30° to the vertical. If the jet strikes at the centre of gravity, 100 mm along the plate from the hinge, find the *horizontal* force which must be applied 150 mm along the plate from the hinge to keep it at that angle.

(44) A 15 mm diameter nozzle is supplied with water at a total head of 30 m. 97% of this head is converted to velocity head to produce a jet which strikes tangentially a vane that is curved back on itself through 165°. Calculate the force on the vane in the direction of the jet (the x-direction) and in the y-direction.

(45) If the jet in Problem 44 is slowed down to 80% of its original value in flowing over the vane, calculate the new x and y forces.

4 Dynamics

Summary

This chapter deals with movement. In the first part the movement is considered without taking into account any forces. This is a subject called kinematics and it is important for analysing the motion of vehicles, missiles and engineering components which move backwards and forwards, by dealing with displacement, speed, velocity and acceleration. These quantities are defined when we look at uniform motion in a straight line. This subject is extended to look at the particular case of motion under the action of gravity, including trajectories. This chapter also looks at how the equations of motion in a straight line can be adapted to angular motion. Finally in the first half the subject of relative velocity is covered as this is very useful in understanding the movement of the individual components in rotating machinery.

In the second part of this chapter we consider the situation where there is a resultant force or moment on a body and so it starts to move or rotate. This topic is known as dynamics and the situation is described by Newton's laws of motion. Once moving forces are involved, we need to look at the mechanical work that is being performed and so the chapter goes on to describe work, power and efficiency. Newton's original work in this area of dynamics was concerned with something called momentum and so this idea is also pursued here, covering the principle of conservation of momentum. The chapter extends Newton's laws and the principle of conservation of momentum to rotary motion, and includes a brief description of d'Alembert's principle which allows a dynamic problem to be converted into a static problem.

Objectives

By the end of this chapter the reader should be able to:

- define displacement, speed, velocity and acceleration;
- use velocity–time graphs and the equations of motion to analyse linear and rotary movement;
- understand motion due to gravity and the formation of trajectories;
- calculate the velocity of one moving object relative to another;
- define the relationships between mass, weight, acceleration and force;
- apply Newton's laws of motion to linear and rotary motion;
- calculate mechanical work, power and efficiency;
- understand the principle of conservation of momentum;
- understand d'Alembert's principle.

4.1 Introduction to kinematics

Kinematics is the name given to the study of movement where we do not need to consider the forces that are causing the movement. Usually this is because some aspect of the motion has been specified. A good example of this is the motion of a passenger lift where the maximum acceleration and deceleration that can be applied during the starting and stopping phases are limited by what is safe and comfortable for the passengers. If we know what value this acceleration needs to be set at, then kinematics will allow a design engineer to calculate such things as the time and the distance that it will take the lift to reach its maximum speed.

Before we can get to that stage, however, we need to define some of the quantities that will occur frequently in the study of movement.

Displacement is the distance moved by the object that is being considered. It is usually given the symbol s and is measured in metres (m). It can be the total distance moved from rest or it can be the distance travelled during one stage of the motion, such as the deceleration phase.

Speed is the term used to describe how fast the object is moving and the units used are metres/second (m/s). Speed is very similar to *velocity* but velocity is defined as a speed *in a particular direction*. This might seem only a slight difference but it is very important. For example, a car driving along a winding road can maintain a constant speed but its velocity will change continually as the driver steers and changes direction to keep the car on the road. We will see in the next section that this means that the driver will have to exert a force on the steering wheel which could be calculated. Often we take speed and velocity as the same thing and use the symbols v or u for both, but do not forget that there is a distinction. If the object is moving at a constant velocity v and it has travelled a distance of s in time t then the velocity is given by

$$v = s/t \tag{4.1.1}$$

Alternatively, if we know that the object has been travelling at a constant velocity v for a time of t then we can calculate the distance travelled as

$$s = vt \tag{4.1.2}$$

Acceleration is the rate at which the velocity is changing with time and so it is defined as the change in velocity in a short time, divided by the short time itself. Therefore the units are metres per second (the units of velocity) divided by seconds and these are written as metres per second^2 (m/s^2). Acceleration is generally given the symbol a. Usually the term acceleration is used for the rate at which an object's speed is increasing, while deceleration is used when the speed is decreasing. Again, do not forget that a change in velocity could also be a directional change at a constant speed.

Having defined some of the common quantities met in the study of kinematics, we can now look at the way that these quantities are linked mathematically.

Velocity is the rate at which an object's displacement is changing with time. Therefore if we were to plot a graph of the object's displacement s against time t then the value of the slope of the line at any point would be the magnitude of the velocity (i.e. the speed). In Figure 4.1.1, an object is starting from the origin of the graph where its displacement is zero at time zero. The line of the graph is straight here, meaning that the

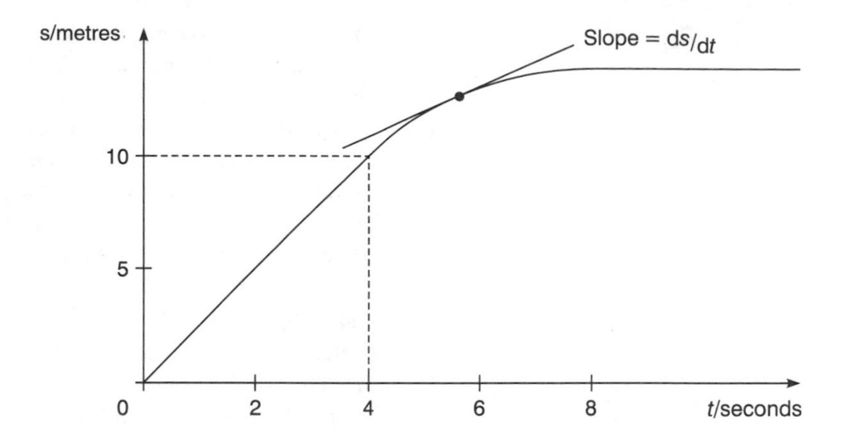

Figure 4.1.1 *A displacement–time graph*

displacement increases at a constant rate. In other words, the speed is constant to begin with and we could measure it by working out the slope of the straight line portion of the graph. The line is straight up to the point where the object has moved by 10 m in 4 s and so the speed for this section is 10 m/4 s = 2.5 m/s. Beyond this section the line starts to become curved as the slope decreases. This shows that the speed is falling even though the displacement is still increasing. We can therefore no longer measure the slope at any point by looking at a whole section of the line as we did for the straight section. We need to work on the instantaneous value of the slope and for this we must adopt the sort of definition of slope that is used in calculus. The instantaneous slope at any point on the curve shown in this graph is ds/dt. This is the rate at which the displacement is changing, which is the velocity and so

$$v = ds/dt \tag{4.1.3}$$

In fact the speed goes on decreasing up to the maximum point on the graph where the line is parallel to the time axis. This means that the slope of the line is zero here and so the object's speed is also zero. The object has come to rest and the displacement no longer changes with time.

This kind of graph is known as a displacement–time graph and is quite useful for analysing the motion of a moving object. However, an even more useful diagram is something known as a velocity–time graph, such as the example shown in Figure 4.1.2.

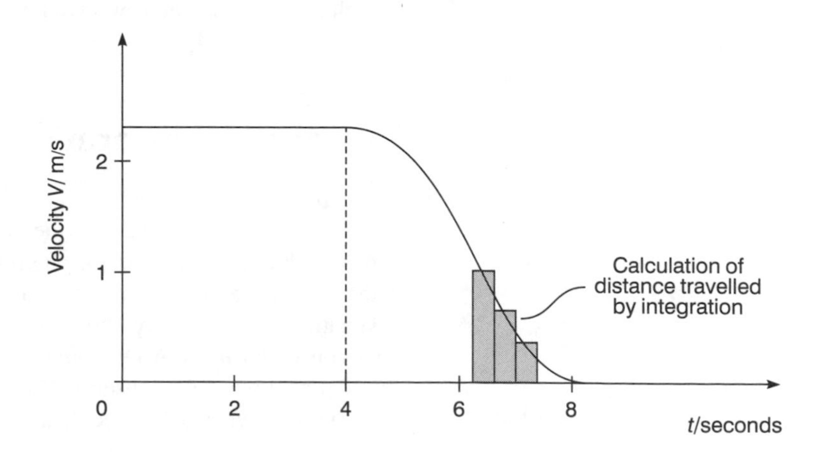

Figure 4.1.2 *A velocity–time graph*

Here the velocity magnitude (or speed) is plotted on the vertical axis against time. The same object's motion is being considered as in Figure 4.1.1 and so the graph starts with the steady velocity of 2.5 m/s calculated above. After a time of 4 s the velocity starts to fall, finally becoming zero as the object comes to rest. The advantage of this type of diagram is that it allows us to study the acceleration of the object, and that is often the quantity that is of most interest to engineers. The acceleration is the rate at which the magnitude of the velocity changes with time and so it is the slope of the line at any point on the velocity–time graph. Over the first portion of the motion, therefore, the acceleration of the object is zero because the line is parallel to the time axis. The acceleration then becomes negative (i.e. it is a deceleration) as the speed starts to fall and the line slopes down.

In general we need to find the slope of this graph using calculus in the way that we did for the displacement–time curve. Therefore

$$a = \mathrm{d}v/\mathrm{d}t \tag{4.1.4}$$

Since $v = \mathrm{d}s/\mathrm{d}t$ we can also write this as

$$a = \mathrm{d}(\mathrm{d}s/\mathrm{d}t)/\mathrm{d}t = \mathrm{d}^2 s/\mathrm{d}t^2 \tag{4.1.5}$$

There is one further important thing that we can get from a velocity–time diagram and that is the distance travelled by the object. To understand how this is done, imagine looking at the object just for a very short time as if you were taking a high-speed photograph of it. The object's velocity during the photograph would effectively be constant because the exposure is so fast that there is not enough time for any acceleration to have a noticeable effect. Therefore from Equation 4.1.2 the small distance travelled during the photograph would be the velocity at that time multiplied by the small length of time it took to take the photograph. On the velocity–time diagram this small distance travelled is represented by the area of the very narrow rectangle formed by the constant velocity multiplied by the short time, as indicated on the magnified part of the whole velocity–time diagram shown here. If we were prepared to add up all the small areas like this from taking a great many high-speed photographs of all the object's motion then the sum would be the total distance travelled by the object. In other words the area underneath the line on a velocity–time diagram is equal to the total distance travelled. The process of adding up all the areas from the multitude of very narrow rectangles is known as integration.

Uniform acceleration

Clearly the analysis of even a simple velocity–time diagram could become quite complicated if the acceleration is continually varying. In this textbook the emphasis is on understanding the basic principles of all the subjects and so from now on we are going to concentrate on the situation where the acceleration is a constant value or at worst a series of constant values. An example of this is shown in Figure 4.1.3.

An object is being observed from time zero, when it has a velocity of u, to time t when it has accelerated to a velocity v. The acceleration a is

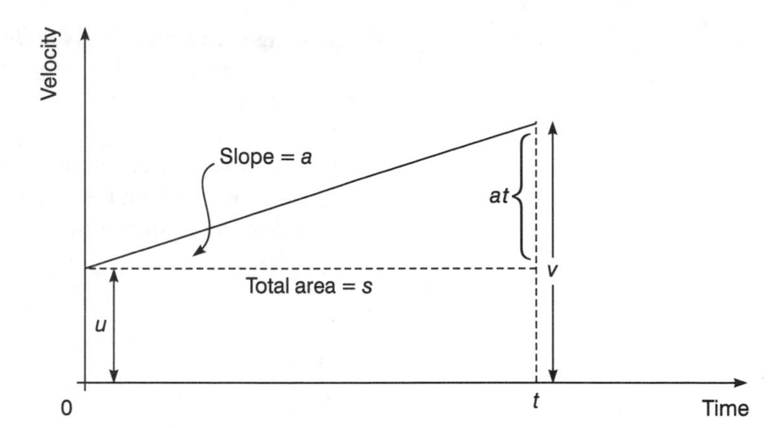

Figure 4.1.3 *Constant acceleration on a velocity–time graph*

the slope of the line on the graph and it is clearly constant because the line between the start and the end of the motion is straight. Now the slope of a line, in geometrical terms, is the amount by which it rises between two points, divided by the horizontal distance between the two points. Therefore considering that the line rises by an amount $(v - u)$ on the vertical axis while moving a distance of $(t - 0)$ along the horizontal axis of the graph, the slope a is given by

$$a = (v - u)/(t - 0)$$

or

$$a = (v - u)/t$$

Multiplying through by t gives

$$at = v - u$$

and finally

$$v = u + at \tag{4.1.6}$$

This is the first of four equations which are collectively known as the *equations of uniform motion in a straight line*. Having found this one it is time to find the other three. As mentioned above, the area under the line on the graph is the distance travelled s and the next stage is to calculate this.

We can think of the area under the line as being made up of two parts, a rectangle representing the distance that the object would have travelled if it had continued at its original velocity of u for all the time, plus the triangle which represents extra distance travelled due to the fact that it was accelerating.

The area of a rectangle is given by height times width and so the distance travelled for constant velocity is ut.

The area of a triangle is equal to half its height times its width. Now the width is the time t, as for the rectangle, and the height is the increase in velocity $(v - u)$. However, it is more useful to note that this velocity increase is also given by at from Equation (4.1.6) as this brings time into the calculation again. The triangle area is then $at^2/2$.

Adding the two areas gives the total distance travelled

$$s = ut + at^2/2 \quad (4.1.7)$$

This is the second equation of uniform motion in a straight line. We could have worked out the area under the line on the graph in a different way, however, as shown in Figure 4.1.4.

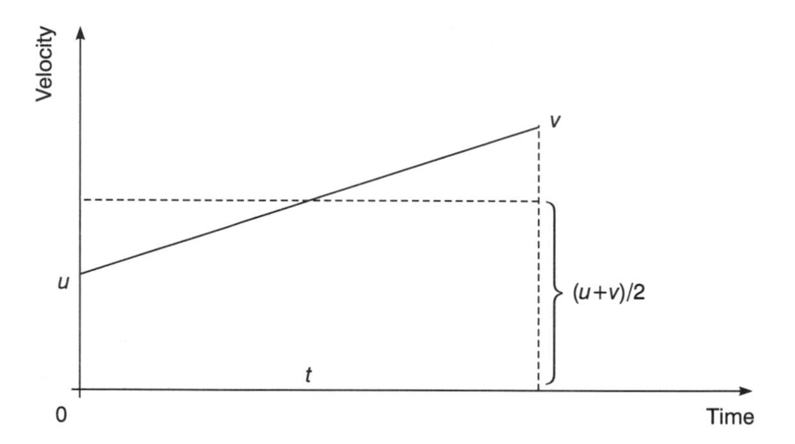

Figure 4.1.4 *Mean velocity on a velocity–time graph*

The real area, as calculated above, has now been replaced by a simple rectangle of the same width but with a mean height which is half way between the two extreme velocities, u and v. This is a mean velocity, calculated as $(u + v)/2$. Therefore the new rectangular area, and hence the distance travelled, is given by

$$s = t(u + v)/2 \quad (4.1.8)$$

This is the third equation of uniform motion in a straight line and is the last one to involve time t. We need another equation because sometimes it is not necessary to consider time explicitly. We can achieve this by taking two of the other equations and eliminating time from them.

From the first equation we find that

$$t = (v - u)/a$$

This version of t can now be substituted into the third equation to give

$$s = ((v - u)/a)\ ((u + v)/2)$$

This can be simplified to give finally

$$v^2 = u^2 + 2as \quad (4.1.9$$

This is the fourth and final equation of uniform motion in a straight line. Other substitutions and combinations are possible but these four equations are enough to solve an enormous range of problems involving movement in a line.

Example 4.1.1

An electric cart is to be used to transport disabled passengers between the two terminals at a new airport. The cart is capable of accelerating at a uniform rate of 1.8 m/s^2 and tests have shown that a uniform deceleration of 2.4 m/s^2 is comfortable for the passengers. Its maximum speed is 4 m/s. Calculate the fastest time in which the cart can travel the distance of 350 m between the terminals.

The first step in solving this problem is to draw a velocity–time diagram, as shown in Figure 4.1.5

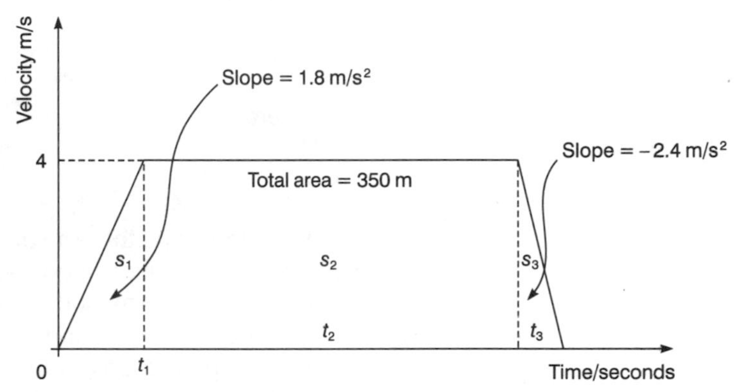

Figure 4.1.5 *Velocity–time graph for the electric cart*

Starting with the origin of the axes, the cart starts from rest with velocity zero at time zero. It accelerates uniformly, which means that the first part of the graph is a straight line which goes up. This acceleration continues until the maximum velocity of 4 m/s is reached. We must assume that there can be a sudden switch from the acceleration phase to the constant velocity phase and so the next part of the graph is a straight line which is parallel to the time axis. This continues until near the end of the journey, when again there is a sudden change to the last phase which is the deceleration back to rest. This is represented by a straight line dropping down to the time axis to show that the cart comes to rest again. We cannot put any definite values on the time axis yet but we can write in that the area under the whole graph line is equal to the distance travelled and therefore has a value of 350 m. Lastly it must be remembered that all the four equations of motion developed above apply to movement with uniform acceleration. The acceleration can have a fixed positive or negative value, or it can be zero, but it must not change. Therefore the last thing to mark on the diagram is that we need to split it up into three portions to show that we must analyse the motion in three sections, one for each value of acceleration.

The overall problem is to calculate the total time so it is best to start by calculating the time for the acceleration and deceleration phases. In both cases we know the initial and final velocities and the acceleration so the equation to use is

$$v = u + at$$

For the start

$$4 = 0 + 1.8t_1$$

Therefore

$$t_1 = 2.22\,\text{s}$$

Similarly for the end

$0 = 4 - 2.4t_3$ (note that deceleration is negative)

Therefore

$$t_3 = 1.67\,\text{s}$$

We cannot find the time of travel at constant speed in this way and so we must ask ourselves what we have not so far used out of the information given in the question. The answer is that we have not used the total distance yet. Therefore the next step is to look at the three distances involved.

We can find the distance travelled during acceleration using

$v^2 = u^2 + 2as$ or $s = ut + at^2/2$

since we have already calculated the time involved. However, as a point of good practice, it is better to work with the original data where possible and so the first of these two equations will be used.

For the acceleration

$$16 = 0 + 2 \quad 1.8\,s_1$$

Therefore

$$s_1 = 4.44\,\text{m}$$

Similarly for deceleration

$$0 = 16 - 2 \quad 2.4\,s_3$$
$$s_3 = 3.33\,\text{m}$$

The distance involved in stopping and starting is

$$s_1 + s_3 = 7.77\,\text{m}$$

and so

$$s_2 = 350 - 7.77$$
$$= 342.22\,\text{m}$$

Since this distance relates to constant speed, the time taken for this portion of the motion is simply

$$t_2 = 342.22\,\text{m}/4\,\text{m/s}$$
$$= 85.56\,\text{s}$$

Hence the total time of travel is (85.56 s + 2.22 s + 1.67 s)

= <u>89.45 s</u>

Note that, within reason, the route does not have to be a straight line. It must be a fairly smooth route so that the accelerations involved in turning corners are small, but otherwise we can apply the equations of uniform motion in a straight line to some circuitous motion.

Motion under gravity

One of the most important examples of uniform acceleration is the acceleration due to gravity. If an object is allowed to fall in air then it accelerates vertically downwards at a rate of approximately 9.81 m/s^2. After falling for quite a time the velocity can build up to such a point that the resistance from the air becomes large and the acceleration decreases. Eventually the object can reach what is called a terminal velocity and there is no further acceleration. An example of this is a free-fall parachutist. For most examples of interest to engineers, however, the acceleration due to gravity can be taken as constant and continuous. They are worthwhile considering, therefore, as a separate case.

The first thing to note is that the acceleration due to gravity is the same for any object, which is why heavy objects only fall at the same rate as light ones. In practice a feather will not fall as fast as a football, but that is simply because of the large air resistance associated with a feather which produces a very low terminal velocity. Therefore we generally do not need to consider the mass of the object. Problems involving falling masses therefore do not involve forces and are examples of kinematics which can be analysed by the equations of uniform motion developed above. This does not necessarily restrict us to straight line movement, as we saw in the worked example, and so we can look briefly at the subject of trajectories.

Example 4.1.2

Suppose a tennis ball is struck so that it leaves the racquet horizontally with a velocity of 25 m/s and at a height above the ground of 1.5 m. How far will the ball travel horizontally before it hits the ground?

Clearly the motion of the ball will be a curve, known as a trajectory, because of the influence of gravity. At first sight this does not seem like the sort of problem that can be analysed by the equations of uniform motion in a straight line. However, the ball's motion can be resolved into horizontal motion and vertical motion, thus allowing direct application of the four equations that are at our disposal. The horizontal motion is essentially at a constant velocity because the air resistance on a tennis ball is low and so there will be negligible slowing during the time it takes for the ball to reach the ground. The vertical motion is a simple application of falling under gravity with constant acceleration (known as g) since again the resistance will be low and there is no need to consider any terminal velocity. Therefore it is easy to calculate the time taken for the ball to drop by the vertical height of 1.5 m. We know the distance of travel s (1.5 m), the starting velocity u (0 m/s in a vertical sense) and the acceleration a or g (9.81 m/s^2) and so the equation to use is

$$s = ut + at^2/2$$

Therefore

$$1.5 = 0 + 9.81t^2/2$$

$$t = 0.553\ \text{s}$$

This value can be thought of not just as the time for the ball to drop to the ground but also as the time of flight for the shot. To find the distance travelled along the ground all that needs to be done is to multiply the horizontal velocity by the time of flight.

$$\text{Distance travelled} = 25 \quad 0.553$$
$$= 13.8\ \text{m}$$

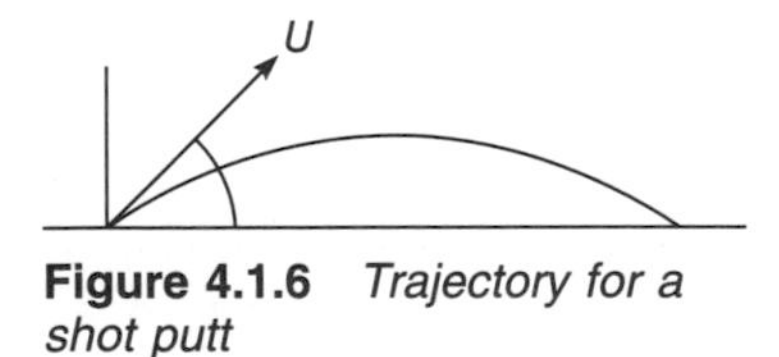

Figure 4.1.6 *Trajectory for a shot putt*

One of the questions that arises often in the subject of trajectories, particularly in its application to sport, is how to achieve the maximum range for a given starting speed. An example of this would be the Olympic event of putting the shot (Figure 4.1.6), where the athlete might wonder whether it is better to aim high and therefore have a long time of flight or to aim low and concentrate the fixed speed of the shot into the horizontal direction.

If we suppose the shot starts out with a velocity of u at an angle of θ to the ground, then the initial velocity in the vertical direction is $u \sin \theta$ upwards and the constant horizontal velocity is $u \cos \theta$. The calculation proceeds like the example above but care has to be taken with the signs because the shot goes upwards in this example before dropping back to the ground. We can take the upwards direction as positive and so the acceleration g will be negative.

By starting with the vertical travel it is possible to work out the time of flight. For simplicity here it is reasonable to neglect the height at which the shot is released and so the shot starts and ends at ground level. In a vertical sense, therefore, the overall distance travelled is zero and the shot arrives back at the ground with the same speed but now in a negative direction. The time of flight t_f is therefore twice the time it takes for the shot to reach its highest point and momentarily have a zero vertical velocity.

Therefore using

$$v = u + at$$

$$0 = u \sin\theta - gt_f/2$$

$$t_f = (2u \sin\theta)/g$$

The horizontal distance travelled is equal to velocity multiplied by time

$$s = (u \cos\theta)(2u \sin\theta)/g$$

The actual distance travelled is not of interest, it is simply the angle required to produce a *maximum* distance that is needed. This depends on finding the maximum value of $\cos\theta \sin\theta$ for angles between zero (throwing it horizontally) and a right angle (throwing it vertically upwards – very dangerous!). We could do this with calculus but it is just as easy to use common sense which tells us that a sine curve and a cosine curve are identical but displaced by 90° and therefore the product will be a maximum for $\theta = 45°$. In other words, if air resistance can be neglected, it is best to launch an object at 45° to the ground if maximum range is required.

Angular motion

Before leaving the equations of motion and moving on to dynamics, it is necessary to consider angular motion. This includes motion in a circle and rotary motion, both of which are important aspects of the types of movement found in machinery. At first sight it seems that the equations of motion cannot be applied to angular motion: the velocity is continually changing because the direction changes as the object revolves and there is no distance travelled for an object spinning on the spot. However, it turns out that they can be applied directly if a few substitutions are made to find the equivalents of velocity, acceleration and distance.

First the distance travelled needs to be replaced by the angle θ through which the object, such as a motor shaft, has turned. Note that this must be measured in radians, where 1 radian equals 57.3°.

Next the linear velocity needs to be replaced by the angular velocity ω. This is measured in radians per second, abbreviated to rad/s.

Finally the linear acceleration must be replaced by the angular acceleration α which is measured in rad/s^2.

The four equations of motion for uniform angular acceleration then become

$$\omega_2 = \omega_1 + \alpha t \qquad (4.1.10)$$

$$\theta = \omega_1 t + \alpha t^2/2 \qquad (4.1.11)$$

$$\theta = t(\omega_1 + \omega_2)/2 \qquad (4.1.12)$$

$$\omega_2^2 = \omega_1^2 + 2\alpha\theta \qquad (4.1.13)$$

Example 4.1.3

A motor running at 2600 rev/min is suddenly switched off and decelerates uniformly to rest after 10 s. Find the angular deceleration and the number of rotations to come to rest.

For the first part of this problem, the initial and final angular velocities are known, as is the time taken for the deceleration. The equation to use is therefore:

$$\omega_2 = \omega_1 + \alpha t$$

First it is necessary to put the initial angular velocity into the correct units. One revolution is equal to 2π radians and so 2600 revolutions equals 16 337 radians. This takes place in 1 minute, or 60 seconds and so the initial angular velocity of 2600 rev/min becomes 272.3 rad/s.

Hence

$$0 = 272.3 - \alpha 10$$

$$\alpha = \underline{27.23 \text{ rad/s}^2}$$

For the second part the equation to use is

$$\begin{aligned} \theta &= t(\omega_1 + \omega_2)/2 \\ &= 10(272.3 + 0)/2 \\ &= \underline{1361 \text{ rad}} \end{aligned}$$

This is not the final answer as it is the number of revolutions that is required. There are 2π radians in one revolution and so this answer needs to be divided by 2   3.141 59, giving a value of 216.7 revolutions.

Motion in a circle

Angular motion can apply not just to objects that are rotating but also to objects that are moving in a circle. An example of this is a train moving on a circular track, as happens in some underground systems around the world. The speed of such a train would normally be expressed in conventional terms as a linear velocity in metres per second (m/s). It could also be expressed as the rate at which an imaginary line from the centre of the track circle to the train is sweeping out an angle like the hands on a clock, i.e. as an angular velocity. The link between the two is

Angular velocity = linear velocity/radius

$$\omega = v/r \qquad (4.1.14)$$

Returning to the original definition of acceleration as either a change in the direction or the magnitude of a velocity, it is apparent that the train's velocity is constantly changing and so the train is accelerating even though it maintains a constant speed. The train's track is constantly turning the train towards the centre and so the acceleration is inwards. The technical term for this is centripetal acceleration. It is a linear acceleration (i.e. with units of m/s^2) and its value is given by

$$a = v^2/r = r\omega^2 \tag{4.1.15}$$

The full importance of this section will become apparent in the next section, on dynamics, where the question of the forces that produce these accelerations is considered. For the time being just imagine that you are in a car that is going round a long bend. Unless you deliberately push yourself away from the side of the car on the outside of the bend and lean towards the centre, thereby providing the inwards force to accompany the inward acceleration, you will find that you are thrown outwards. Your body is trying to go straight ahead but the car is accelerating towards the geometrical centre of the bend.

Example 4.1.4

A car is to be driven at a steady speed of 80 km/h round a smooth bend which has a radius of 30 m. Calculate the linear acceleration towards the centre of the bend experienced by a passenger in the car.

This is a straightforward application of Equation (4.1.15) but the velocity needs to be converted to the correct units of m/s.

Velocity of the car is $v = 80\text{ km/h} = 80\,000/(60 \quad 60)\text{ m/s}$

$= 22.222\text{ m/s}$

Linear acceleration $a = v^2/r = 22.222^2/30 = \underline{16.5\text{ m/s}^2}$

Clearly this is a dramatic speed at which to go round this bend as the passenger experiences an acceleration which is nearly twice that due to gravity. In fact the car would probably skid out of control!

Problems 4.1.1

(1) A car is initially moving at 6 m/s and with a uniform acceleration of 2.4 m/s^2. Find its velocity after 6 s and the distance travelled in that time.

(2) A ball is falling under gravity ($g = 9.81\text{ m/s}^2$). At time $t = 0$ its velocity is 8 m/s and at the moment it hits the ground its velocity is 62 m/s. How far above the ground was it at $t = 0$? At what time did it hit the ground? What was its average velocity?

(3) A train passes station A with velocity 80 km/hour and acceleration 3 m/s^2. After 30 s it passes a signal which

instructs the driver to slow down to a halt. The train then decelerates at 2.4 m/s^2 until it stops. Find the total time and total distance from the station A when it comes to rest.

(4) A telecommunications tower, 1000 m high, is to have an observation platform built at the top. This will be connected to the ground by a lift. If it is required that the travel time of the lift should be 1 minute, and its maximum velocity can be 21 m/s, find the uniform acceleration which needs to be applied at the start and end of the journey.

(5) A set of buffers is being designed for a railway station. They must be capable of bringing a train uniformly to rest in a distance of 9 m from a velocity of 12 m/s. Find the uniform deceleration required to achieve this, as a multiple of g, the acceleration due to gravity.

(6) A motor-bike passes point A with a velocity of 8 m/s and accelerates uniformly until it passes point B at a velocity of 20 m/s. If the distance from A to B is 1 km determine:

(a) the average speed;
(b) the time to travel from A to B;
(c) the value of the acceleration.

(7) A light aeroplane starts its take off from rest and accelerates at 1.25 m/s^2. It suddenly suffers an engine failure and decelerates back to a halt at 1.875 m/s^2, coming to rest after a total distance of 150 m along the runway. Find the maximum velocity reached, and the total time taken.

(8) A car accelerates uniformly from rest at 2 m/s^2 to a velocity of V_1. It then continues to accelerate up to a velocity of 34 m/s, but at 1.4 m/s^2. The total distance travelled is 364 m. Calculate the value of V_1, and the times taken for the two stages of the journey.

(9) A large airport is installing a 'people mover' which is a sort of conveyor belt for shifting passengers down long corridors. The passengers stand on one end of the belt and a system of sliding sections allows them to accelerate at 2 m/s^2 up to a maximum velocity of V_{max}. A similar system at the other end brings them back to rest at a deceleration of 3 m/s^2. The total distance travelled is 945 m and the total time taken is 60 s. Find V_{max} and the times for the three stages of the journey.

(10) A car accelerates uniformly from 14 m/s to 32 m/s in a time of 11.5 s. Calculate:

(a) the average velocity;
(b) the distance travelled;
(c) the acceleration.

(11) The same car then decelerates uniformly to come to rest in a distance of 120 m. Calculate:

(a) the deceleration;
(b) the time taken.

(12) A train passes a signal at a velocity of 31 m/s and with an acceleration of 2.4 m/s^2. Find the distance it travels in the next 10 s.

(13) A piece of masonry is dislodged from the top of a tower. By the time it passes the sixth floor window it is travelling at 10 m/s. Find its velocity 5 s later.

(14) If the total time for the masonry in Problem 13 to reach the ground is 9 s, calculate:

(a) the height of the tower;
(b) the height of the sixth floor window;
(c) the masonry's velocity as it hits the ground.

(15) A tram which travels backwards and forwards along Blackpool's famous Golden Mile can accelerate at 2.5 m/s^2 and decelerate at 3.1 m/s^2. Its maximum velocity is 35 m/s. Calculate the shortest time to travel from the terminus at the start of the line to the terminus at the other end, a distance of 1580 m, if:

(a) the tram makes no stops on the way;
(b) the tram makes one stop of 10 s duration.

4.2 Dynamics – analysis of motion due to forces

In the chapter on statics we shall consider the situation where all the forces and moments on a body add up to zero. This is given the name of equilibrium and in this situation the body will remain still; it will not move from one place to another and it will not rotate. In the second half of this chapter we shall consider the situation where there is a resultant force or moment on the body and so it starts to move or rotate. This topic is known as dynamics and the situation is described by Newton's laws of motion which are the subject of most of the first part of this section, where linear motion is introduced.

Most of this section is devoted to the work of Sir Isaac Newton and is therefore known as Newton's laws of motion. When Newton first published these laws back in the seventeenth century he caused a great deal of controversy. Even the top scientists and mathematicians of the day found difficulty in understanding what he was getting at and hardly anyone could follow his reasoning. Today we have little difficulty with the topic because we are familiar with concepts such as gravity and acceleration from watching astronauts floating around in space or satellites orbiting the earth. We can even experience them for ourselves directly on the roller coaster rides at amusement parks.

Newton lived in the second half of the seventeenth century and was born into a well-to-do family in Lincolnshire. He proved to be a genius at a very early age and was appointed to a Professorship at Cambridge University at the remarkably young age of 21. He spent his time investigating such things as astronomy, optics and heat, but the thing for which he is best remembered is his work on gravity and the laws of motion. For this he not only had to carry out an experimental study of forces but also he had to invent the subject of calculus.

According to legend Newton came up with the concept of gravity by studying a falling apple. The legend goes something like this.

While Newton was working in Cambridge, England became afflicted by the Great Plague, particularly in London. Fearing that this would soon spread to other cities in the hot summer weather, Newton decided to head back to the family's country estate where he soon found that there was little to do. As a result he spent a great deal of time sitting under a Bramley apple tree in the garden. He started to ponder why the apples were hanging downwards and why they eventually all fell to earth, speeding up as they fell. When one of the apples suddenly fell off and hit him on the head he had a flash of inspiration and realized that it was all to do with forces.

There is some truth in the legend and certainly by the time it was safe for him to return to Cambridge he had formulated the basis of his masterwork. After many years of subsequent work Newton was eventually able to publish his ideas in a finished form. Basically he stated that:

- Bodies will stay at rest or in steady motion unless acted upon by a resultant force.
- A resultant force will make a body accelerate in the direction of the force.
- Every action due to a force produces an equal and opposite reaction.

These are known as Newton's first, second and third laws of motion respectively.

The first law of motion will be covered fully in Chapter 5 on Statics and so we shall concentrate on the other two. We shall start by looking at the relationship between force, acceleration, mass and gravity. For the time being, however, let us just finish this section by noting that the unit of force, the newton, is just about equal to the weight of one Bramley apple!

Newton's second law of motion

Let us go back to the legend of Newton and the apple. From the work on statics we will find that the apple stays on the tree as long as the apple stalk is strong enough to support the weight of the apple. As the apple grows there will come a point when the weight is too great and so the stalk will break and the apple falls. The quantity that is being added to the apple as it grows is *mass*. Sometimes this is confused with weight but there have now been many examples of fruits and seeds being grown inside orbiting spacecraft where every object is weightless and would float if not anchored down.

Mass is the amount of matter in a body, measured in kilograms (kg).

The reason that the apple hangs downwards on the tree, and eventually falls downwards, is that there is a force of attraction between the earth and any object that is close to it. This is the gravitational force and is directed towards the centre of the earth. We experience this as a vertical, downward force. The apple is therefore pulled downwards by the effect of gravity, commonly known as its weight. Any apple that started to grow upwards on its stalk would quickly bend the stalk over and hang down due to this gravitational effect. As the mass of the apple increases due to growth, so does the weight until the point is reached where the weight is just greater than the largest force that can be tolerated by the stalk. As a result the stalk breaks.

Now let us look at the motion of the apple once it has parted company with the tree. It still has mass, of course, as this is a fundamental property to do with its size and the amount of matter it contains. Therefore it must also have a weight, since this is the force due to gravity acting on the apple's mass. Hence there is a downwards force on the apple which makes it start to move downwards and get faster and faster. In other words, as we found in the first section of this chapter, it accelerates. What Newton showed was that the force and mass for any body were linked to this acceleration for any kind of motion, not just falling under gravity.

What Newton stated as his Second Law of Motion may be written as follows.

A body's acceleration *a* is proportional to the resultant force *F* on the body and inversely proportional to the mass *m* of the body.

This seems quite complicated in words but is very simple mathematically in the form that we shall use.

$$F = ma \qquad (4.2.1)$$

Let us use this first to look at the downward acceleration caused by gravity. The gravitational force between two objects depends on the two masses. For most examples of gravity in engineering one of the objects is the earth and this has a constant mass. Therefore the weight of an object, defined as the force due to the earth's gravity on the object, is proportional to the object's mass. We can write this mathematically as

$$W = mC$$

where C is a constant.

Now from Newton's second law this force W must be equal to the body's mass m multiplied by the acceleration a with which it is falling.

$$W = ma$$

By comparison between these two equations, it is clear that the acceleration due to gravity is constant, irrespective of the mass of the object. In other words, all objects accelerate under gravity at the same rate no matter whether they are heavy or light. The acceleration due to gravity has been determined experimentally and found to have a value of $9.81\,\text{m/s}^2$. It is given a special symbol g to show that it is constant.

The weight of an object is therefore given by

$$W = mg$$

and this is true even if the object is not free to fall and cannot accelerate.

Sometimes this relationship between weight and mass can be confusing. If you were to go into a shop and ask for some particular amount of cheese to be cut for you then you would ask for a specified *weight* of cheese and it would be measured out in kilograms. As far as scientists and engineers are concerned you are really requesting a *mass* of cheese; a weight is a force and is measured in newtons.

Before we pursue the way in which this study of dynamics is applied to engineering situations, we must not forget that force is a vector quantity, i.e. it has direction as well as magnitude. Consider the case of two astronauts who are manoeuvring in the cargo bay of a space shuttle with the aid of thruster packs on their backs which release a jet of compressed gas to drive them along (Figure 4.2.1).

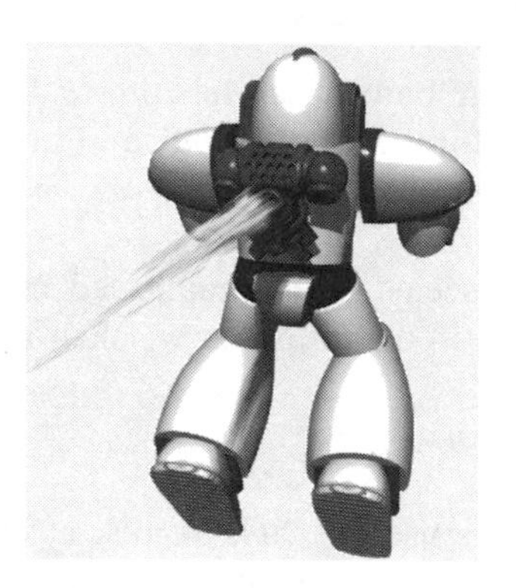

Figure 4.2.1 *Two astronauts with jet packs*

Both packs are the same and deliver a force of 20 N. The first astronaut has a mass of 60 kg and has the jet nozzle turned so it points directly to his left. The second astronaut has a mass of 80 kg and has the nozzle pointing directly backwards. Let us calculate the accelerations produced.

Applying Newton's second law to the first astronaut

$$F = m_1 a_1$$

$$20 = 60a_1$$

$$a_1 = 0.333\,\text{m/s}^2$$

This acceleration will be in the direction of the force, which in turn will be directly opposed to the direction of the nozzle and the gas flow. The astronaut will therefore move to the right.

For the second astronaut

$$F = m_2 a_2$$

$$20 = 80a_2$$

$$a_2 = 0.250\,\text{m/s}^2$$

This acceleration will be away from the jet flow and so it will be directly forward.

This example is very simple because there is no resistance to prevent the acceleration occurring and there are no other forces acting on the astronauts. In general we need to consider friction and we need to calculate the resultant force on a body, as in this next example.

Example 4.2.1

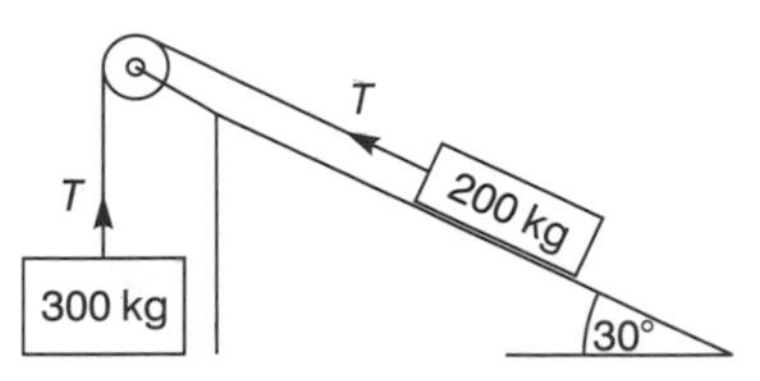

Figure 4.2.2 *The ramp and pulley system*

A large mass is being used to raise a smaller mass up a ramp with a cable and pulley system, as shown in Figure 4.2.2. If the coefficient of friction between the small mass and the ramp is 0.3, find the distance travelled by the large mass in the first 25 s after the system is released from rest.

This is an application of Newton's second law. We need to find the acceleration in order to calculate the distance travelled using the equations of uniform motion that were introduced in the chapter on kinematics. We already know the masses involved and so the remaining step is to calculate the resultant force on the large mass. Clearly the large mass has a weight which will act vertically down and this is given by

$$W_1 = 300 \quad 9.81\,\text{N}$$
$$= 2943\,\text{N}$$

This is opposed by the tension force T in the cable which acts upwards and so the resultant downwards force on the large mass is

$$2943 - T$$

The acceleration is therefore given by

$$2943 - T = 300a$$

We can also calculate this acceleration by looking at the other mass, since the two masses are linked and must have the same acceleration. The same tension force will also act equally on the other side of the pulley, which is assumed to be frictionless. There are three forces acting on the smaller mass along the plane of the ramp: the tension force pulling up the slope, the component of the body's weight acting down the slope and the friction force opposing any motion. The resultant force on the smaller mass is therefore

$$T - (200 \quad 9.81 \quad \sin 30°) - (0.3 \quad 200 \quad 9.81 \quad \cos 30°)$$
$$= T - 981 - 509.7$$
$$= T - 1490.7$$

The acceleration is given by

$$T - 1490.7 = 200a$$

We therefore have two equations which contain the two unknowns T and a and so we can solve them simultaneously by adding them to eliminate T.

$$2942 - 1490.7 = 500a$$
$$a = 2.90\,\text{m/s}^2$$

Since we know the acceleration and the time of travel, we can find the distance travelled using the following formula

$$s = ut + at^2/2$$

$$= 0 + 2.90 \quad 25^2/2$$

$$= \underline{906\,\text{m}}$$

Rotary motion

In practice most machines involve rotary motion as well as linear motion. This could be such examples as electric motors, gears, pulleys and internal combustion engines. Therefore if we need to calculate how fast a machine will reach full speed (in other words calculate the acceleration of all its components) then we must consider rotary acceleration, and the associated torques, as well as linear acceleration. Fortunately Newton's second law of motion applies equally well to rotary motion provided we use the correct version of the formula.

Suppose we are looking at the acceleration of a solid disc mounted on a shaft and rotated by a pull cord wrapped around its rim. We cannot use the conventional form of Newton's second law, $F = ma$, because although there is a linear force being applied in the form of the tension in the cord, the acceleration is definitely not in a line. We therefore cannot identify an acceleration a in units of m/s^2. Furthermore some of the mass of the disc is close to the axle and not moving from one point to another, only turning on the spot, while some of the mass is moving very quickly at the rim. Now in the section on kinematics we saw how the equations of uniform motion could be adapted for rotary motion by substituting the equivalent rotary quantity instead of the linear term. We found that linear distance travelled s could be replaced by angle of rotation θ, while linear acceleration a could be replaced by the angular acceleration α, where $\alpha = a/r$. It is also quite clear that if we are looking at rotary motion then we should be using the torque $\tau = Fr$ instead of force. All that remains is to find the quantity that will be the equivalent of mass in rotary motion.

The mass of an object is really the resistance that the object offers to being moved in a straight line even when there is no friction. It can also be called the inertia of the object; an object with a large mass will accelerate much slower than one with a low mass under the action of an identical force, as we saw in the example of the two astronauts. We are therefore looking for the resistance that an object offers to being rotated in the absence of friction. This must be a combination of mass and shape because a flywheel where most of the mass is concentrated in the rim, supported by slender spokes, will offer much more resistance than a uniform wheel of the same mass. The quantity we are seeking is something known as the *mass moment of inertia* or simply as the *moment of inertia*, I, which has units of kilograms times metres squared (kg m^2). For a uniform disc, which is the most common shape found in engineering objects such as pulleys, this is given by the formula

$$I = mr^2/2$$

We are now in a position to write Newton's second law in a form suitable for rotary motion

$$\tau = I\alpha \tag{4.2.2}$$

With this equation we can solve problems involving real engineering components that utilize rotary motion and then go on to consider combinations of linear and rotary components.

One of the most common rotary motion devices is the flywheel. This is a massive wheel which is mounted on a shaft to deliberately provide a great deal of resistance to angular acceleration. The purpose of the device is to ensure a smooth running speed, especially for something that is being driven by a series of pulses such as those coming from an internal combustion engine.

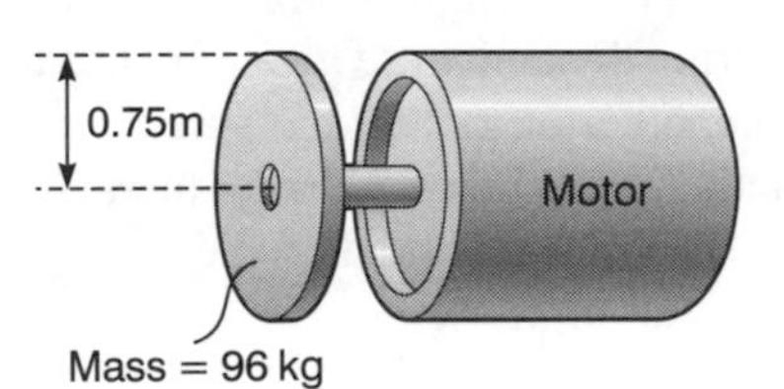

Figure 4.2.3 *The flywheel and motor*

Example 4.2.2

An electric motor is being used to accelerate a flywheel from rest to a speed of 3000 rev/min in 40 s (Figure 4.2.3). The flywheel is a uniform disc of mass 96 kg and radius 0.75 m. Calculate the angular acceleration required and hence find the output torque which must be produced by the motor.

The first part of this problem is an application of kinematics to rotary motion where we know the time, the initial velocity and the final velocity, and we need to find the acceleration. The equation to use is therefore

$$\omega_2 = \omega_1 + \alpha t$$

The start velocity is zero because the flywheel begins from rest. The final angular velocity is

$$3000\ \text{rev/min} = 3000 \quad 2\pi/60 = 314.2\ \text{rad/s}$$

Therefore

$$314.2 = 0 + 40\alpha$$

$$\alpha = \underline{7.854\ \text{rad/s}^2}$$

Now that the angular acceleration is known it is possible to calculate the torque or couple which is necessary to produce it, once the moment of inertia is calculated. The flywheel is a uniform disc so the moment of inertia is given by

$$I = mr^2/2 = 96 \quad 0.75^2/2 = 27\ \text{kg m}^2$$

Therefore the driving torque is given by

$$\tau = I\alpha$$

$$= 27 \quad 7.854$$

$$= \underline{212\ \text{N m}}$$

This general method can be applied to much more complicated shapes such as pulleys which may be regarded as a series of uniform discs on a common axis. Because all the components are all on the same axis, the individual moments of inertia may all be added to give the value for the single shape. Sometimes the moment of inertia of a rotating object is given directly by the manufacturer but there is also another standard way in which they can quote the value for designers. This is in terms of something called the *radius of gyration*, k. With this the moment of inertia of a body is found by multiplying the mass of the body by the square of the radius of gyration such that

$$I = mk^2 \tag{4.2.3}$$

Work, energy and power

So far in this treatment of dynamics we have discussed forces in terms of the accelerations that they can produce but in engineering terms that is only half the story. Forces in engineering dynamics are generally used to move something from one place to another or to alter the speed of a piece of machinery, particularly rotating machinery. Clearly there is some work involved and so we need to be able to calculate the energy that is expended. We start with a definition of work.

$$\begin{aligned} \text{Mechanical work} &= \text{force} \quad \text{distance moved in the direction of the force} \\ &= Fd \end{aligned} \tag{4.2.4}$$

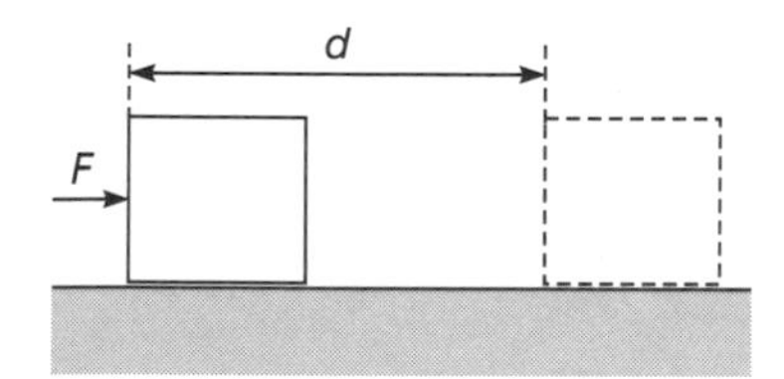

Figure 4.2.4 *A block being moved*

To understand this, suppose we have a solid block resting on the floor and a force of F is required to move it slowly against the frictional resistance offered by the floor (Figure 4.2.4).

The force is applied for as long as it takes for the mass to be moved by the required distance d. Commonsense tells us that the force must be applied in the direction in which the block needs to move. The work performed is then Fd, which is measured in joules (J). This work is expended as heat which is soon dissipated and is very difficult to recover.

Now suppose we want to lift the same block vertically by a height of h. A force needs to be applied vertically upwards to just overcome the weight mg of the block (remember that weight is a vertical downward force equal to the mass multiplied by the acceleration with which the block would fall down due to gravity if it was unsupported). The mechanical work performed in lifting the block to the required height is therefore

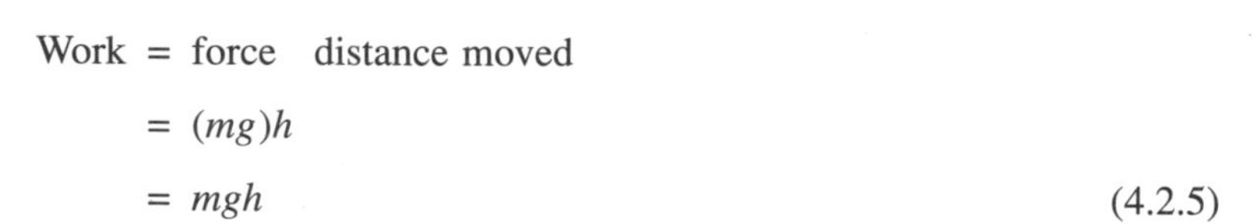

$$\begin{aligned} \text{Work} &= \text{force} \quad \text{distance moved} \\ &= (mg)h \\ &= mgh \end{aligned} \tag{4.2.5}$$

This work is very different from that performed in the first example where it was to overcome friction. Here the work is effectively stored because the block could be released at any time and would fall back to the floor, accelerating as it went. For this reason the block is said to have a *potential energy* of mgh when it is at a height of h because that amount of work has been carried out at some time in raising it to that level. If the block was now replaced by an item of machinery of the same mass which had been raised piece by piece and then assembled at the top, the potential energy associated with it would still be the same. The object does not have to have been physically raised in one go for the potential energy to be calculated using this equation.

Let us look at another example of potential energy in order to understand better where the name comes from. Suppose you went on a helter-skelter at the fair. You do a great deal of work climbing up the stairs to the top and some of it goes into biological heat. Most of it, however, goes into mechanical work which is stored as potential energy mgh. The reason for the name is that you have the potential to use that energy whenever you are ready to sit on the mat and slide down the chute. Once you are sitting on the mat at the top then you do not need to exert any more force or expend any more energy. You just let yourself go and you accelerate down the chute to the bottom, getting faster all the time. If you had wanted to go that fast on the level then you would have had to use up a great deal of energy in sprinting so clearly there is energy associated with the fact that you are going fast. What has happened is that the potential energy you possessed at the top of the helter-skelter has been converted into *kinetic energy*. This is the energy associated with movement and is given by the equation

$$\text{Kinetic energy (k.e.)} = mv^2/2 \qquad (4.2.6)$$

This equation would allow you to calculate your velocity (or strictly your speed) at the bottom of the helter-skelter if there were no friction between the mat on which you have to sit and the chute. In practice there is always some friction and so part of the potential energy is lost in the form of heat. Note that we are not disobeying the conservation of energy law, it is simply that friction converts some of the useful energy into a form that we cannot easily use or recover. This loss of energy in overcoming friction is also a problem in machinery where frictional losses have to be taken into account even though special bearings and other devices may have been used. The way that this loss is included in energy calculations is to introduce the idea of *mechanical efficiency* to represent how close a piece of mechanical equipment is to being ideal in energy terms.

$$\text{Mechanical efficiency} = (\text{useful energy out/energy in}) \quad 100\%$$

Machines are always less than 100% efficient and so we never get as much energy out as we put in.

So far we have not included time directly in any of this discussion about work and energy but clearly it is important. Going back to the helter-skelter example, if you were to run up the stairs to the top then you would not be surprised to arrive out of breath. The energy you would have expended would be the same since you end up at rest at the same height, but clearly the fact that you have done it in a much shorter time has put a bigger demand on you. The difference is that in running up the stairs you are performing work faster and therefore operating at a greater *power*, which is defined as follows

Power = rate of doing work
= (force distance moved)/time
= force (distance moved/time)
= force velocity

Therefore

$$P = Fv \quad (4.2.7)$$

The units of power are *joules/second* which are also known as *watts* (W).

This gives us the opportunity to come up with an alternative definition of efficiency. Sometimes it is not possible to measure the total amount of work put into a machine or the total amount of work produced because the machine may be in continuous use. In those circumstances it is possible to think in terms of the rate at which energy is being used and work is being performed. Therefore

Efficiency = (power out/power in) 100%

This is the definition of mechanical efficiency that is often of most use in the case of rotating machinery so it is appropriate that we should look at how any of these equations could be applied to rotary motion. Using the idea developed in the earlier sections of replacing force F by torque τ, distance moved s by angle of rotation θ, mass m by moment of inertia I and velocity v by angular velocity ω

$$\text{Work} = \tau\theta \quad (4.2.8)$$

$$\text{Power} = \tau\omega \quad (4.2.9)$$

$$\text{Kinetic energy} = I\omega^2/2 \quad (4.2.10)$$

Example 4.2.3

A farm tractor has an engine with a power output of 90 kW. The tractor travels at a maximum steady speed to the top of a 600 m hill in a time of 4 min 20 s. If the mass of the tractor is 2.4 tonnes, what is the efficiency of the tractor's drive system?

This is an example of the use of the following equation

Efficiency = (power out/power in) 100%

The power being put into the drive system of the tractor (i.e. the gearbox, transmission and wheels) is the output power of 90 000 W from the engine. The output power of the drive system can be found from the work involved in raising the tractor up the hill within a given time.

Work done $= Fd$

$= mgh$

$= 2.4 \quad 1000 \quad 9.81 \quad 600$

$= 14\,126\,400$ J

This work is performed in a time of 260 s and so the output power (i.e. the rate of doing work) is

Output power = 14 126 400 J/260 s

= 54 332 W

Therefore the mechanical efficiency is

Output power/input power = (54 332/90 000) 100%

= 60.4%

Momentum

When Newton was investigating forces and motion he carried out a series of experiments which involved rolling balls of different masses down inclined planes. This was designed mainly to test his second law of motion concerning forces and accelerations but as part of the experiment he also let the balls collide and measured the result of the collisions. From this he came up with the *principle of conservation of momentum*. To understand this, let us first look at the concept of momentum.

Suppose you were asked to rate how easy it was to catch the following balls:

- A tennis ball tossed to you by hand.
- A ten-pin bowling ball tossed to you, again by hand.
- A ten-pin bowling ball thrown at you.

You would probably rate the first one the easiest and the last one the most difficult. This is because the tennis ball has small mass and small velocity, whereas the bowling ball in the last case has large mass and large velocity. We found earlier in this chapter that we could think of the mass by itself as inertia, the resistance of an object to being moved, but clearly the mass and the velocity together contribute to the 'unstoppability' of a moving object. This property of a moving body is known as its momentum, given by

Momentum = mass velocity

$= mv$

What Newton found in his experiments was that the total momentum of any two balls before a collision was equal to their total momentum after the collision. If we applied this to the third situation described above, but with you standing on a skateboard, then Figure 4.2.5 shows what would happen.

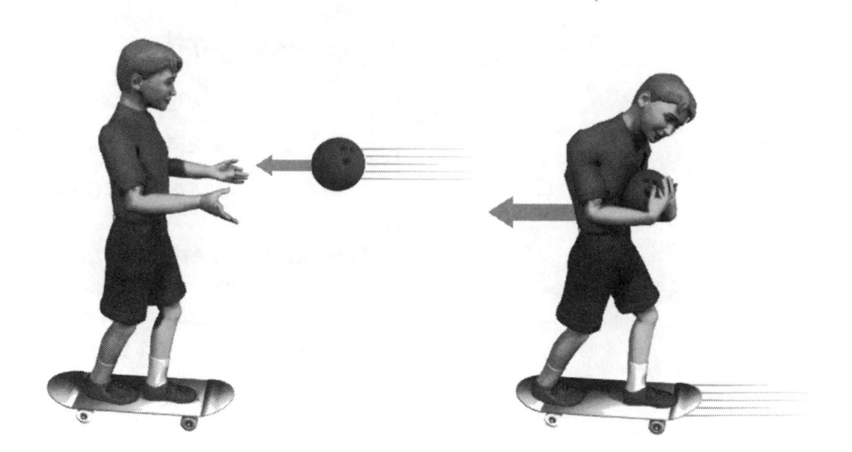

Figure 4.2.5 *Catching a bowling ball on a skateboard*

Although you were initially at rest on the skateboard, the act of catching the fast-moving bowling ball would push you backwards slowly. The initial momentum would be mV for the ball but zero for yourself as you are at rest. The final momentum would be equal to the total mass of yourself and the ball $(M + m)$ multiplied by the new, small, common velocity v. The principle of conservation of momentum tells us that

$$mV = (M + m)v$$

This principle applies to all collisions, however, not just those where the two bodies join together (such as catching the ball). In general the principle of conservation of momentum for two bodies colliding may be stated as

$$m_1 u_1 + m_2 u_2 = m_1 v_1 + m_2 v_2 \tag{4.2.11}$$

where u refers to velocities before the impact and v refers to velocities after the impact.

This equation is very widely used as it does not only cover the case of two objects colliding, it can handle the situation where two objects are initially united and subsequently separate. Examples of this are shells being fired from guns, stages of a space rocket separating and fuel being ejected from a jet engine. Imagine the special case of two objects that are initially united and are at rest. It could be you on the skateboard again, this time holding the bowling ball. The total momentum would be zero because you are at rest. Now suppose that you throw the ball forward. This would give the ball some forward momentum but, since the total momentum must remain zero, this can only happen if you start to move backwards on the skateboard with equal backwards momentum. What has happened is that you have given the ball an *impulse*, you have given it forward momentum by pushing it. But at the same time the push has given you an impulse in the opposite direction and it is this that has caused you to move back with equal negative momentum. This is a direct example of Newton's third law in action. One statement of this law is that

for every action there is an equal, opposite reaction.

Example 4.2.4

A loaded railway truck of total mass 8600 kg is travelling along a level track at a velocity of 5 m/s and is struck from behind by an empty truck of mass 3500 kg travelling at 8 m/s. The two trucks become coupled together during the collision. Find the velocity of the trucks after the impact.

We use the following equation

$$m_1 u_1 + m_2 u_2 = m_1 v_1 + m_2 v_2$$

which becomes

$$(8600 \quad 5) + (3500 \quad 8) = (8600 + 3500) \quad V$$

where V is the common final velocity.

Therefore

$$V = (43\,000 + 28\,000)/12\,100$$
$$= \underline{5.87\text{ m/s}}$$

This is quite a simple example in two respects: the trucks join together so there is only a single final velocity and they are both initially moving in the same direction. In general, care has to be taken to make it clear in which direction the objects are moving as it is not always so obvious as in this example. If the two objects were moving towards each other in opposite directions initially then a positive direction would have to be chosen so that one of the objects could be given a positive velocity and the other one a negative velocity. If the two bodies do not start out united or do not join together then there is a major problem in that there are too many unknowns to find from this single equation and another equation is required.

Before we leave this example of the two railway trucks, therefore, it is worthwhile looking at the energy involved. There are only two forms of energy which are of interest in dynamics: kinetic energy and potential energy. For trains on level tracks we can neglect the potential energy as the height of the trucks does not alter. That leaves the kinetic energy to be calculated according to the formula

$$\text{Kinetic energy (ke)} = mv^2/2$$

The initial total kinetic energy is therefore

$$(8600 \quad 25/2) + (3500 \quad 64/2) = 219\,500\text{ J}$$

The final kinetic energy is

$$(8600 + 3500) \quad 5.87^2/2 = 208\,464\text{ J}$$

Clearly some energy was lost in the impact between the two trucks and so the principle of conservation of momentum does not say anything

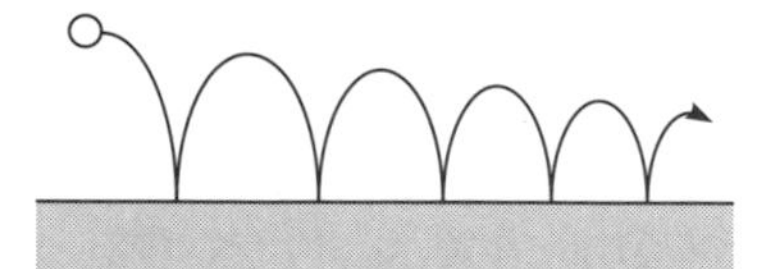

Figure 4.2.6 *A steel ball bouncing on a hard surface*

about energy being conserved. The question of energy in impacts is left to something called the *coefficient of restitution*. To understand this, look at the steel ball bouncing on a hard surface in Figure 4.2.6.

The maximum height of the ball at the top of each bounce gets smaller and smaller as energy is lost in each impact with the hard surface. This shows that the velocity after each impact is less in magnitude than the velocity before the impact. The ratio of the magnitudes of these two velocities is known as the coefficient of restitution e which clearly always has a value of less than one. For simplicity it is best to define e in terms of speed since then we do not have to take into account the fact that the ball changes direction at each bounce.

$$\text{Coefficient of restitution} = e = \frac{\text{relative speed of separation}}{\text{relative speed of approach}}$$

The idea of using relative speeds is to make it applicable to cases where both bodies are moving before and after the impact, unlike the simple case above where the hard surface is fixed.

Example 4.2.5

A ball of mass 12 kg is moving at a speed of 2 m/s along a smooth surface. It is hit from behind by a ball of the same diameter but a mass of only 4 kg travelling in the same direction with a speed of 8 m/s. The coefficient of restitution between the two balls is 0.75. Calculate the final velocities of the two balls.

First we apply the principle of conservation of momentum

$$(12 \quad 2) + (4 \quad 8) = 12v_1 + 4v_2$$

Therefore

$$14 = 3v_1 + v_2$$

Next we apply the coefficient of restitution

$$0.75 = (v_1 - v_2)/(8 - 2)$$

Therefore

$$(v_1 - v_2) = 4.5$$

We now have two equations relating the two final velocities and so we can solve them simultaneously or simply make a substitution. Finally this gives us

$v_1 = 4.625$ m/s

$v_2 = -0.125$ m/s (i.e. it moves backwards after the impact)

Angular momentum

Before we leave the topic of momentum, we should again consider how the equations that we have learned for linear motion apply to rotary motion. We have already seen that the linear velocity v in m/s can be replaced by the angular velocity ω in rad/s, such that $v = r\omega$. We have also seen that the equivalent of mass m in kg in rotational terms is the moment of inertia I in kgm^2. It follows from this that the rotational equivalent of momentum mv is the angular momentum $I\omega$.

The principle of conservation of angular momentum then becomes

$$I_1\omega_1 + I_2\omega_2 = I_1\Omega_1 + I_2\Omega_2$$

where the symbol Ω is used for the final angular velocities.

Problems with angular momentum are solved in exactly the same way as those for linear momentum except that there is rarely any need to use a coefficient of restitution. This is because problems generally involve a single object which changes its shape and therefore its moment of inertia. An example would be a telecommunications satellite which is needed to spin in its final position in space. Usually this is achieved by spinning the satellite while it is still attached to its anchor points in the cargo bay of a space shuttle. The difficulty is that the satellite will unfold very long solar panels once it is released and this will increase its moment of inertia. The engineers in charge of the project will solve the problem by calculating the final momentum with the panels unfolded as well as the initial moment of inertia with the panels folded away. Using the principle of conservation of angular momentum and knowing the final angular velocity required, it is straightforward to calculate the initial angular velocity that must be reached before the satellite can be released.

Example 4.2.6

A telecommunications satellite has a moment of inertia of 82 kg m^2 when it is in its final configuration with its solar panels extended. It is required to rotate at a rate of 0.5 rev/s in this configuration. When the satellite is still in the cargo bay of a shuttle vehicle and the panels are still folded away, it has a moment of inertia of 6.8 kg m^2. To what rotational speed should it be accelerated before being released and unfolded.

The initial angular momentum = 6.8 ω

The final angular momentum = 82 0.5 2 π (converting from revs to rads)

The initial and final angular momentum are equal, therefore:

ω = 82 0.5 2 π/6.8 = 37.9 rad/s (or 6 rev/s)

d'Alembert's principle

The final section in this chapter is really more to do with statics, surprisingly, and concerns a way of tackling dynamics problems that was devised by a French scientist called d'Alembert. At the time of its introduction it was very difficult to understand but nowadays we are all familiar with some of its applications in space travel. Suppose we consider the case of astronauts being launched into space on a space shuttle. During the launch when the shuttle is vertical and accelerating rapidly up through the atmosphere with the rockets at full blast, the astronauts feel as though they have suddenly become extremely heavy; they cannot raise their arms from the arms of the seat, they cannot raise their heads from the headrests and the skin on their faces is pulled back tight because it is apparently so heavy. We know that all this is because the shuttle is accelerating so fast that the astronauts are subjected to what are known as 'g-forces', the sort of forces you feel on a roller coaster as it goes through the bottom of a tight curve. However, if you were in the space shuttle and the windows were blacked out so that you could not see any movement, you could be excused for thinking that the astronauts really were being subjected to some large extra force acting in the opposite direction to the acceleration. For an astronaut of mass m experiencing an acceleration of a directly upwards, the extra imaginary force would be $F = ma$ directly down, according to Newton's second law. From that point on, any engineering dynamics problem concerning the astronaut, such as working out the reaction loads on the seat anchor points, could be carried out as a statics calculation with this imaginary force included. This is d'Alembert's principle.

Problems 4.2.1

(1) A car of mass 1200 kg accelerates at a rate of 2.8 m/s^2. Calculate the resultant driving force which must be acting on it.

(2) A force of 5.4 kN is used to accelerate a fork-lift truck of mass 1845 kg. Calculate the acceleration produced.

(3) A man exerts a horizontal force of 300 N to push a trolley. If the trolley accelerates at 0.87 m/s^2, calculate its mass.

(4) A car of mass 1.4 tonnes rests on a 30° slope. If it were allowed to roll down the hill, what would be its acceleration? Assume that the resistance to motion is equivalent to a force of 1000 N acting up the hill.

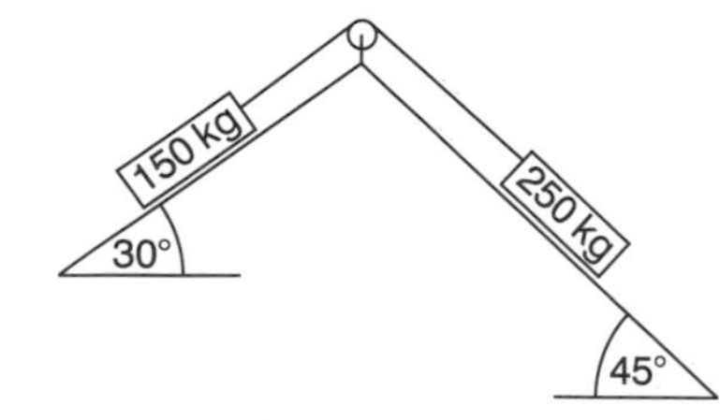

Figure 4.2.7

(5) Two masses are connected by a light cable passing over a frictionless pulley (Figure 4.2.7). If they are released from rest calculate their initial acceleration:

(a) if they slide freely on the slopes;
(b) if they have a coefficient of friction of 0.3.

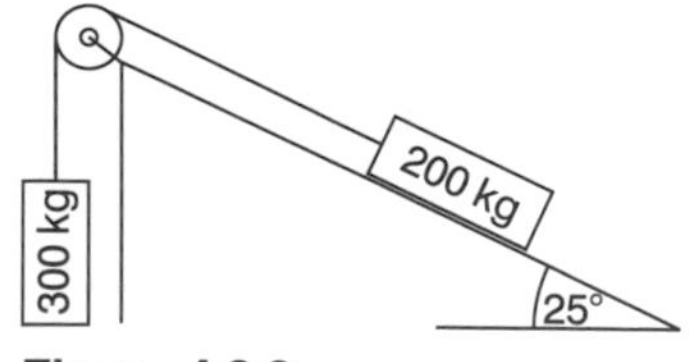

Figure 4.2.8

(6) Two masses are connected as shown in Figure 4.2.8 by a light cable passing over a frictionless pulley. If the coefficient of friction between the mass and the incline is 0.28 find the distance that the large mass falls in the first 30 s after the system is released from rest.

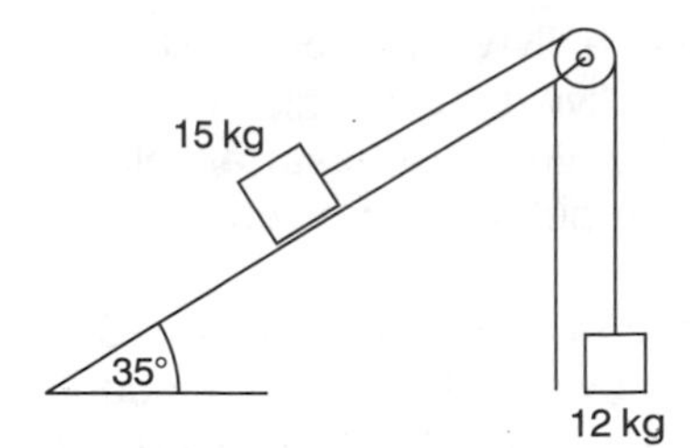

Figure 4.2.9

(7) Calculate the acceleration of the two bodies shown in Figure 4.2.9 if the coefficient of friction on the slope is 0.24.

(8) Calculate the acceleration if the coefficient of friction on the slope shown in Figure 4.2.10 is 0.32:

(a) if the body is already moving down the slope;
(b) if the system is released from rest.

(9) Calculate the acceleration of the two bodies shown in Figure 4.2.11:

(a) if they are already moving from right to left;
(b) if they are released from rest.

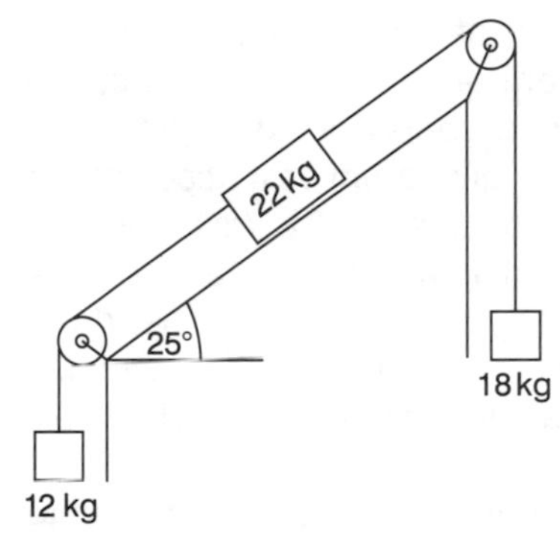

Figure 4.2.10

(10) A harrier jump-jet has a mass of 12 500 kg when fully laden and can take off vertically with an acceleration of 4 m/s^2. It then switches to forward flight, while maintaining constant thrust, and initially experiences an air resistance force of 20 kN. Calculate the horizontal acceleration and hence find the distance travelled in the first 15 s of horizontal flight.

(11) A helicopter can accelerate vertically upwards at a rate of 3.5 m/s^2 when it is unloaded and has a mass of 2250 kg. Calculate the maximum extra load it could support.

(12) A lunar landing module of mass 5425 kg, including crew and fuel, etc., is designed for a maximum deceleration of 5.6 m/s^2 when using half thrust on its rocket motors during a landing on the moon. While on the moon 100 kg of fuel is used, 200 kg of rubbish are jettisoned and 450 kg of equipment are left behind. Calculate the acceleration on take-off at full power.

($g_{moon} = 1/6\, g_{earth}$)

(13) By the time the module in Problem 12 has reached orbit around the moon, it has used another 675 kg of fuel. Calculate the acceleration it could then achieve at three-quarters thrust.

(14) A lift cage has a mass of 850 kg and the tension in the cable when the lift is moving upwards at a steady velocity with six people inside it, each of mass 150 kg, is 20 kN. Assuming that friction is constant, calculate the tension in the cable when the same lift containing only four of the people accelerates upwards from rest at a rate of 2 m/s^2.

(15) An electric motor with a moment of inertia 118.3 kg m^2 experiences a torque of 249 N m. Calculate the angular acceleration of the motor under no-load conditions, and hence find the time to reach a speed of 3600 rpm from rest, assuming the acceleration remains uniform.

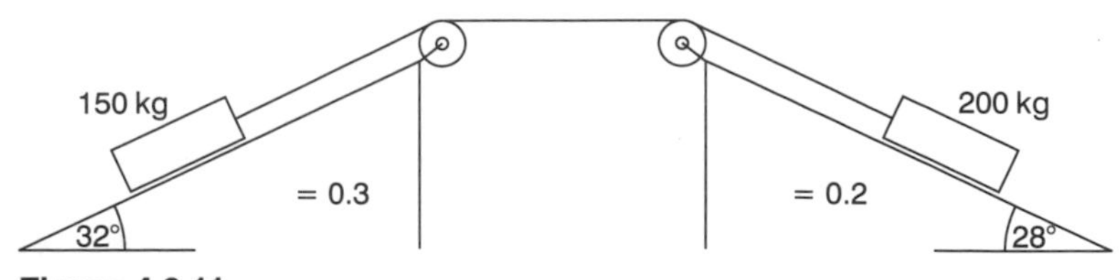

Figure 4.2.11

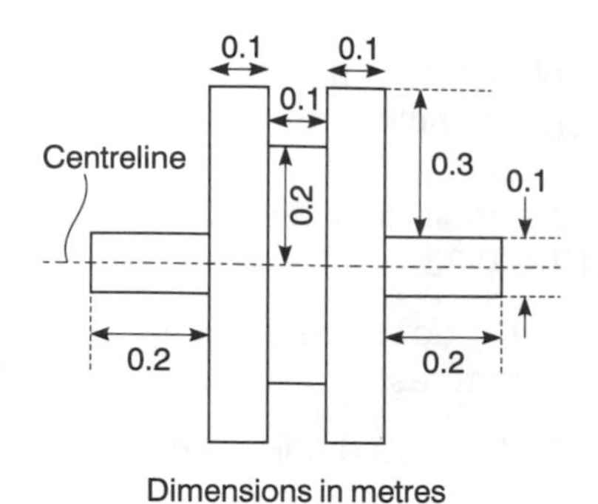

Figure 4.2.12

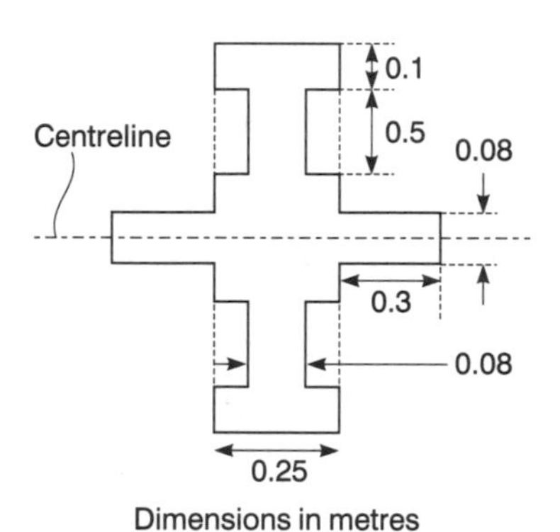

Figure 4.2.13

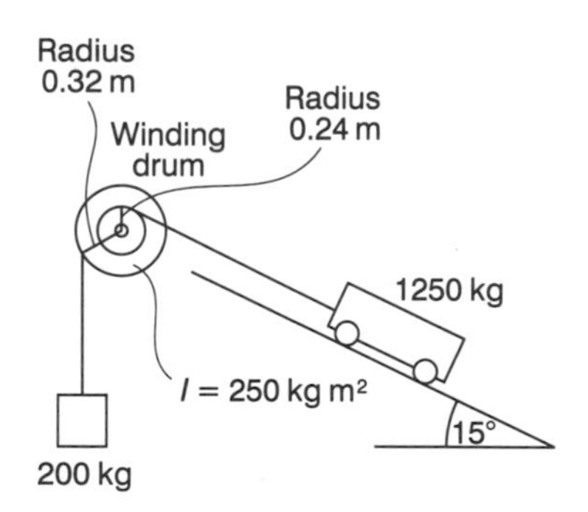

Figure 4.2.14

(16) A motor is used to accelerate a flywheel from rest to a speed of 2400 rpm in 32 s. The flywheel is a disc of mass 105 kg and radius 0.85 m. Calculate the angular acceleration required and hence find the output torque which must be produced by the motor.

(17) A locomotive turntable is a horizontal disc of diameter 20 m and mass 105 tonnes, which is evenly distributed. It needs to reach its maximum speed of 1 rpm within 20 s when it is unloaded and the resistance due to friction is equivalent to a force of 102 N acting tangentially at the edge of the turntable. Calculate the driving torque required.

(18) A pulley on a shaft (Figure 4.2.12) can be considered as a number of discs mounted on the same axis. If the density of the material is 7800 kg/m^3, calculate the moment of inertia of the pulley and shaft, and hence find the torque required to accelerate it from rest to a speed of 5400 rpm in 12 s.

(19) A flywheel is constructed as shown in Figure 4.2.13. By considering it as a number of discs, both real and imaginary, calculate its moment of inertia. Hence calculate the time it will take to reach a speed of 4800 rpm if the torque which can be applied to give a uniform acceleration is 206 N m. Density of the material is 8200 kg/m^3.

(20) A winding drum with a built-in flywheel is attached to a suspended mass of 110 kg by a light cord. The drum diameter is 0.2 m and the moment of inertia of the assembly is 59 kg m^2. Calculate the distance that the mass will fall in 9 s if it is released from rest.

(21) For the same arrangement, what torque must be applied to the drum to accelerate the mass upwards from rest at a rate of 2 m/s^2.

(22) A truck is hauled up a slope by a cable system which is driven by an electric motor (Figure 4.2.14); but also has a counterweight as shown. If the resistance to motion of the truck is 800 N, calculate the torque that the motor must apply at the winding drum in order to pull the truck up the slope a distance of 10 m from rest in 8 s.

(23) The waterside at Niagara Falls can be reached by a cable car which moves up and down the side of the gorge (Figure 4.2.15). If the motor which drives it were to fail, calculate the time it would take to move 50 m up the slope, and find the braking force which would need to be applied to the car subsequently to bring it back to rest at the top of the slope.

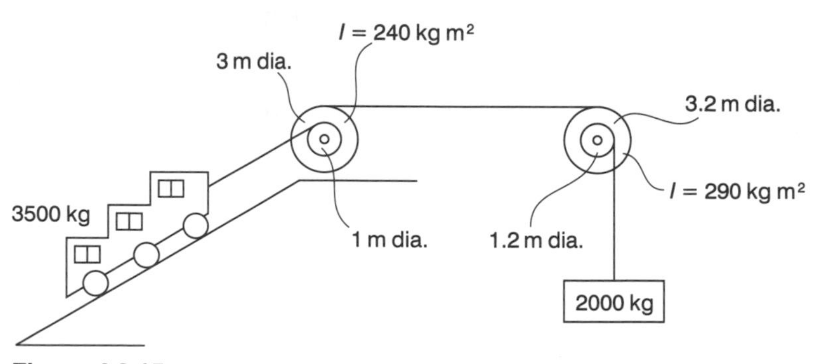

Figure 4.2.15

(24) Calculate the work required to move a packing case along the floor by 18.5 m if the force required to overcome friction is 150 N.

(25) How much work is required to raise a 26 kg box by 16 m?

(26) How much kinetic energy does a car of mass 1.05 tonnes have at a speed of 80 km/hour?

(27) A lump of lead, of mass unknown, is dropped from a tower. Because of air resistance the conversion of its potential energy to kinetic energy is only 86% efficient. Calculate its velocity after it has fallen by 9 m.

(28) A crossbow exerts a steady force of 400 N over a distance of 0.15 m when firing a bolt, of mass 0.035 kg. If the crossbow is 76% efficient, calculate the velocity of the bolt as it leaves the bow.

(29) An electric motor of 60 W power drives a winch which can raise a 10 kg mass by 15 m in 29.9 s. Calculate the efficiency of the winch.

(30) A railway wagon has a velocity of 20 m/s as it reaches the start of an inclined track (angle of elevation 1.2°). How far will it travel up the inclined track, if all resistance can be neglected and it is allowed to freewheel until it comes to rest?

(31) For the same situation as Problem 30, what will be the velocity of the wagon after it has travelled 200 m up the slope?

(32) If the wagon is only 75% efficient at converting kinetic energy to potential energy when freewheeling, calculate its velocity after it has travelled 300 m up the incline.

(33) A car has an engine with a power output of 46 kW. It can travel at a steady speed to the top of a 500 m high hill in 3 min 42 s. How efficient is the car if its mass is 1.45 tonnes?

(34) A 24 kg shell is fired from a 4 m long vertical gun barrel, and reaches a velocity of 450 m/s as it leaves the gun. Calculate:

(a) the constant force propelling it along the barrel;
(b) the maximum height above ground level reached by the shell;
(c) the speed at which it strikes the ground on return.

Assume an average constant resistance due to the air of 45 N once the shell is out of the gun.

(35) The resistance to motion of a car of mass 850 kg can be found using the formula Resistance = $1.2\,V^2$ newtons, where V is the car's velocity in m/s. Calculate the power required from the car's engine to produce speeds of 5, 15 and 30 m/s:

(a) along the straight horizontal track;
(b) ascending a slope of $\sin^{-1} 0.1$.

In all cases also calculate the total work done by the tractive force over a distance of 1 km.

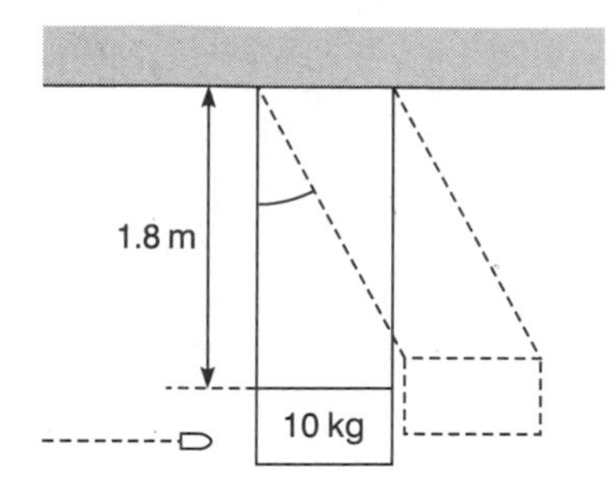

Figure 4.2.16

(36) A bullet is fired into a sand box suspended on four wires as shown in Figure 4.2.16, and becomes embedded in the sand. The sand box has a mass of 10 kg; the bullet has a mass of 0.01 kg. Only 1% of the bullet's energy goes towards moving the box; the rest is dissipated in the sand. If $\theta = 18°$ for the furthest point of the swing, estimate the velocity of the bullet just before it strikes the box.

(37) Water of density 1000 kg/m^3 is to be raised by a pump from a reservoir through a vertical height of 12 m at a rate of 5 m^3/s. Calculate the required power of the pump, neglecting any energy loss due to friction.

(38) Water emerges from a fire hose in a 50 mm diameter jet at a speed of 20 m/s. Calculate the power of the pump required to produce this jet and estimate the vertical height the jet would reach.

density of water = 1000 kg/m^3;
flow rate in kg/s = jet area velocity density

(39) A body of mass m is at rest at the top of an inclined track of slope 30°. It is released from rest and moves down the slope without friction. Calculate its speed after it has moved 20 m by using the energy method and by using the equation for motion in a straight line.

(40) The body from Problem 39 now starts from rest and moves down another 30° slope, but this time against a friction force of 20 N. Calculate its speed after it has moved 20 m down this slope, again using both methods.

(41) A body of mass 30 kg is projected up an inclined plane of slope 35° with an initial speed of 12 m/s. The coefficient of friction between body and plane is 0.15. Calculate using the energy method:

(a) how far up the slope it travels before coming to rest;
(b) how fast it is travelling when it returns to its initial position.

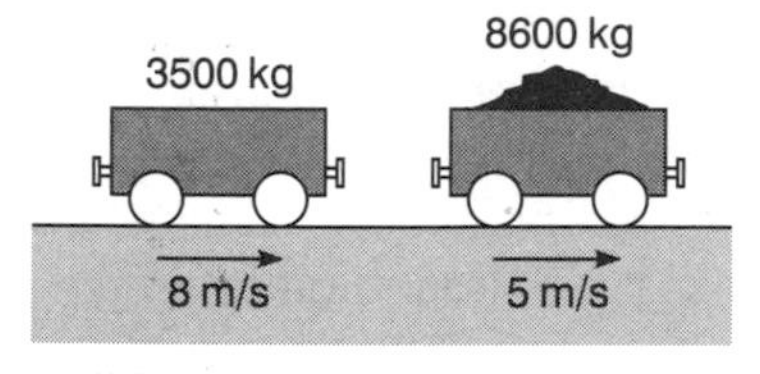

Figure 4.2.17

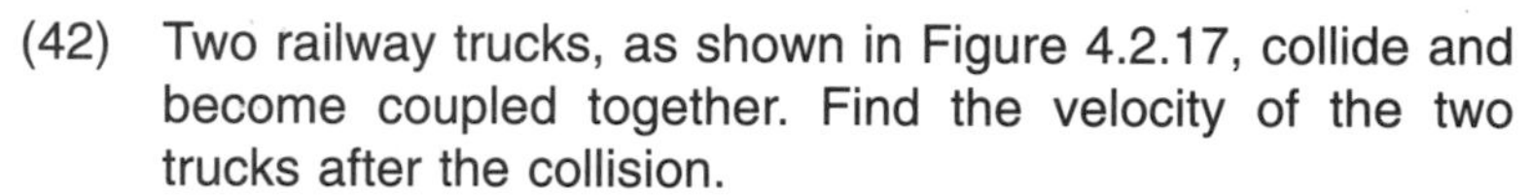

(42) Two railway trucks, as shown in Figure 4.2.17, collide and become coupled together. Find the velocity of the two trucks after the collision.

(43) Two spacecraft, moving in the same direction, are about to undertake a docking manoeuvre (Figure 4.2.18). Find the final velocity of the two craft if the manoeuvring rockets fail and the docking becomes more like a collision.

(44) A large gun is mounted on a railway wagon which is travelling at 5 m/s. The total mass is 3850 kg. The gun fires a 5 kg shell horizontally at a velocity of 1050 m/s back along the track. Find the new velocity of the wagon, which you may assume to be freewheeling.

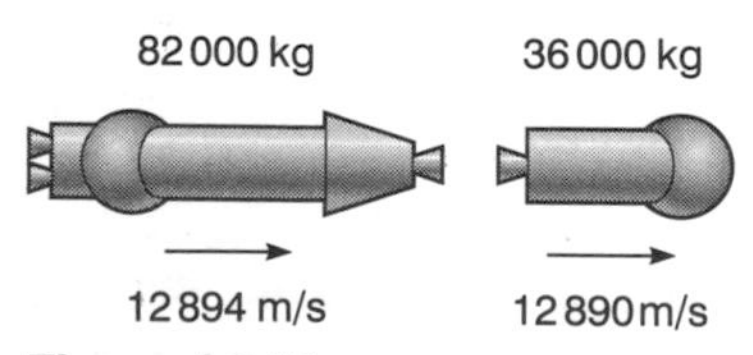

Figure 4.2.18

(45) Two railway wagons are approaching each other from opposite directions, one with mass 1200 kg moving at a velocity of 2 m/s and the other with mass 1400 kg and velocity 3 m/s. If the wagons lock together at collision, find their common velocity finally.

(46) A man of mass 82 kg is skating at 5 m/s carrying his partner, a woman of mass 53 kg, as part of an ice dancing

movement. He then throws her forward so that she moves off at 5.9 m/s. Find the man's velocity just after the throw.

(47) A ball of mass 4 kg and moving at 8 m/s collides with a ball of mass 12 kg moving in the opposite direction at 2 m/s. If the coefficient of restitution between them is 0.75, find the final velocities.

(48) A railway wagon of mass 120 kg and moving at a velocity of 8 m/s runs into the back of a second wagon of mass 180 kg moving in the same direction at 6 m/s. If the coefficient of restitution is 0.8, find the final velocities.

(49) A satellite of mass 40 kg and travelling at a speed of 5 m/s accidentally collides with a larger satellite of mass 80 kg. It rebounds and moves in the opposite direction at a speed of 5 m/s. If the coefficient of restitution between the two satellites is 0.8, find the initial velocity of the larger satellite.

5 Statics

Summary

Mechanics is the study of state of rest or motion of bodies under the action of forces; *statics* is that aspect of it involving the study of bodies at rest. Newton's first law states that bodies stay at rest or in steady motion unless acted on by a resultant force. Thus when we have bodies at rest there is no resultant force and thus the forces acting on a body at rest must cancel each other out to leave no resultant, they are said to be in equilibrium. Engineers are interested in statics because it enables them to design structures such as bridges or buildings and determine the conditions necessary for them to remain standing and not fall down.

The chapter starts with a revision of the term vector and the conditions for equilibrium and then analyses situations involving both point and distributed forces. The main part of the chapter is concerned with the analysis of structures, this including trusses, frames, beams, chains and cables and a discussion of stress and strain. The chapter concludes with a discussion of systems involving friction and a consideration of the virtual work principle and its use in the analysis of mechanical systems. To bring together the various aspects of the chapter there is a case study of the problems engineers have faced over the centuries of bridging gaps, whether they be buildings with the gaps between walls being bridged or bridges over rivers.

Objectives

By the end of this chapter, the reader should be able to:

- analyse equilibrium conditions involving both point and distributed forces;
- analyse the forces involved in trusses and frames using the method of joints and the method of sections;
- determine tensile, compressive and shear stresses and strains and determines the stresses arising from temperature changes with restrained systems;
- determine bending moments, shear forces, stresses and deflections for beams subject to bending;
- analyse the forces involved in structures using chains and cables;
- use the virtual work principle in the analysis of structures.

5.1 Equilibrium

Mechanical systems deal with two kinds of quantities – scalars and vectors; forces are vector quantities. *Scalar quantities*, e.g. mass, can be fully defined by just a number and so if we add two masses, say 10 g and 20 g, then we just add the two numerical values, i.e. 10 + 20 = 30 g. However, with forces we need to know both size and direction before we can consider the outcome of 'adding' two forces. We cannot just say the result of adding a force of 100 N to a force of 300 N will be a force of 400 N, this can only be the case if both the 100 N and 300 N forces were acting in the same direction. If they were in opposite directions the result would be 200 N in the direction of the 300 N force. Quantities for which both the size and direction have to be specified are termed *vectors*.

Thus we need to know how to add and subtract vector quantities in order to be able to analyse the forces acting on structures and determine the conditions for equilibrium. Equilibrium is at the heart of all the problems in statics – when there are a number of forces acting on a structure, what are the conditions necessary for them to cancel each other out so that the structure remains at rest? This is what this section is about.

Vectors

Before considering the conditions for equilibrium, here are some basic points about vectors. To represent vector quantities on a diagram we use arrows with the length of the arrow chosen to represent the size of the vector and the direction of the arrow the direction of the vector. Figure 5.1.1 shows a representation of two vectors of the same size but in opposite directions. To indicate in texts when we are referring to a vector quantity, rather than a scalar quantity or just the size of a vector quantity, it is common practice to use a bold letter such as **a**. The negative of this vector, i.e. –**a**, is the vector of the same size as **a** but in the opposite direction.

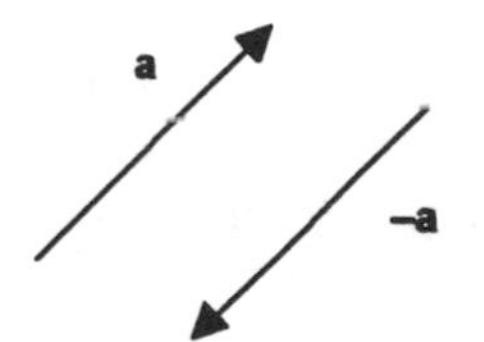

Figure 5.1.1 *Vectors can be represented by arrows, the arrow head indicating direction and the length of line the size*

When we want to refer to the sum of two vector quantities we can write **a** + **b**. This is a vector equation and so we cannot obtain the sum by the simple process of adding two numbers representing the sizes of the vectors. If we have two vectors of the same size but acting in opposite directions then the sum will be zero, think of two equal and opposed tug-of-war teams, and so when we are referring to the vector acting in directly the opposite direction we write –**a** so that **a** + (–**a**) = 0.

If we want to refer just to the size (the term magnitude is often used) of **a** we can write it as just *a* or, to clearly indicate that it is the size of a vector, as $|\mathbf{a}|$.

Adding vectors

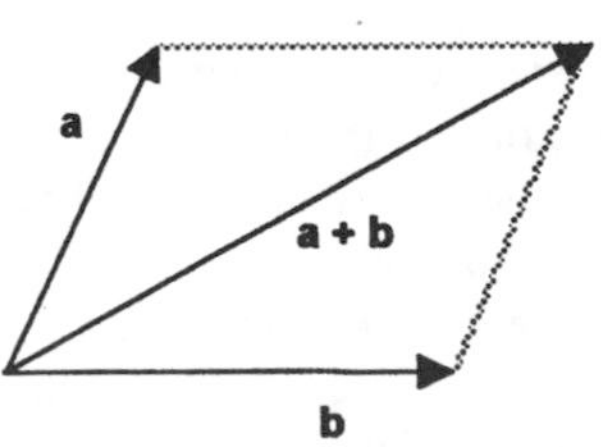

Figure 5.1.2 *Parallelogram rule which gives the diagonal as the sum, i.e. resultant, of a pair of vectors*

The sum of two vectors can be obtained by the *parallelogram rule*. This can be stated as: if we place the tails of the arrows representing the two vectors $\mathbf{F}_1$ and $\mathbf{F}_2$ together and complete a parallelogram, then the diagonal of that parallelogram drawn from the junction of the two tails represents the sum of the vectors $\mathbf{F}_1$ and $\mathbf{F}_2$. Figure 5.1.2 illustrates this. The procedure for drawing the parallelogram is:

(1) Select a suitable scale for drawing lines to represent the vectors.
(2) Draw an arrowed line to represent the first vector.
(3) From the start of the first line, i.e. its tail end, draw an arrowed line to represent the second vector.
(4) Complete the parallelogram by drawing lines parallel to the vector lines.
(5) The sum of the two vectors, i.e. the *resultant*, is the line drawn as the diagonal from the start point, the direction of the resultant being outwards from the start point.

Figure 5.1.3 *Two forces in equilibrium: equal in size but in opposite directions*

Forces in equilibrium

If when two or more forces act on an object there is no resultant force (the object remains at rest or moving with a constant velocity – Newton's first law), then the forces are said to be in *equilibrium*. This means that the sum of the force vectors is zero.

For two forces to be in equilibrium (Figure 5.1.3) they have to:

- Be equal in size.
- Have lines of action which pass through the same point (such forces are said to be *concurrent*).
- Act in exactly opposite direction.

For three forces to be in equilibrium they have to:

- All be in the same plane (such forces are said to be *coplanar*).
- Have lines of action which pass through the same point (such forces are said to be concurrent).
- Give no resultant force in any direction. We can use the parallelogram rule to determine the resultant of two forces and hence determine the size and direction of the third force that has to be applied for there to be no resultant force.

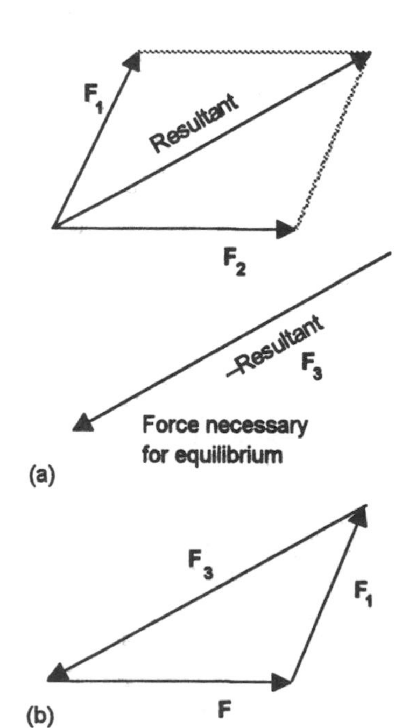

Figure 5.1.4 *Equilibrium of three forces: (a) the force necessary for equilibrium is equal in size but in opposite direction to the resultant; (b) three forces in equilibrium form the sides of a triangle*

If three forces are in equilibrium, the resultant of two of the forces must be the same size and in the opposite direction to the third force (Figure 5.1.4(a)). Thus an alternative to drawing the parallelogram and then reversing the direction of the resultant is to use the *triangle rule*; this is just half the parallelogram (Figure 5.1.4(b)). If the three forces are represented on a diagram in magnitude and direction by the vectors $\mathbf{F}_1$, $\mathbf{F}_2$ and $\mathbf{F}_3$ then their arrow-headed lines when taken in the order of the forces must form a triangle if the three forces are to be in equilibrium. If the forces are not in equilibrium, a closed triangle is not formed. Drawing the triangle of forces involves the following steps:

(1) Select a suitable scale to represent the sizes of the forces.
(2) Draw an arrow-headed line to represent one of the forces.
(3) Take the forces in the sequence they occur when going, say, clockwise. Draw the arrow-headed line for the next force so that it starts with its tail end from the arrowed end of the first force.
(4) Then draw the arrow-headed line for the third force, starting with its tail end from the arrow end of the second force.
(5) If the forces are in equilibrium, the arrow end of the third force will coincide with the tail end of the first arrow-headed line to give a closed triangle.

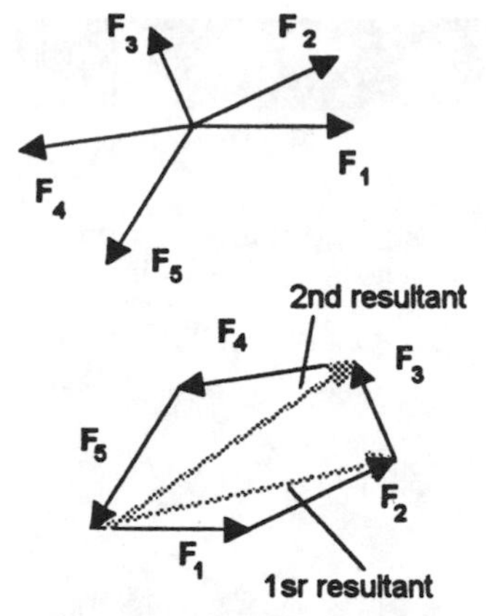

Figure 5.1.5 *Polygon rule: the forces when taken in sequence give a closed polygon*

The triangle law can be used to determine whether three forces are in equilibrium and if we have more than three forces we use the *polygon rule*. Suppose we have vectors $\mathbf{F_1}$, $\mathbf{F_2}$, $\mathbf{F_3}$, $\mathbf{F_4}$ and $\mathbf{F_5}$. If we take the vectors in sequence and draw the arrows tail to head, when there is equilibrium we complete a closed polygon shape (Figure 5.1.5). If they are not in equilibrium, the shape is not closed. All we are effectively doing is using the parallelogram rule to determine the resultant of the first two forces and then the second resultant when the first resultant is added to the next force and so on until we end up with the resultant of all but the last force needed to give equilibrium and this must be in the opposite direction to this last resultant.

Reaction forces

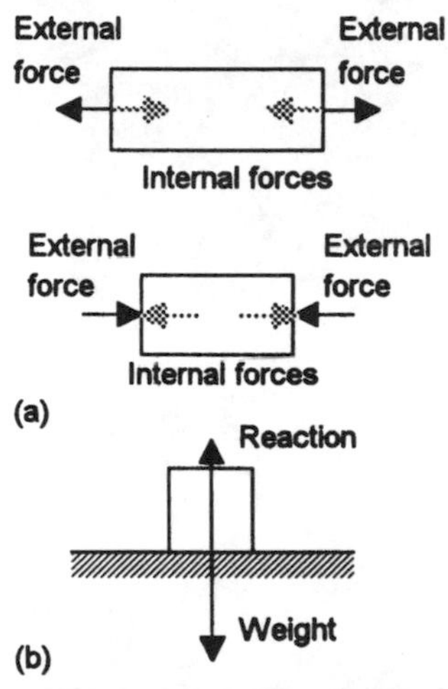

Figure 5.1.6 *(a) Applying external forces to objects gives rise to resisting internal forces; (b) the resisting, reactive, force arising from the weight of an object resting on the ground*

When external forces are applied to an object from outside, internal forces are induced in the object to counteract the externally applied forces and maintain equilibrium (Figure 5.1.6(a)). Think of the forces you apply to stretch a rubber band. Your externally applied forces result in internal forces being set up in the rubber band. So when you let go of the band these internal forces pull the band back to its original unstretched state. Internal forces are produced in reaction to the application of external forces and so are often called the *reactive forces* or *reactions*.

When an object is at rest on the ground, the weight of the object must be counterbalanced by an opposing force to give equilibrium. This counterbalance force is the reaction force of the ground (Figure 5.1.6(b)). If you think of the ground as being like a sheet of rubber, then the weight of the object causes it to become deformed and stretched and so exert the reactive force.

Free-body diagram

If a system is in equilibrium then each element in that system is in equilibrium. Thus if we isolate an element from the system we can consider the equilibrium of just those forces acting on it. The term *free-body diagram* is used for a diagram which shows just the forces involved with an element. For example, Figure 5.1.7(a) shows a system involving a load suspended by means of ropes from two fixed points. We might draw the free-body diagram for point A, the junction of the three ropes, and show just the three forces involved at that point (Figure 5.1.7(b)); alternatively we might draw the free-body diagram for point B, where the single rope connects to the load, and show just the two forces involved at that point (Figure 5.1.7(c)).

The procedure for drawing a free-body diagram is thus:

(1) Imagine the particle to be isolated from the rest of the system.
(2) Draw a diagram of this free particle and show all the forces that act on it, putting on all the external loads as well as the reactive forces.

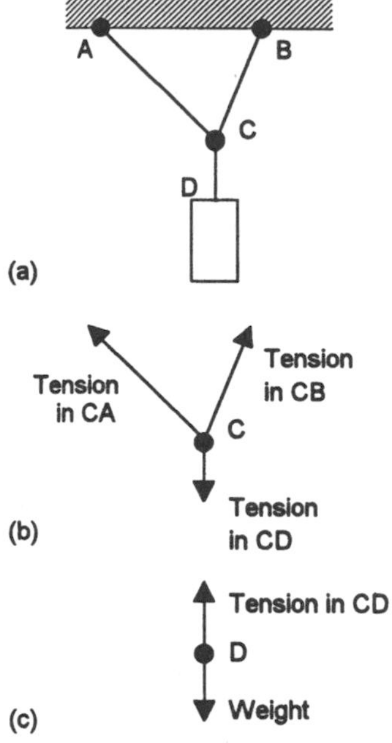

Figure 5.1.7 *(a) Load supported by ropes from two fixed points; (b) and (c) the free body diagrams for points C and D*

Mathematics in action

Sine and cosine rules

Problems involving the parallelogram and triangle rules can be solved graphically or by calculation. Useful rules that can be used in such calculations are the sine rule and the cosine rule. The *sine* rule can be stated as:

$$\frac{a}{\sin A} = \frac{b}{\sin B} = \frac{c}{\sin C}$$

where a is the length of the side of the triangle opposite the angle A, b the length of the side opposite the angle B and c the length of the side opposite the angle C (Figure 5.1.8). The *cosine rule* can be stated as: the square of a side is equal to the sum of the squares of the other two sides minus twice the product of those sides multiplied by the cosine of the angle between them and so:

$$a^2 = b^2 + c^2 - 2bc \cos A$$

$$b^2 = a^2 + c^2 - 2ac \cos B$$

$$c^2 = a^2 + b^2 - 2ab \cos C$$

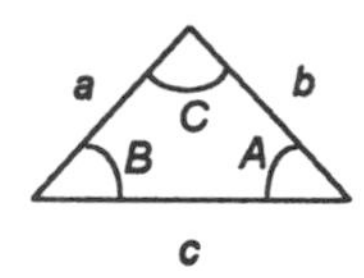

Figure 5.1.8 *Triangle*

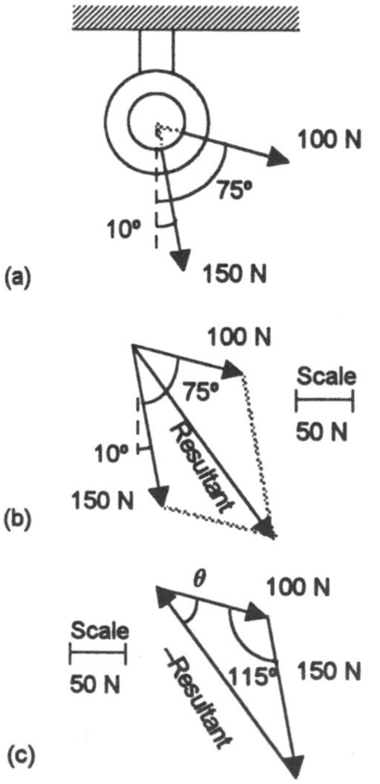

Figure 5.1.9 *Example 5.1.1*

Example 5.1.1

A screw eye is subject to two forces of 150 N and 100 N in the directions shown in Figure 5.1.9(a). Determine the resultant force.

One method of determining the resultant is to draw the parallelogram of forces to scale, as in Figure 5.1.9(b), and determine the length of the diagonal and its angle. Another possibility is to use the triangle of forces to determine the force needed to give equilibrium, the resultant will then be in the opposite direction but the same size. Figure 5.1.9(c) shows the triangle of forces drawn to scale.

We can use the cosine rule to calculate the size of the resultant. For Figure 5.1.9(c), the cosine rule gives:

$$(\text{Resultant})^2 = 100^2 + 150^2 - 2 \times 100 \times 150 \cos 115°$$

Hence the size of the resultant is 212.6 N. The angle θ between the resultant and the 100 N force can be determined by the sine rule:

$$\frac{150}{\sin \theta} = \frac{212.6}{\sin 115°}$$

Hence θ = 39.7° and so it is at 75° – 39.7 = 35.3° to the vertical.

Example 5.1.2

A rigid rod, of negligible mass, supporting a load of 250 N is held at an angle of 30° to the vertical by a tie rope at 75° to the vertical, as shown in Figure 5.1.10(a). Draw the free-body diagram for point A where the rope is attached to the rod and hence determine the tension in the rope and the compression in the rod.

Figure 5.1.10(b) shows the free-body diagram for point A. The triangle of forces (Figure 5.1.10(c)) can be drawn by first drawing the line for the 250 N force. The compression in the rod will be a force of unknown size at an angle of 30° to the vertical 250 N force and so a line can be drawn in this direction. The tension in the tie rope will be a force of unknown size at an angle of 75° to the vertical 250 N force and so a line can be drawn in this direction. The intersection of these two lines gives the completion of the triangle and hence enables the sizes of the forces to be determined. This can be done by a scale diagram or by calculation using the sine rule.

$$\frac{250}{\sin(180° - 30° - 75°)} = \frac{T}{\sin 30°} = \frac{C}{\sin 75°}$$

Hence T = 129.4 N and C = 250 N.

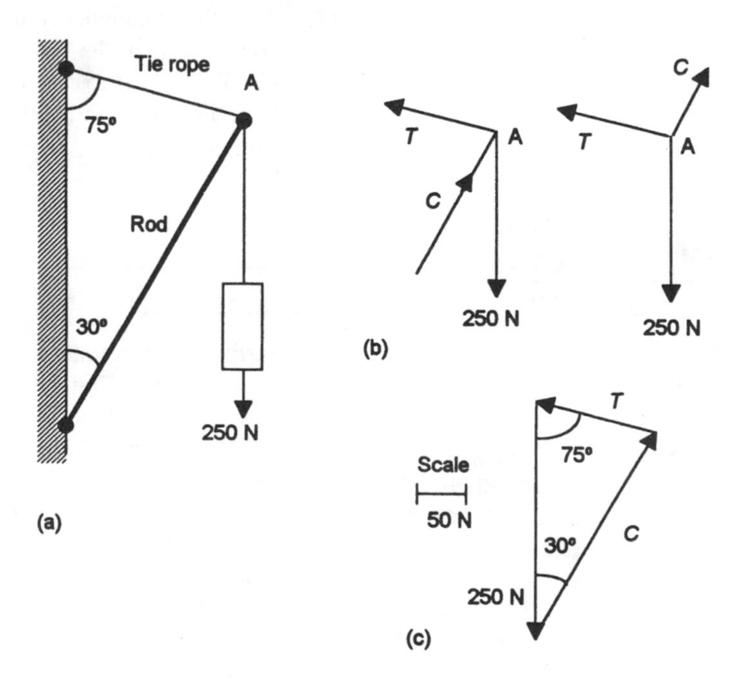

Figure 5.1.10 *Example 5.1.2*

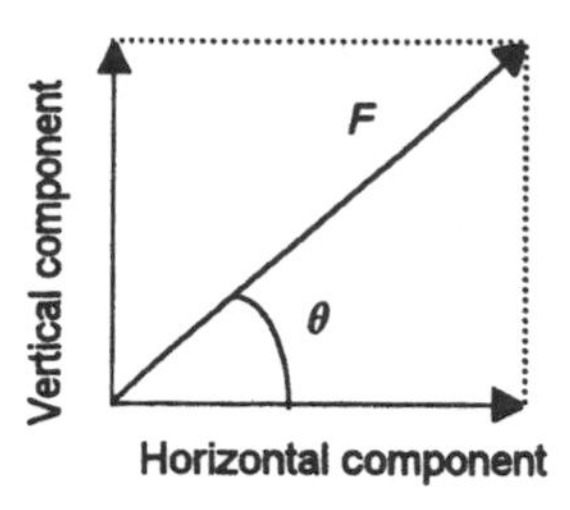

Figure 5.1.11 *Resolving a force F into vertical and horizontal components*

Resolving forces

A single force can be replaced by two forces at right angles to each other. This is known as *resolving* a force into its *components* and involves using the parallelogram of forces and starting with the diagonal of the parallelogram and then finding the two forces which would 'fit' the sides of such a parallelogram (Figure 5.1.11). Thus, if we have a force F, then its components are:

$$\text{Horizontal component} = F\cos\theta \qquad (5.1.1)$$

$$\text{Vertical component} = F\sin\theta \qquad (5.1.2)$$

Pythagoras' theorem gives:

$$F^2 = (\text{horizontal component})^2 + (\text{vertical component})^2 \qquad (5.1.3)$$

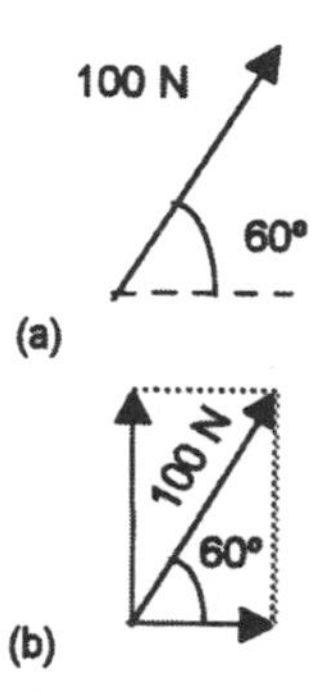

Figure 5.1.12 *Example 5.1.3*

Example 5.1.3

Determine the horizontal and vertical components of the 100 N force shown in Figure 5.1.12(a).

Figure 5.1.12(b) shows the parallelogram which enables the two components to be obtained. The horizontal component is 100 cos 60° = 50 N and the vertical component is 100 sin 60° = 86.6 N.

Resultant using components

The procedure for finding the resultant of a number of forces by dealing with their components involves:

(1) Resolve all the forces into two right-angled directions.
(2) Add the components in the horizontal direction and add the components in the vertical direction.
(3) Use Pythagoras to determine the size of resultant of these summed horizontal and vertical components and tan θ = (vertical components)/(horizontal components) to obtain its angle to the horizontal.

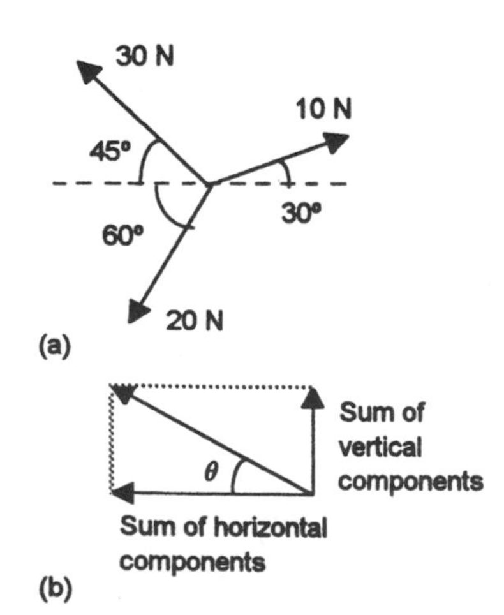

Figure 5.1.13 *Example 5.1.4*

Example 5.1.4

Determine the resultant of the system of concurrent coplanar forces shown in Figure 5.1.13(a).

The sum of the horizontal components is 30 cos 45° + 20 cos 60° − 10 cos 30° = 22.6 N to the left. The sum of the vertical components is 30 sin 45° + 10 sin 30° − 20 sin 60° = 8.9 N upwards. The resultant (Figure 5.1.13(b)) is thus given by Pythagoras as $\sqrt{(22.6^2 + 8.9^2)}$ = 24.3 N in a direction given by tan θ = 8.9/22.6 and so at an angle of 21.5° upwards from the horizontal.

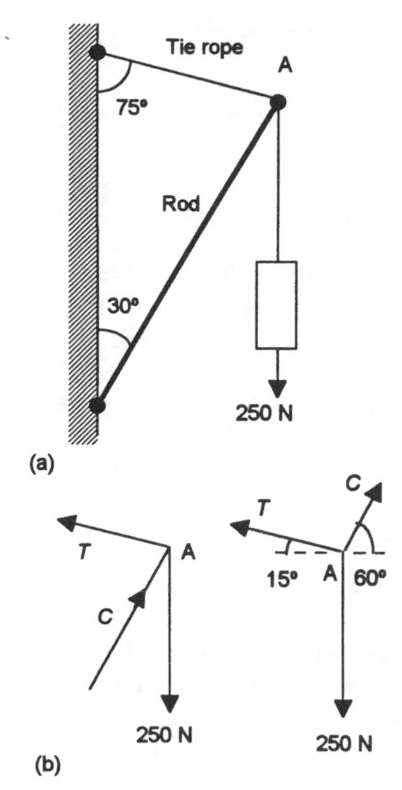

Figure 5.1.14 *Example 5.1.5*

Equilibrium conditions with components

A body is in equilibrium if there is no resultant force acting on it. Thus for a body under the action of a number of concurrent forces, for there to be no resultant force we must have:

sum of horizontal components of all the forces = 0 (5.1.4)

sum of vertical components of all the forces = 0 (5.1.5)

Example 5.1.5

A rigid rod, of negligible mass, supporting a load of 250 N is held at an angle of 30° to the vertical by a tie rope at 75° to the vertical, as shown in Figure 5.1.14(a). Determine the tension in the rope and the compression in the rod. (This is a repeat of Example 5.1.2.)

Figure 5.1.14(b) shows the free-body diagram for point A. For equilibrium, the horizontal components are $T \cos 15° - C \cos 60° = 0$, hence $T = 0.518C$, and the vertical components are $T \sin 15° + C \sin 60° - 250 = 0$. Hence, $T = 250$ N and $C = 129.5$ N.

Moment of a force

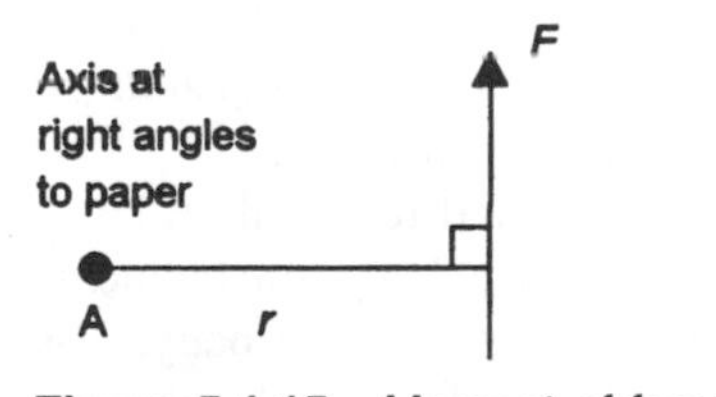

Figure 5.1.15 *Moment of force F is Fr*

In considering the equilibrium of structures we have to consider not only that there is no resultant force but also that the way the forces act on it there is no way that they will cause it to rotate. For this reason we need to consider the moments of the forces. The *moment of a force* about the axis at A in Figure 5.1.15 of the force is defined as being the product of the size of the force F and its perpendicular distance r from the axis to the line of action of the force (Figure 5.1.15):

$$\text{Moment} = Fr \tag{5.1.6}$$

The SI unit of the moment is the newton metre (N m). Moment is a vector quantity and its direction is given as being along the axis about which the moment acts and in the direction along that axis that a right-handed screw would rotate.

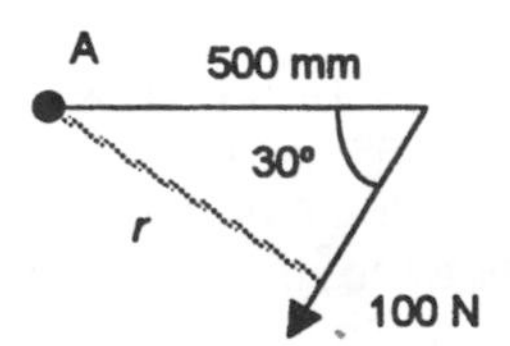

Figure 5.1.16 *Example 5.1.6*

Example 5.1.6

Determine the moment of the force shown in Figure 5.1.16 about the axis through A.

The perpendicular distance r of the line of action of the force from the pivot axis is $r = 500 \sin 30° = 250$ mm and so the moment is $100 \times 0.250 = 25$ N m. The direction of the moment about the axis is clockwise.

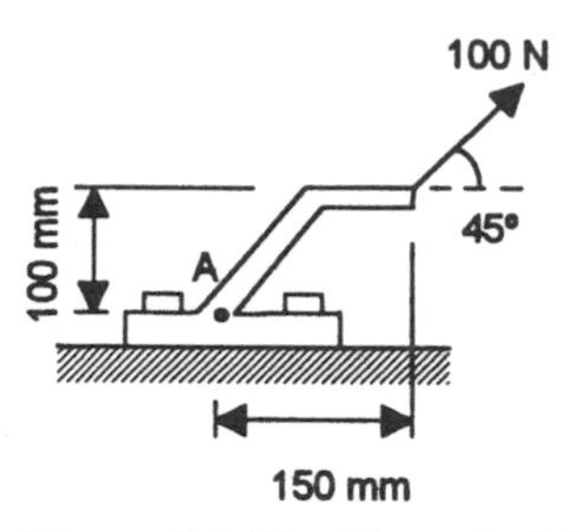

Figure 5.1.17 *Example 5.1.7*

Example 5.1.7

Determine the moment of the force shown in Figure 5.1.17 about the axis through A.

While we can work out the perpendicular distance between the line of action of the force and the axis through A, a procedure which is often simpler is to resolve the force into two components and then find the moment of each component about A. The components of the force are 100 sin 45° = 70.7 N vertically and 100 cos 45° = 70.7 N horizontally. The moment of the horizontal component about A is 70.7×0.100 = 7.07 N m in a clockwise direction and the moment of the vertical component about A is 70.7×0.150 = 10.61 N m in an anticlockwise direction. Thus the moment of the force is 10.61–7.07 = 3.54 N m in an anticlockwise direction.

Couple

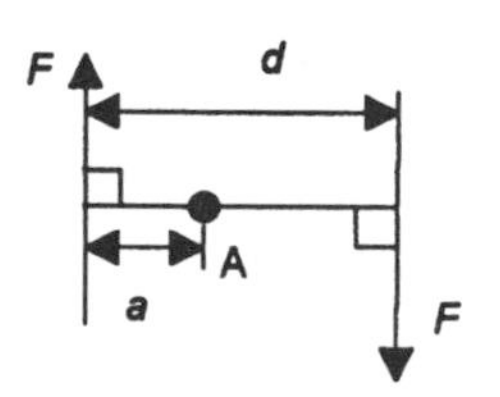

Figure 5.1.18 *A couple consists of two equal and oppositely directed forces and has a moment of Fd*

A couple consists of two equal and oppositely directed parallel forces (Figure 5.1.18). Their vector sum is zero but they will produce a tendency to rotate and hence a moment. For example, the combined moment of the two forces about an axis at right angles to the plane in which they are located at, say, A is $Fa + F(d - a) = Fd$.

$$\text{Moment of couple} = Fd \tag{5.1.7}$$

where d is the perpendicular distance between the lines of action of the forces. The moment of a couple is thus a constant, regardless of the location of the axis about which the moments are taken, and equal to the product of the force size and the distance between the forces.

A couple acting on a body produces rotation and for equilibrium to occur must be balanced by another couple of equal and opposite moment.

When there are a number of coplanar forces acting on a body, it is possible to simplify the situation by replacing them by a single resultant force and a single resultant couple.

- The resultant force acting on the body can be obtained by combining the sum of the vertical components of all the forces and the sum of all the horizontal components of the forces.
- The resultant couple is the sum of the moments of the forces about any point in the plane plus the sum of all the couples in the system.

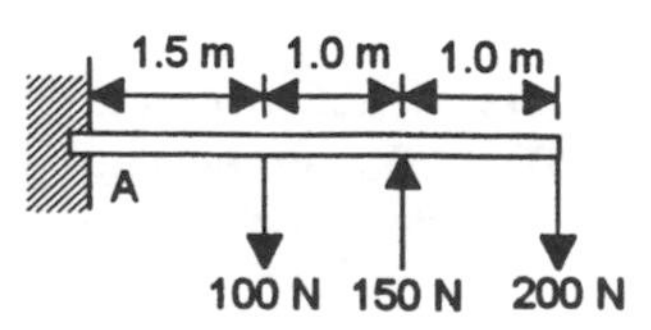

Figure 5.1.19 *Example 5.1.8*

Example 5.1.8

Replace the system of forces shown acting on the beam in Figure 5.1.19 by an equivalent single force.

The resultant force = 100 – 150 + 200 = 150 N in a downward direction. The moments of the forces about A = $100 \times 1.5 - 150 \times 2.5 + 200 \times 3.5$ = 475 N m in a clockwise direction. To give this moment, the resultant force must be a distance of 475/150 = 3.2 m from A.

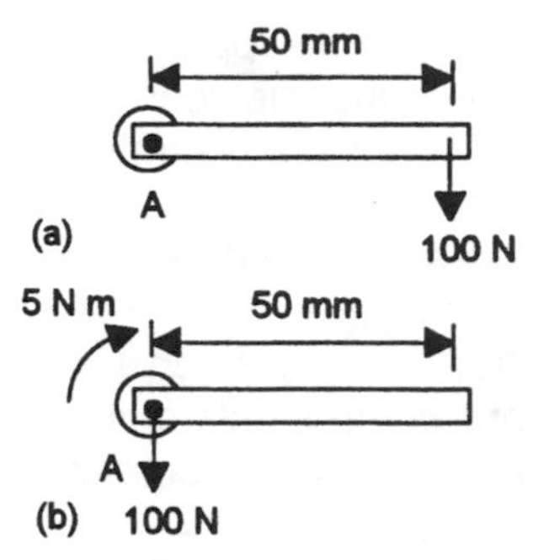

Figure 5.1.20 *Example 5.1.9*

Example 5.1.9

Replace the force shown acting on the control handle in Figure 5.1.20(a) by an equivalent single force and a single couple acting at point A.

The resultant force is 100 N downwards and so the equivalent system must have a downwards directed force of 100 N. The 100 N force has a moment about A of $100 \times 0.050 = 5$ N m in a clockwise direction. Thus moving the 100 N force to A requires the addition of a clockwise couple of 5 N m. Hence the equivalent system is as shown in Figure 5.1.20(b).

Equilibrium of a rigid body

If an object fails to rotate about some axis then the turning effect of one force must be balanced by the opposite direction turning effect of another force. Thus, when there is no rotation, the algebraic sum of the clockwise moments about an axis must equal the algebraic sum of the anticlockwise moments about the same axis; this is known as the *principle of moments*. Thus the general conditions for equilibrium of a body can be stated as:

- There is no resultant force acting on the body; this can be restated as the sum of the vertical components of all the forces must be zero and the sum of all the horizontal components of the forces must be zero.
- The sum of the moments about any point in the plane must be zero.

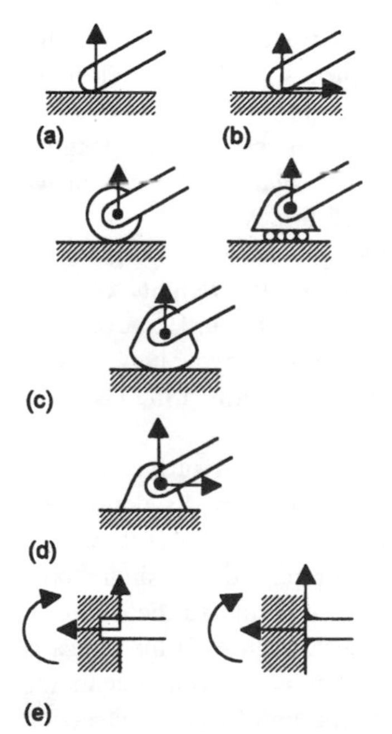

Figure 5.1.21 *Reactions at supports: (a) smooth surfaces; (b) rough surfaces; (c) roller, rocker or ball support; (d) freely hinged pin; (e) built-in support*

In tackling problems involving supported systems we need to consider the reactive forces at supports. When the smooth surfaces of two bodies are in contact (Figure 5.1.21(a)), the force exerted by one on the other is at right angles to the surfaces and thus the reactive force is at right angles. When the surfaces are rough (Figure 5.1.21(b)), the force exerted by one surface on the other need not be at right angles and thus the reactive force may not be at right angles and so, in general, will have a normal and a tangential component. A roller, rocker or ball support (Figure 5.1.21(c)) just transmits a compressive force normal to the supporting surface and so there is just a reactive force normal to the supporting surface. A freely hinged pinned support (Figure 5.1.21(d)) is capable of supporting a force in any direction in the plane normal to the pin axis and so there can be a reactive force in any such direction; this is normally represented by its two right-angled components. A built-in or fixed support (Figure 5.1.21(e)) is capable of supporting an axial force, a transverse force and a couple and thus there are two such reactive forces and a reactive couple.

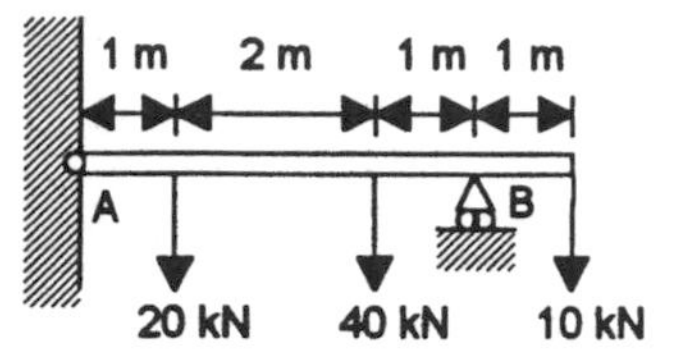

Figure 5.1.22 *Example 5.1.10*

Example 5.1.10

Figure 5.1.22 shows a beam which has a pinned support at A and rests on a roller support at B. Determine the reactive forces at the two supports due to the three concentrated loads shown if the beam is in equilibrium.

The support at A is pinned and so there can be vertical and horizontal components of a reactive force V_A and H_A. The roller support at B provides only a vertical reactive force V_B. For equilibrium, the sum of the horizontal components of the forces must be zero and so we must have $H_A = 0$. The sum of the vertical components should also be zero and so $V_A + V_B = 20 + 40 + 10 = 70$ kN. The sum of the moments about any axis must be zero; thus if we take moments about A (in order to eliminate one of the unknown forces) we have $20 \times 1 + 40 \times 3 - 4V_B + 10 \times 5 = 0$ and so $V_B = 47.5$ kN. Since $V_A + V_B = 70$ kN then $V_A = 70 - 47.5 = 22.5$ kN.

Distributed forces

So far in this section, the forces have been considered as acting at points. However, we might be considering a structure such as a bridge which is built up from a large number of pieces, each of which has a weight and so we have a large number of weight forces acting over the length of structure. Forces which are applied to or act over large lengths, areas or volumes of a body are termed *distributed forces*. Here we consider how we can handle such distributed forces.

The most commonly encountered distributed force is the weight of an object; the weight of a body is made up of the weights of each constituent particle and so does not act at a single point. However, it is possible to replace all the weight forces of an object by a single weight force acting at a particular point, this point being termed the *centre of gravity*.

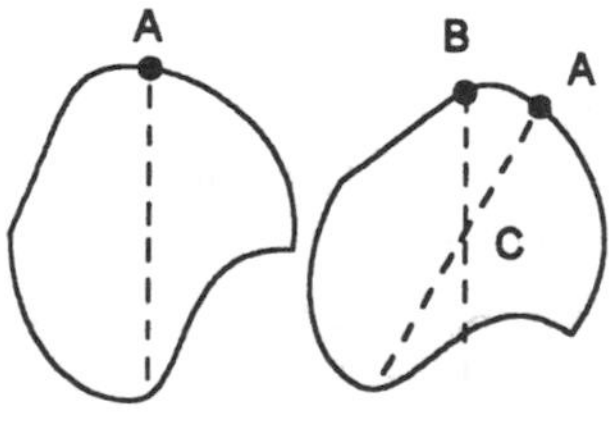

If an object is suspended from some point A on its surface (Figure 5.1.23), equilibrium will occur when the moments of its constituent particles about the axis through A balance. When this occurs we can think of all the individual weight forces being replaced by a single force with a line of action vertically passing through A and so the centre of gravity to lie somewhere on the vertical line through A. If the object is now suspended from some other point B, when equilibrium occurs the centre of gravity will lie on the vertical line through B. The intersection of these lines gives the location C of the centre of gravity.

Figure 5.1.23 *The centre of gravity of an object lies vertically below a point of suspension when the object is free to move about the point of suspension and take up its equilibrium position*

Consider an object to be made up of a large number of small elements:

segment 1: weight δw_1 a distance x_1 from some axis

segment 2: weight δw_2 a distance x_2 from the same axis

segment 3: weight δw_3 a distance x_3 from the same axis

and so on for further segments.

Note that the symbol δ is used to indicate that we are dealing with a small bit of the quantity defined by the symbol following the δ.

The total moment of the segments about the axis will be the sum of all the moment terms, i.e. $\delta w_1 x_1 + \delta w_2 x_2 + \delta w_3 x_3 + \ldots$. If a single force W, i.e. the total weight, acting through the centre of gravity is to be used to replace the forces due to each segment then we must have a moment about the axis of Wx, where x is the distance of the centre of gravity from the axis, such that Wx = sum of all the moment terms due to the segments. Hence:

$$x = \frac{\delta w_1 x_1 + \delta w_2 x_2 + \delta w_3 x_3 + \ldots}{W} = \frac{\Sigma x \, \delta w}{W}$$

If we consider infinitesimally small elements, i.e. $\delta w \to 0$, then we can write this as:

$$x = \frac{\int x \, \mathrm{d}w}{W} \qquad (5.1.8)$$

Note that as the bits being summed become vanishingly small we replace the summation sign by the integral sign and δ by d.

For symmetrical homogeneous objects, the centre of gravity is the geometrical centre. Thus, for a sphere the centre of gravity is at the centre of the sphere. For a cube, the centre of gravity is at the centre of the cube. For a rectangular cross-section homogeneous (i.e. one whose properties does not vary) beam, the centre of gravity is halfway along its length and in the centre of the cross-section

Centroids

We often have situations involving objects cut from uniform sheets of material and so having a constant weight per square metre or objects entirely made from the same material and thus having a constant weight per cubic metre throughout. In such situations where there is a constant weight per unit area or unit volume, we can modify our centre of gravity definition and eliminate the weight term from it. We then talk of the centroid of an object.

When we have a uniform sheet of material with the constant weight per square metre w, then $W = wA$ and $\delta w = w \, \delta A$ and so:

$$x = \frac{\int x \, \mathrm{d}w}{W} = \frac{\int xw \, \mathrm{d}A}{wA} = \frac{\int x \, \mathrm{d}A}{A} \qquad (5.1.9)$$

For a body of volume V and the constant weight per unit volume w, then $W = wV$ and $\delta w = w \, \delta V$ and so:

$$x = \frac{\int x \, \mathrm{d}w}{W} = \frac{\int xw \, \mathrm{d}V}{wV} = \frac{\int x \, \mathrm{d}V}{V} \qquad (5.1.10)$$

If we have a body of constant cross-sectional area and weight per unit volume, then $W = wAL$ and $\delta w = wA \, \delta AL$ and so:

$$x = \frac{\int x \, \mathrm{d}w}{W} = \frac{\int x \, wA \, \mathrm{d}L}{wAL} = \frac{\int x \, \mathrm{d}L}{L} \qquad (5.1.11)$$

The term *centroid* is used for the 'geometrical centre' of an object defined by the above equations; this being because it is independent of the weight of the body and depends only on geometrical properties.

The centroid of a circular area is at its centre, of a rectangular area at its centre, of a triangular area at one-third the altitude from each of the sides and of a semicircular area $4r/3\pi$ above the base diameter along the radius at right angles to the base.

Mathematics in action

Centroids of common shapes

Many objects can be considered to be constituted from a number of basic shapes, e.g. rectangles, triangles and hemispheres, and a knowledge of the centroids of such shapes enables the centroids of the more complex shapes to be determined. Here we look at how we can derive the position of centroids, considering the examples of a triangular area, a hemisphere and a circular arc.

(1) *Centroid of a triangular area*
For the triangle shown in Figure 5.1.24, consider a small strip of area $\delta A = x\,\delta y$. By similar triangles $x/(h-y) = b/h$ and so $x\,\delta y = [b(h-y)/h]\,\delta y$. The total area $A = \frac{1}{2}bh$. Hence, the y co-ordinate of the centroid is:

$$\bar{y} = \frac{2}{bh}\int_0^h y\frac{b(h-y)}{h}\,dy = \tfrac{1}{3}h \tag{5.1.12}$$

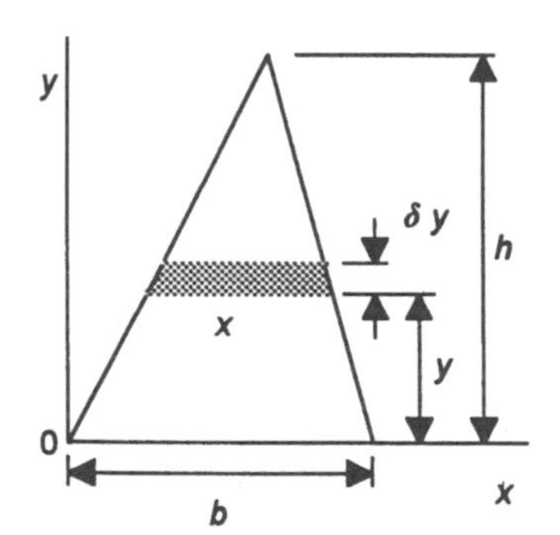

Figure 5.1.24 *Triangular area*

The centroid is located at one-third the altitude of the triangle. The same result is obtained if we consider the location with respect to the other sides. Since the point of intersection of the lines drawn from each apex to area the midpoint of the opposite side (the term median is used) meet at one-third the altitude of the triangle, the centroid is at the intersection of the medians.

(2) *Centroid of a hemisphere*
The same approach can be used to find the centroid of three-dimensional objects. The object is considered to be a stack of thin two-dimensional slices, each of which can be treated in the manner described above. For the hemisphere of radius r shown in Figure 5.1.25, we take circular slices of thickness δx. The radius r of the slice is given by applying Pythagoras' theorem: $R^2 = r^2 + x^2$. The area of the slice is $\pi r^2 = \pi(R^2 - x^2)$ and so its volume is $dV = \pi(R^2 - x^2)\,\delta x$. Hence, since the volume of a hemisphere is half that of a sphere, the x co-ordinate of the centroid is:

$$\bar{x} = \frac{1}{\frac{2}{3}\pi R^3}\int_0^R x\pi(R^2 - x^2)\,dx = \tfrac{3}{8}R \tag{5.1.13}$$

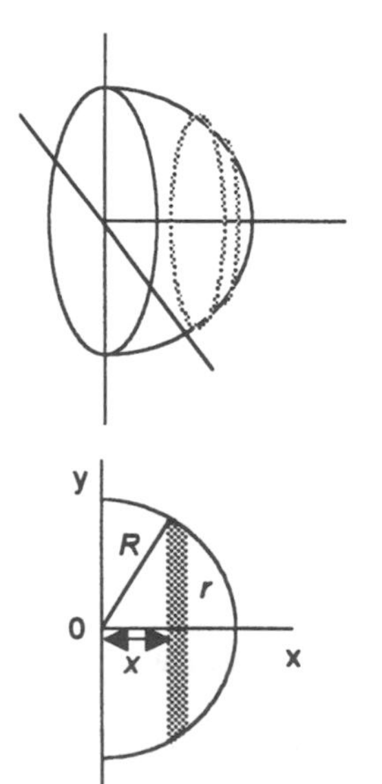

Figure 5.1.25 *Hemisphere*

The centroid is thus located at three-eighths of the radius along the radius at right angles to the flat surface of the hemisphere.

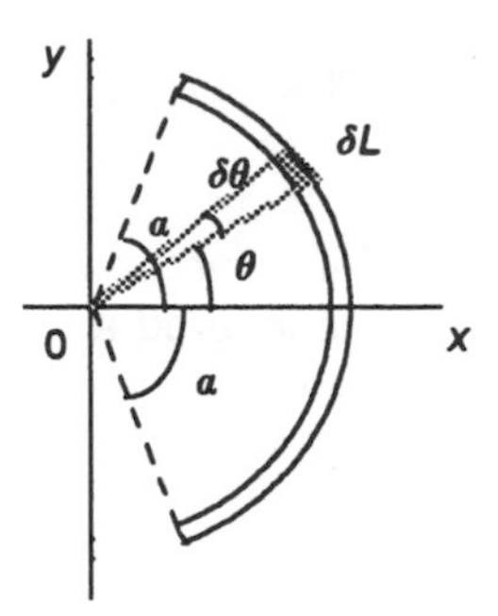

Figure 5.1.26 *Circular arc*

(3) *Centroid of a circular arc wire*
This is a body of constant cross-sectional area and weight per unit volume and so we use Equation (5.1.11). For the circular arc of radius r shown in Figure 5.1.26 we take length segments of length δL. This is an arc of radius r subtending an angle $\delta\theta$ at the centre; hence $\delta L = r\,\delta\theta$. The x co-ordinate of the element is $r \cos \theta$. Hence, the position of the centroid along the x-axis is:

$$\bar{x} = \frac{\int x\, dL}{L} = \frac{1}{2r\alpha} \int_{-\alpha}^{\alpha} (r \cos \theta) r\, d\theta = \frac{r \sin \alpha}{\alpha} \quad (5.1.14)$$

Composite bodies

For objects which can be divided into several parts for which the centres of gravity and weights, or centroids and areas, are known, the centre of gravity/centroid can be determined by the following procedure:

(1) On a sketch of the body, divide it into a number of composite parts. A hole, i.e. a part having no material, can be considered to be a part having a negative weight or area.
(2) Establish the co-ordinate axes on the sketch and locate the centre of gravity, or centroid, of each constituent part.
(3) Take moments about some convenient axis of the weights or areas of the constituent parts and equate the sum to the moment the centre of gravity or centroid of the total weight or area about the same axis.

Example 5.1.11

Determine the position of the centroid for the uniform thickness sheet shown in Figure 5.1.27.

For the object shown in Figure 5.1.27, the two constituent parts are different sized, homogeneous, rectangular objects of constant thickness sheet. An axis is chosen through the centres of the two parts. The centroid of each piece is at its centre. Thus, one piece has its centroid a distance along its centreline from A of 50 mm and the other a distance from A of 230 mm. Taking moments about A gives, for the two parts, $(160 \times 100) \times 50 + (260 \times 60) \times 230$ and this is equal to the product of the total area, i.e. $160 \times 100 + 260 \times 60$, multiplied by the distance of the centroid of the composite from X. Thus:

$$\bar{x} = \frac{160 \times 100 \times 50 + 260 \times 60 \times 230}{160 \times 100 + 260 \times 60} = 138.9\,\text{m}$$

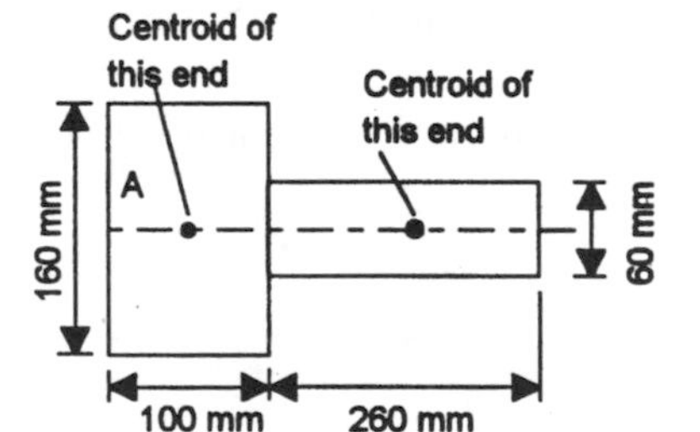

Figure 5.1.27 *Example 5.1.11*

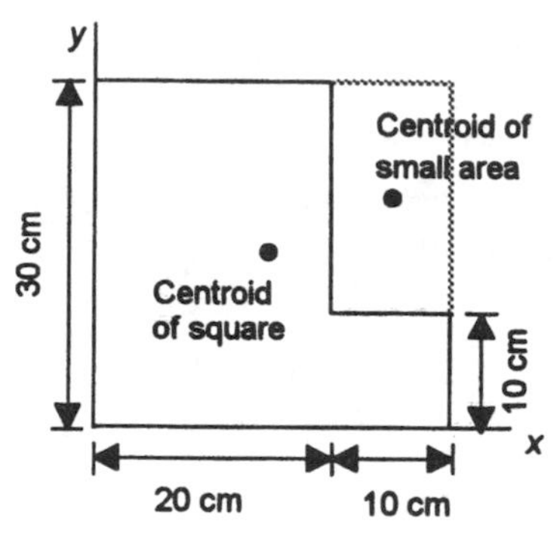

Figure 5.1.28 *Example 5.1.12*

Example 5.1.12

Determine the position of the centroid for the uniform thickness sheet shown in Figure 5.1.28.

The sheets can be considered to have two elements, a large square 30 cm by 30 cm and a negative rectangular area 20 cm by 10 cm. The centroid of each segment is located at its midpoint. Taking moments about the *y*-axis gives:

$$\bar{x} = \frac{30 \times 30 \times 15 - 20 \times 10 \times 25}{30 \times 30 - 20 \times 10} = 12.1 \text{ cm}$$

Taking moments about the *x*-axis gives:

$$\bar{y} = \frac{30 \times 30 \times 15 - 20 \times 10 \times 25}{30 \times 30 - 20 \times 10} = 12.1 \text{ cm}$$

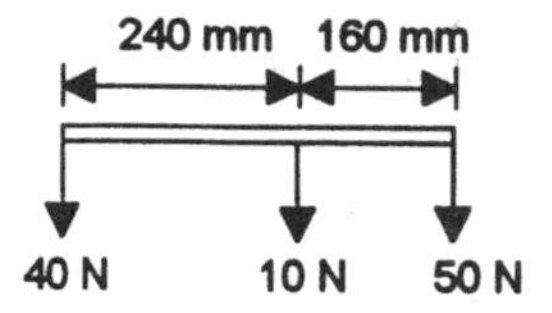

Figure 5.1.29 *Example 5.1.13*

Example 5.1.13

Determine the position of the centre of gravity of the horizontal uniform beam shown in Figure 5.1.29 when it is loaded in the manner shown. Neglect the weight of the beam.

Taking moments about the left-hand end gives $10 \times 240 + 50 \times 400 = 22\,400$ N mm. The total load on the beam is 100 N and thus the centre of gravity is located $22\,400/100 = 22.4$ mm from the left-hand end.

Beams with distributed forces

The weight of a uniform beam can be considered to act at its centre of gravity, this being located at its midpoint. However, beams can also be subject to a distributed load spread over part of their length. An example of this is a floor beam in a building where the loading due to a machine might be spread over just part of the length of the beam. Distributed loading on beams is commonly represented on diagrams in the manner shown in Figure 5.1.30 where there is a uniformly distributed load of 20 N/m over part of it.

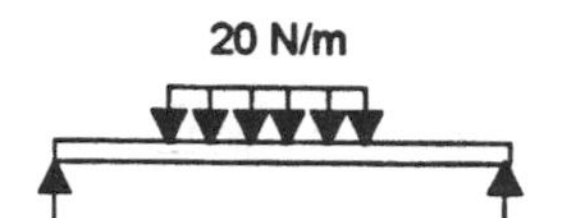

Figure 5.1.30 *Distributed load of 20 N/m over part of the beam is represented by a block of arrows*

Example 5.1.14

Determine the reactions at the supports of the uniform horizontal beam shown in Figure 5.1.31, with its point load and a uniformly distributed load of 40 N/m and the beam having a weight of 15 N.

The weight of the beam can be considered to act at its centre of gravity and thus, since the beam is uniform, this point will be located a distance of 0.6 m from the left-hand end. The distributed load can be considered to act at its

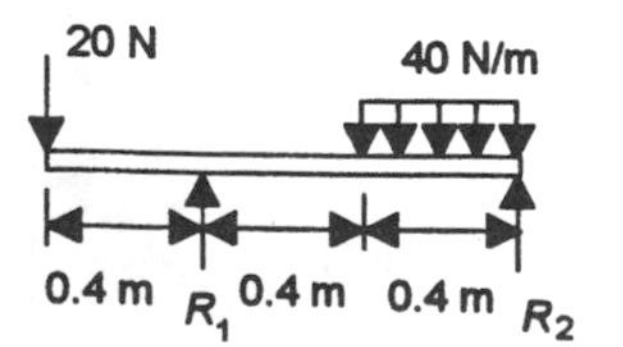

Figure 5.1.31 *Example 5.1.14*

centre of gravity, i.e. the midpoint of the length over which it acts, and so is 16 N at 1.0 m from the left-hand end. Taking moments about the left-hand end gives:

$$0.4R_1 + 1.2R_2 = 15 \times 0.6 + 16 \times 1.0 = 25$$

and $R_1 = 62.5 - 3R_2$. The sum of the vertical forces must be zero and so:

$$R_1 + R_2 = 20 + 15 + 16 = 51$$

Hence:

$$62.5 - 3R_2 + R_2 = 51$$

$R_2 = 5.75\,\text{N}$ and $R_1 = 45.25\,\text{N}$.

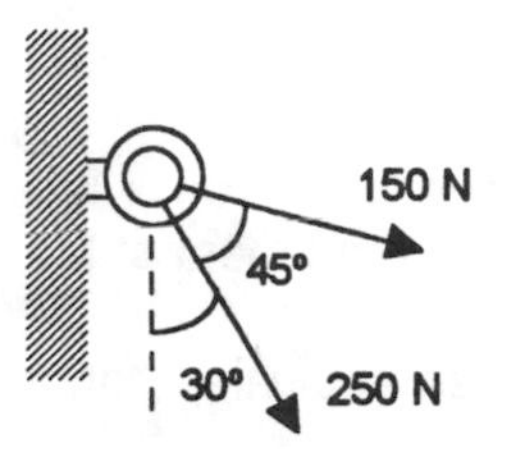

Figure 5.1.32 *Problem 1*

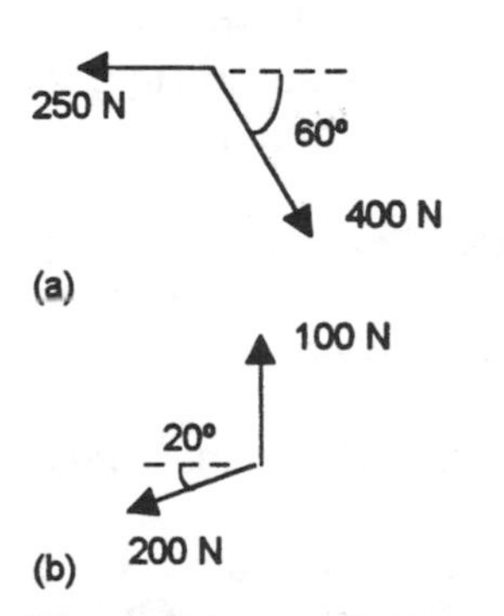

Figure 5.1.33 *Problem 2*

Problems 5.1.1

(1) An eye bolt is subject to two forces of 250 N and 150 N in the directions shown in Figure 5.1.32. Determine the resultant force.

(2) A particle is in equilibrium under the action of the three coplanar concurrent forces. If two of the forces are as shown in Figure 5.1.33(a) and (b), what will be the size and direction of the third force?

(3) Determine the tension in the two ropes attached to the rigid supports in the systems shown in Figure 5.1.34 if the supported blocks are in equilibrium.

(4) A vehicle of weight 10 kN is stationary on a hill which is inclined at 20° to the horizontal. Determine the components of the weight at right angles to the surface of the hill and parallel to it.

(5) A horizontal rod of negligible weight is pin-jointed at one end and supported at the other by a wire inclined at 30° to the horizontal (Figure 5.1.35). If it supports a load of 100 N at its midpoint, determine the tension in the wire.

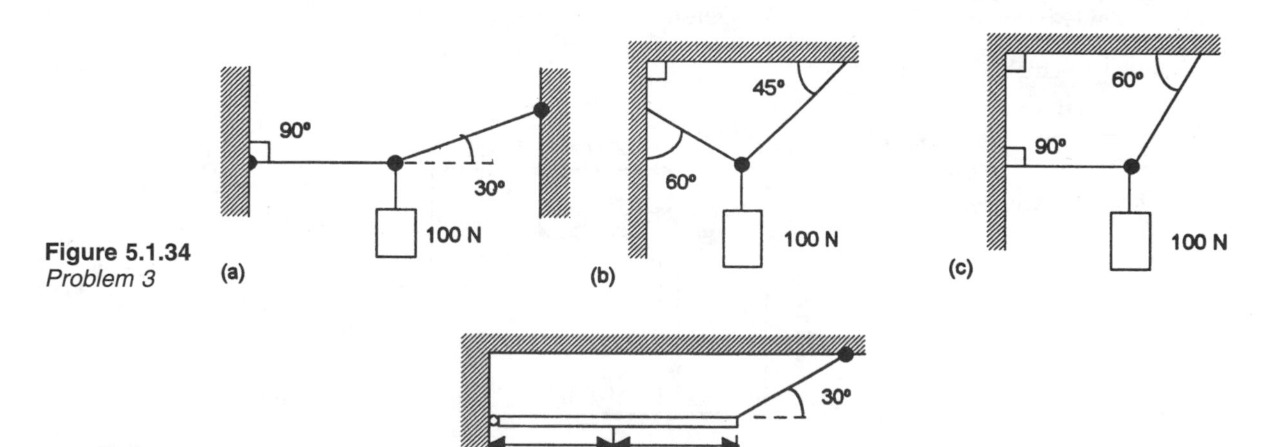

Figure 5.1.34 *Problem 3*

Figure 5.1.35 *Problem 5*

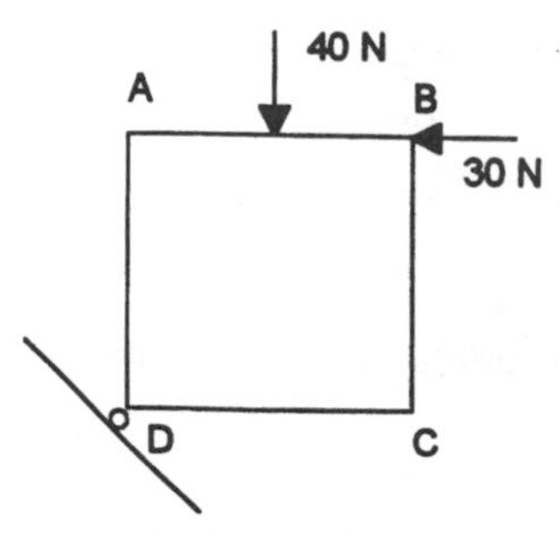

Figure 5.1.36 *Problem 6*

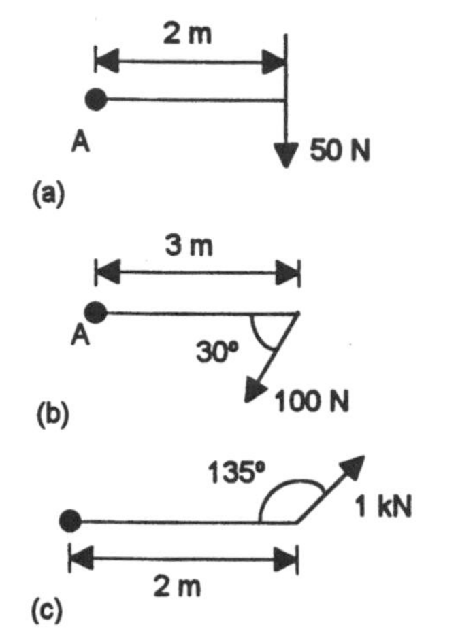

Figure 5.1.37 *Problem 7*

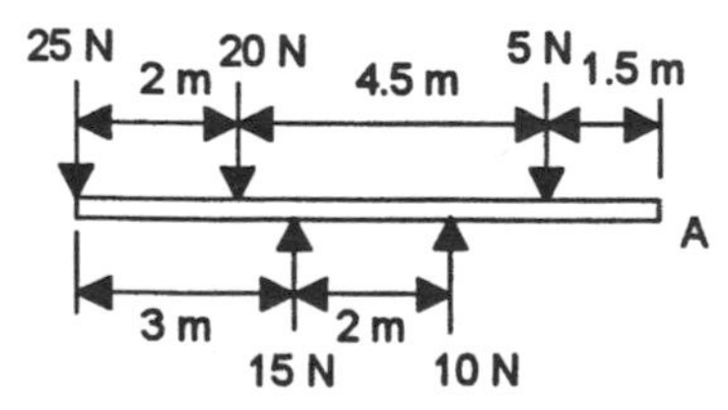

Figure 5.1.38 *Problem 8*

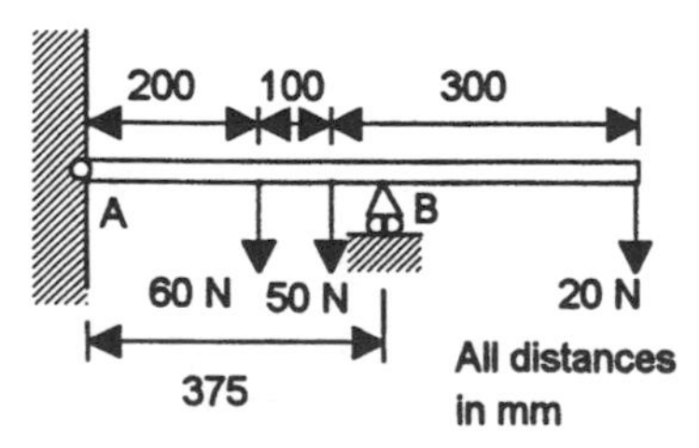

Figure 5.1.39 *Problem 9*

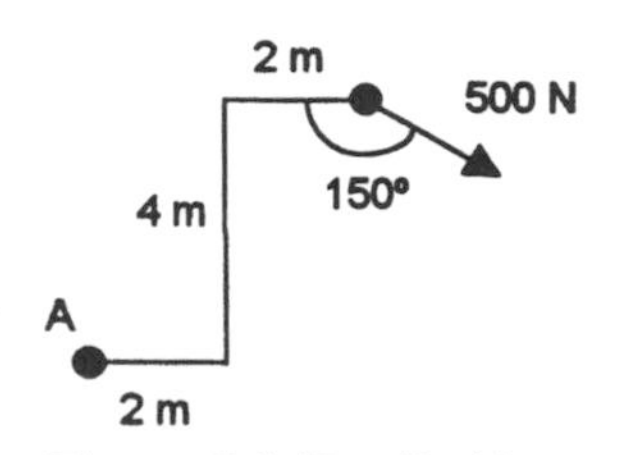

Figure 5.1.40 *Problem 11*

(6) The square plate shown in Figure 5.1.36 is pin-jointed at D and is subject to the coplanar forces shown, the 40 N force being applied at the midpoint of the upper face and the 30 N force along the line BA. Determine the force that has to be applied along the diagonal AC to give equilibrium and the vertical and horizontal components of the reaction force at D.

(7) Determine the magnitude and direction of the moments of the forces shown in Figure 5.1.37 about the axes through point A.

(8) Replace the system of parallel coplanar forces acting on the beam in Figure 5.1.38 by an equivalent resultant force and couple acting at end A.

(9) Figure 5.1.39 shows a beam which has a pinned support at A and rests on a roller support at B. Determine the reactive forces at the two supports due to the three concentrated loads shown if the beam is in equilibrium.

(10) Two parallel oppositely directed forces of size 150 N are separated by a distance of 2 m. What is the moment of the couple?

(11) Replace the force in Figure 5.1.40 with an equivalent force and couple at A.

(12) The system of coplanar forces shown in Figure 5.1.41 is in equilibrium. Determine the sizes of the forces *P*, *Q* and *R*.

(13) For the system shown in Figure 5.1.42, determine the direction along which the 30 N force is to be applied to give the maximum moment about the axis through A and determine the value of this moment.

(14) Figure 5.1.43 shows a beam suspended horizontally by three wires and supporting a central load of 500 N. If the beam is in equilibrium and of negligible weight, determine the tensions in the wires.

(15) Determine the positions of the centroids for the areas shown in Figure 5.1.44.

(16) Determine the location of the centre of gravity of a horizontal uniform beam of negligible weight when it has point loads of 20 N a distance of 2 m, 50 N at 5 m and 30 N at 6 m from the left-hand end.

(17) Determine, by integration, the location of the centroid of a quarter circular area of radius *r*.

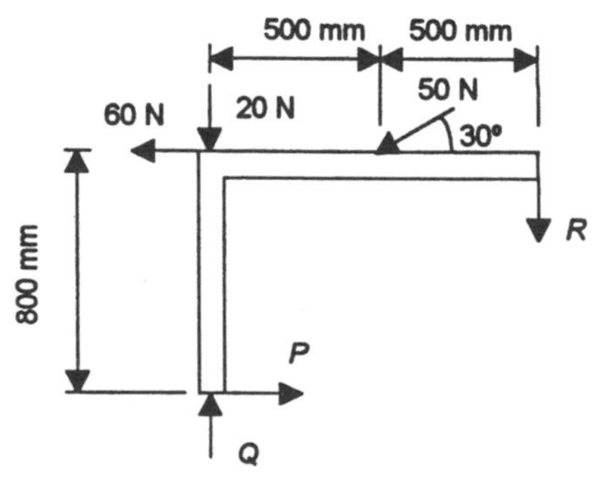

Figure 5.1.41 *Problem 12*

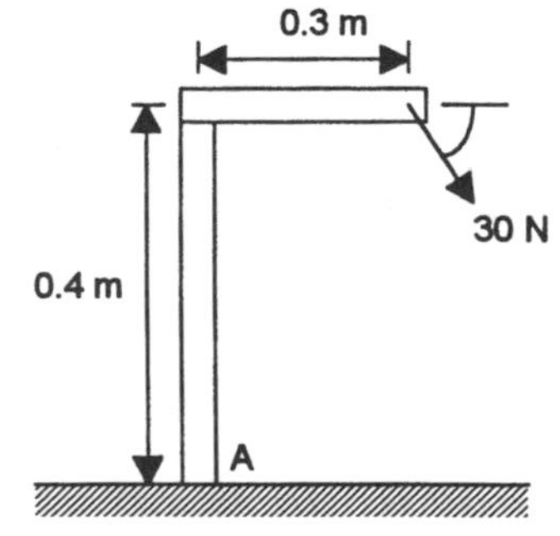

Figure 5.1.42 *Problem 13*

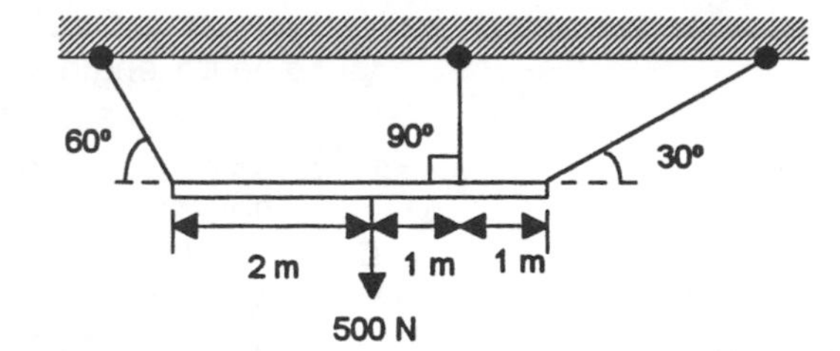

Figure 5.1.43 *Problem 14*

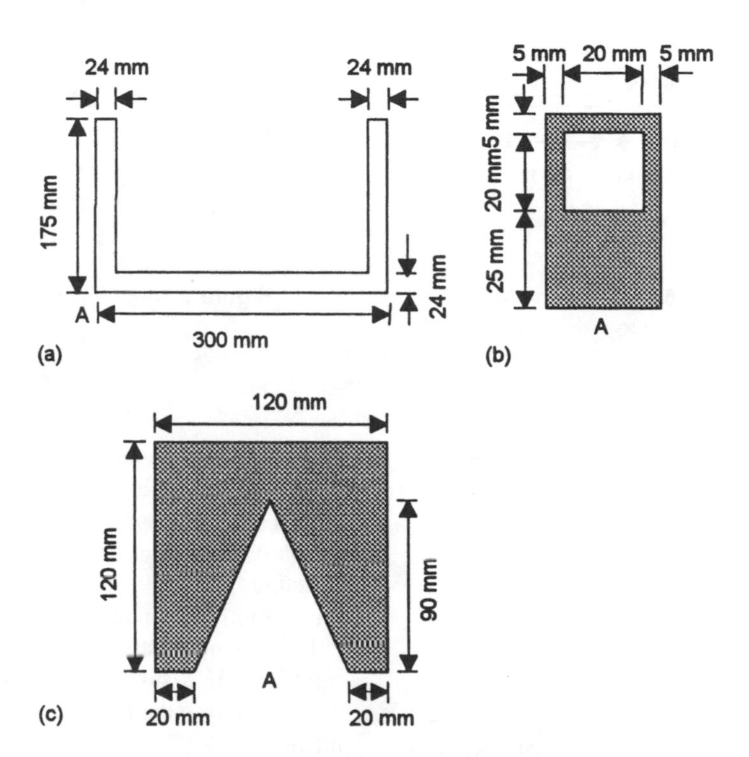

Figure 5.1.44 *Problem 15*

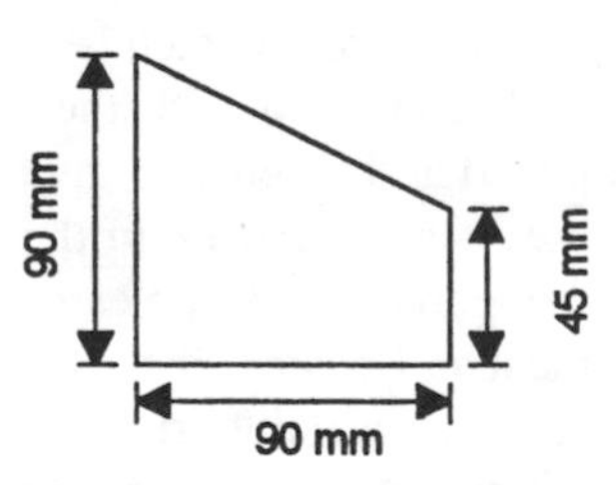

Figure 5.1.45 *Problem 19*

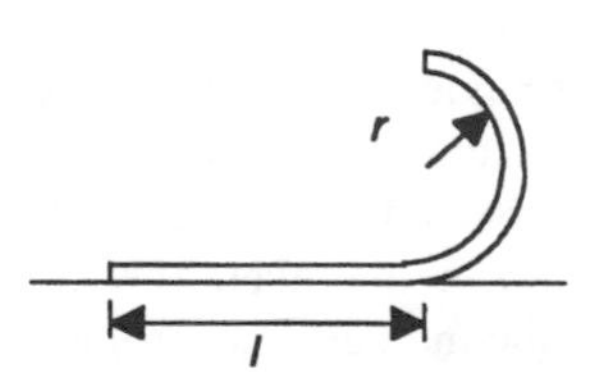

Figure 5.1.46 *Problem 22*

(18) Determine, by integration, the location of the centroid of a hemispherical shell of radius r and negligible thickness.

(19) Determine the position of the centroid for the area shown in Figure 5.1.45.

(20) Determine, by means of integration, the location of the centroid of a wire bent into the form of a quarter circle.

(21) A cylindrical tin can has a base but no lid and is made from thin, uniform thickness, sheet. If the base diameter is 60 mm and the can height is 160 mm, at what point is the centre of gravity of the can?

(22) A thin uniform cross-section metal strip has part of it bent into a semicircle of radius r with a length l tangential to the semicircle, as illustrated in Figure 5.1.46. Show that the strip can rest with the straight part on a horizontal bench if l is greater than $2r$.

(23) A length of wire of uniform cross-section and mass per unit length is bent into the form of a semicircle. What angle will the diameter make with the vertical when the wire loop is suspended freely from one end of its diameter?

(24) A uniform beam of length 5.0 m rests on supports 1.0 m from the left-hand end and 0.5 m from the right-hand end. A

uniformly distributed load of 10 kN/m is spread over its entire length and the beam has a weight of 10 kN. What will be the reactions at the supports?

(25) Determine the reactions at the supports of the uniform horizontal beams shown in Figure 5.1.47.

(26) The tapered concrete beam shown in Figure 5.1.48 has a rectangular cross-section of uniform thickness 0.30 m and a density of 2400 kg/m^3. If it is supported horizontally by two supports, one at each end, determine the reactions at the supports.

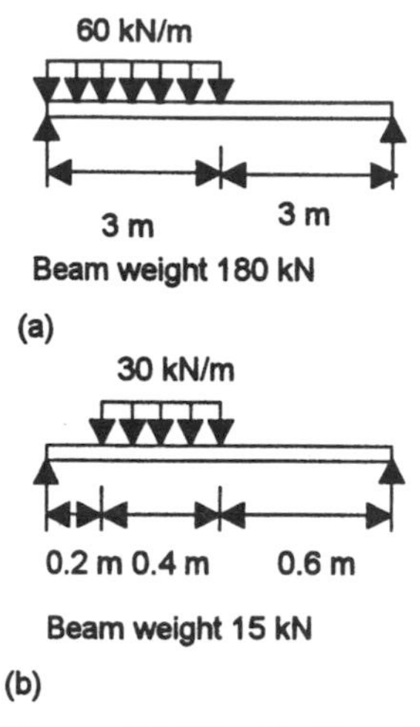

Figure 5.1.47 *Problem 25*

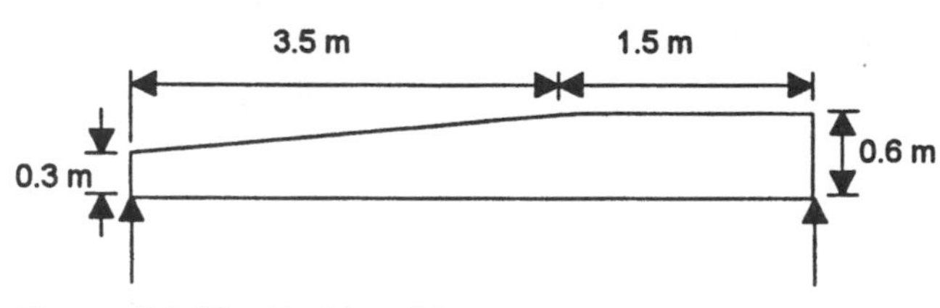

Figure 5.1.48 *Problem 26*

5.2 Structures

This section is about the analysis of structures. The term *structure* is used for an assembly of members such as bars, plates and walls which form a stiff unit capable of supporting loads. We can thus apply the term to complex structures such as those of buildings and bridges or simpler structures such as those of a support for a child's swing or a cantilevered bracket to hold a pub sign (Figure 5.2.1).

In Section 5.1, structures were considered which where either a single rigid body or a system of connected members and free-body diagrams used to analyse the forces acting on individual members or junctions of members.

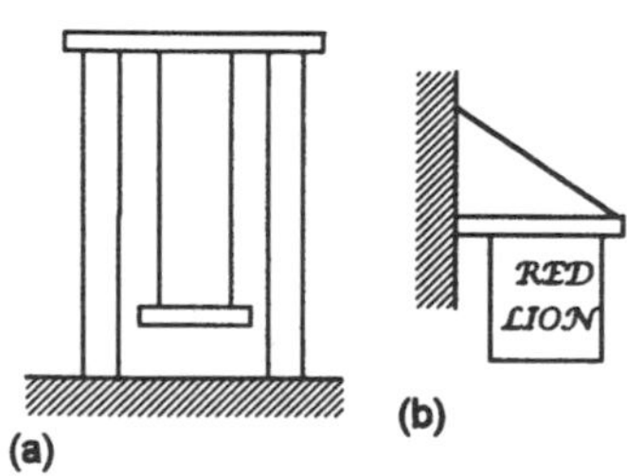

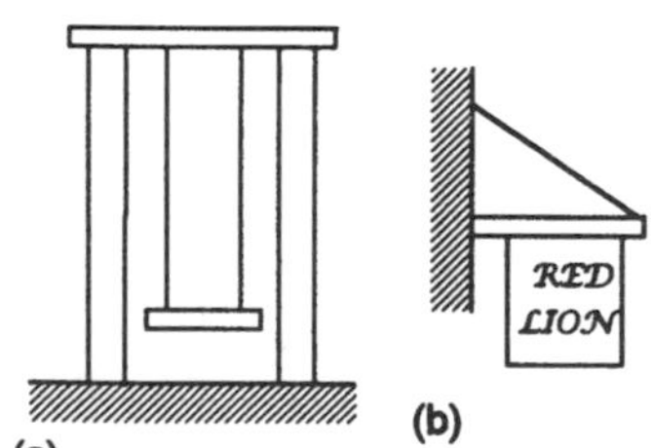

Figure 5.2.1 *Example of structures: (a) a child's swing; (b) a hanging pub sign*

Terms that are encountered when talking of structures are frameworks and trusses. The term *framework* is used for an assembly of members which have sectional dimensions that are small compared with their length. A framework composed of members joined at their ends to give a rigid structure is called a *truss* and when the members all lie in the same plane a *plane truss*. Bridges and roof supports are examples of trusses, the structural members being typically I-beams, bars or channels which are fastened together at their ends by welding, riveting or bolts which behave like pin-jointed connections and permit forces in any direction. Figure 5.2.2 shows some examples of trusses; a truss structure that is used with bridges and one that is used to support the roof of a house.

(a) A Warren bridge truss

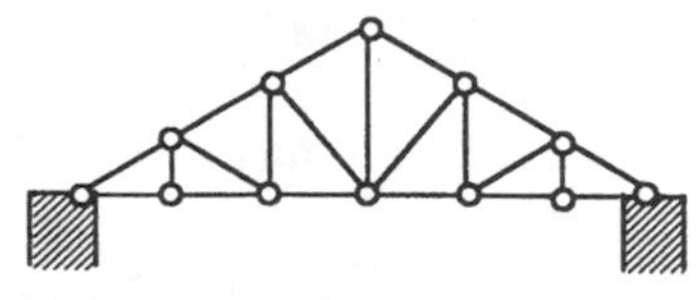

(b) A Howe roof truss

Figure 5.2.2 *Examples of trusses: (a) a bridge; (b) a roof*

Statically determinate structures

The basic element of a plane truss is the triangle, this being three members pin-jointed at their ends to give a stable rigid framework (Figure 5.2.3(a)). Such a pin-jointed structure is said to be *statically determinate* in that the equations of equilibrium for members, i.e. no resultant force in any direction and the sum of the moments is zero, are sufficient to enable all the forces to be determined.

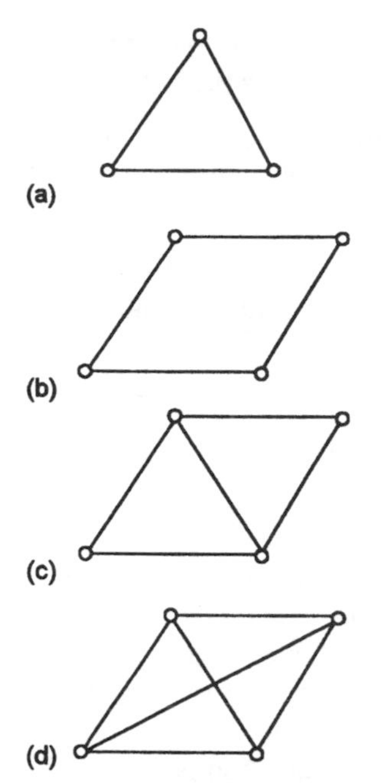

Figure 5.2.3 *Frameworks: (a) stable; (b) unstable; (c) stable; (d) stable with a redundant member*

The arrangement of four members to form a rectangular framework (Figure 5.2.3(b)) gives an unstable structure since a small sideways force can cause the structure to collapse. To make the rectangle stable, a diagonal member is required and this converts it into two triangular frameworks (Figure 5.2.3(c)). The addition of two diagonal members (Figure 5.2.3(d)) gives a rigid structure but one of the diagonal members is 'redundant' in not being required for stability; such a structure is termed *statically indeterminate* since it cannot be analysed by just the equations for equilibrium. Note that the analysis of members in trusses might show that a member is unloaded. An unloaded member must not be confused with a redundant member; it may be that the member becomes loaded under different conditions.

If m is the number of members and j the number of joints, then a stable pin-jointed structure can be produced if:

$$m + 3 = 2j$$

If $m + 3$ is greater than $2j$ then 'redundant' members are present; if $m + 3$ is less than $2j$ then the structure is unstable.

Many structures are designed to include 'redundant' members so that the structure will be 'fail safe' and not collapse if one or more members fail or so that they can more easily cope with different loading conditions. The loading capacity of members subject to compression depends on their length with short members able to carry greater loads; this is because slender members can more easily buckle. Tension members can, however, be much thinner and so compression members tend to be more expensive than tension members. Thus if a truss has to be designed to cope with different loading conditions which can result in a member being subject sometimes to tensile loading and sometimes to compressive loading, it will have to be made thick enough to cope with the compressive loading. One way this can be avoided is to design the truss with a redundant member. Figure 5.2.4 illustrates this with a truss which is designed to withstand wind loading from either the left or the right. The diagonal bracing members are slender cables which can only operate in tension. With the loading from the left, member BD is in tension and member AC goes slack and takes no load. With the loading from the right, member AC is in tension and member BD is slack and takes no load. The truss can thus withstand both types of loading without having a diagonal member designed for compression. Such a form of bracing was widely used in early biplanes to cope with the loading on the wings changing from when they were on the ground to when in the air. Nowadays such bracing is often found in structures subject to wind loading, e.g. water towers.

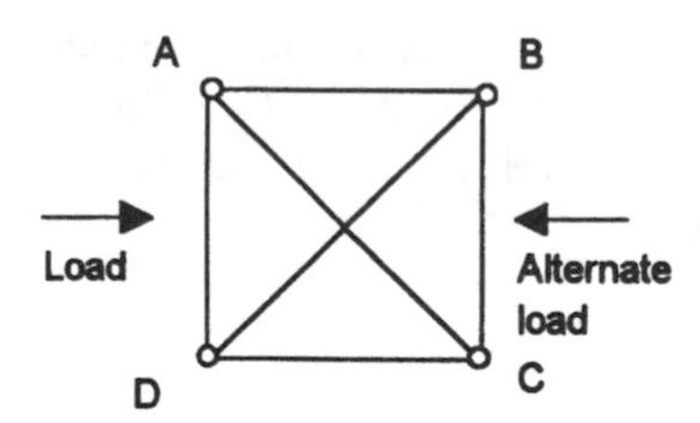

Figure 5.2.4 *Designing with redundancy*

Example 5.2.1

Use the above criteria to determine whether the frameworks shown in Figure 5.2.3 can be stable, are unstable or contain redundant members.

For Figure 5.2.3(a) we have $m = 3$ and $j = 3$, hence when we have $m + 3 = 6$ and $2j = 6$ then the structure can be stable.

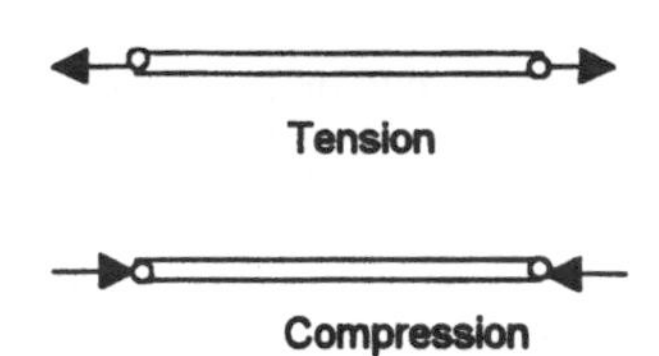

Figure 5.2.5 *Two-force members: equilibrium occurring under the action of just two forces of the same size and acting in the same straight line but opposite directions*

For Figure 5.2.3(b) we have $m = 4$ and $j = 4$, hence when we have $m + 3 = 7$ and $2j = 8$ then the structure is unstable.

For Figure 5.2.3(c) we have $m = 5$ and $j = 4$, hence when we have $m + 3 = 8$ and $2j = 8$ then the structure can be stable.

For Figure 5.2.3(c) we have $m = 6$ and $j = 4$, hence when we have $m + 3 = 9$ and $2j = 8$ then the structure contains a 'redundant' member.

Analysis of frameworks

In the following pages we discuss and apply methods that can be used to analyse frameworks and determine the forces in individual members. For example, we might want to determine the forces in the individual members of the Warren bridge truss of Figure 5.2.2 when there is a heavy lorry on the bridge.

The analysis of frameworks is based on several assumptions:

- Each member can be represented on a diagram as a straight line joining its two end points where external forces are applied; external forces are only applied at the ends of members. The lines represent the longitudinal axes of the members and the joints between members are treated as points located at the intersection of the members. The weight of a member is assumed to be small compared with the forces acting on it.
- All members are assumed to be two-force members; equilibrium occurs under the action of just two forces with the forces being of equal size and having the same line of action but in opposite directions so that a member is subject to either just tension or just compression (Figure 5.2.5). A member which is in tension is called a *tie*; a member that is in compression is called a *strut*.
- All the joints are assumed to behave as pin-jointed and thus the joint is capable of supporting a force in any direction (see Figure 5.1.21(d) and associated text). Welded and riveted joints can usually be assumed to behave in this way.

Bow's notation

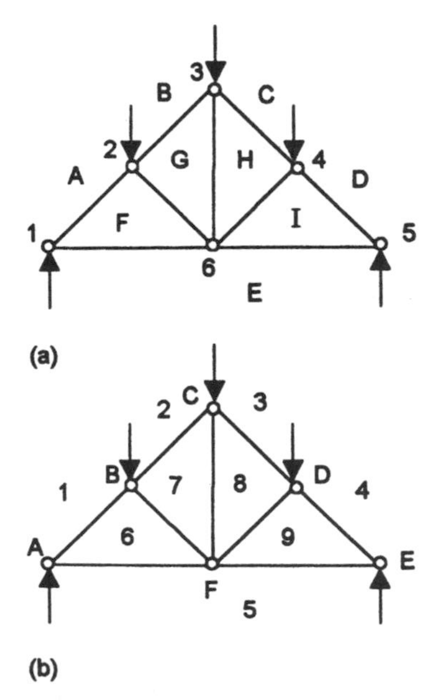

Figure 5.2.6 *Bow's notation, (a) and (b) being two alternative forms; in this book (a) is used*

Bow's notation is a useful method of labelling the forces in a truss. This is based on labelling all the spaces between the members and their external forces and reactions using letters or numbers or a combination of letters and numbers. The advantage of using all letters for the spaces is that the joints can be given numbers, as illustrated in Figure 5.2.6(a); or, conversely, using numbers for the spaces means letters can be used to identify joints (Figure 5.2.6(b)). The internal forces are then labelled by the two letters or numbers on each side of it, generally taken in a clockwise direction. Thus, in Figure 5.2.6(a), the force in the member linking junctions 1 and 2 is F_{AF} and the force in the member linking junctions 3 and 6 is F_{GH}. In Figure 5.2.6(b), the force in the member linking junctions A and B is F_{16} and the force in the member linking junctions C and F is F_{78}.

In the above illustration of Bow's notation, the joints were labelled independently of the spaces between forces. However, the space labelling can be used to identify the joints without the need for independent labelling for them. The joints are labelled by the space letters or numbers surrounding them when read in a clockwise direction. Thus, in Figure 5.2.6(a), junction 1 could be identified as junction AFE and junction 3 as junction BCHG.

Method of joints

Each joint in a structure will be in equilibrium if the structure is in equilibrium, thus the analysis of trusses by the *method of joints* involves considering the equilibrium conditions at each joint in isolation from the rest of the truss. The procedure is:

(1) Draw a line diagram of the framework.
(2) Label the diagram using Bow's notation, or some other form of notation.
(3) Determine any unknown external forces or reactions at supports by considering the truss at a single entity, ignoring all internal forces in truss members.
(4) Consider a junction in isolation from the rest of the truss and the forces, both external and internal, acting on that junction. The sum of the components of these forces in the vertical direction must be zero, as must be the sum of the components in the horizontal direction. Solve the two equations to obtain the unknown forces. Because we only have two equations at a junction, the junctions to be first selected for this treatment should be where there are no more than two unknown forces.
(5) Consider each junction in turn, selecting them in the order which leaves no more than two unknown forces to be determined at a junction.

Example 5.2.2

Determine the forces acting on the members of the truss shown in Figure 5.2.7. The ends of the truss rest on smooth surfaces and its span is 20 m.

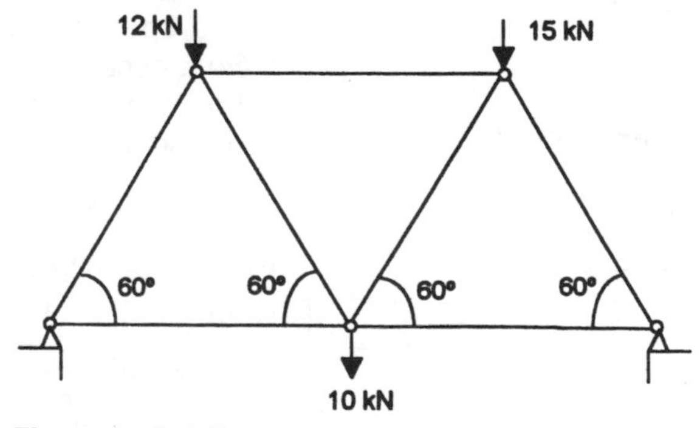

Figure 5.2.7 *Example 5.2.2*

Note that if the span had not been given, we could assume an arbitrary length of 1 unit for a member and then relate other distances to this length. Figure 5.2.8(a) shows Figure 5.2.7 redrawn and labelled using Bow's notation with the spaces being labelled by letters.

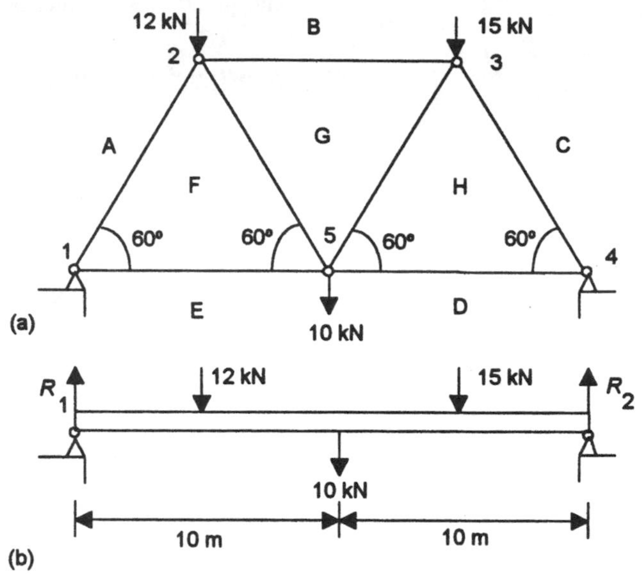

Figure 5.2.8 *Example 5.2.2.*

Considering the truss as an entity we have the situation shown in Figure 5.2.8(b). Because the supporting surfaces for the truss are smooth, the reactions at the supports will be vertical. Taking moments about the end at which reaction R_1 acts gives:

$$12 \times 5 + 10 \times 10 + 15 \times 15 = 20R_2$$

Hence R_2 = 19.25 kN. Equating the vertical components of the forces gives:

$$R_1 + R_2 = 12 + 10 + 15$$

and so R_1 = 17.75 kN.

Figure 5.2.9 shows free-body diagrams for each of the joints in the framework. Assumptions have been made about the directions of the forces in the members; if the forces are in the opposite directions then, when calculated, they will have a negative sign.

For joint 1, the sum of the vertical components must be zero and so:

$$17.75 - F_{AF} \sin 60° = 0$$

Hence F_{AF} = 20.5 kN. The sun of the horizontal components must be zero and so:

$$F_{AF} \cos 60° - F_{FE} = 0$$

Hence F_{FE} = 10.25 kN.

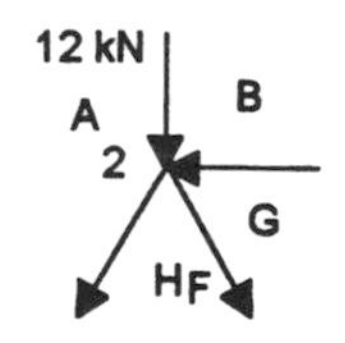

Joint 1, i.e. joint AFE

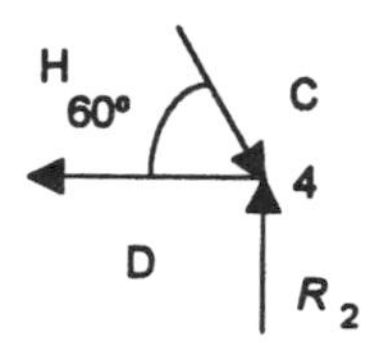

Joint 2, i.e. joint ABGF

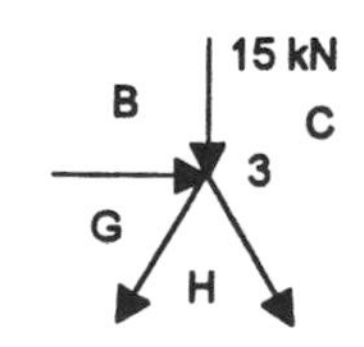

Joint 4, i.e. joint CDH

Joint 3, i.e. joint BCHG

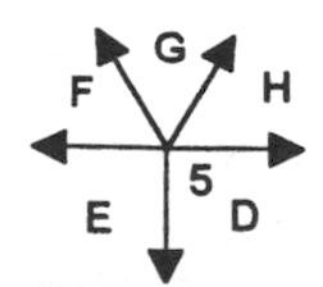

Joint 5, i.e. joint DEFGH

Figure 5.2.9 *Example 5.2.2*

For joint 2, the sum of the vertical components must be zero and so:

$$12 + F_{FG} \sin 60° - F_{AF} \sin 60° = 0$$

With F_{AF} = 20.5 kN, then F_{FG} = 6.6 kN. The sum of the horizontal components must be zero and so:

$$F_{EG} - F_{FG} \cos 60° - F_{AF} \cos 60° = 0$$

Hence F_{BG} = 13.6 kN.

For joint 4, the sum of the vertical components must be zero and so:

$$F_{CH} \sin 60° - R_2 = 0$$

Hence F_{CH} = 22.2 kN. The sum of the horizontal components must be zero and so:

$$F_{DH} - F_{CH} \cos 60° = 0$$

Hence F_{DH} = 11.1 kN.

For joint 3, the sum of the vertical components must be zero and so:

$$15 + F_{GH} \sin 60° - F_{CH} \sin 60° = 0$$

Hence F_{GH} = 4.9 kN. The sum of the horizontal components must be zero and so:

$$F_{GH} \cos 60° + F_{CH} \cos 60° - F_{BG} = 0$$

This is correct to the accuracy with which these forces have already been calculated.

For joint 5, the sum of the vertical components must be zero and so:

$$10 - F_{FG} \sin 60° - F_{GH} \sin 60° = 0$$

This is correct to the accuracy with which these forces have already been calculated. The sum of the horizontal components must be zero and so:

$$F_{FE} + F_{FG} \cos 60° - F_{DH} - F_{GH} \cos 60° = 0$$

This is correct to the accuracy with which these forces have already been calculated.

The directions of the resulting internal forces are such that member AF is in compression, BG is in compression, CH is in compression, DH is in tension, FE is in tension, FG is in tension and GH is in tension. The convention is adopted of labelling tensile forces by positive signs and compressive forces by negative signs, this being because tensile forces tend to increase length whereas compressive forces decrease length. Thus the internal forces in the members of the truss are F_{AF} = −20.5 kN, F_{BG} = −13.6 kN, F_{CH} = −22.2 kN, F_{DH} = +11.1 kN, F_{FE} = +10.25 kN, F_{FG} = +6.6 kN, F_{GH} = +4.9 kN.

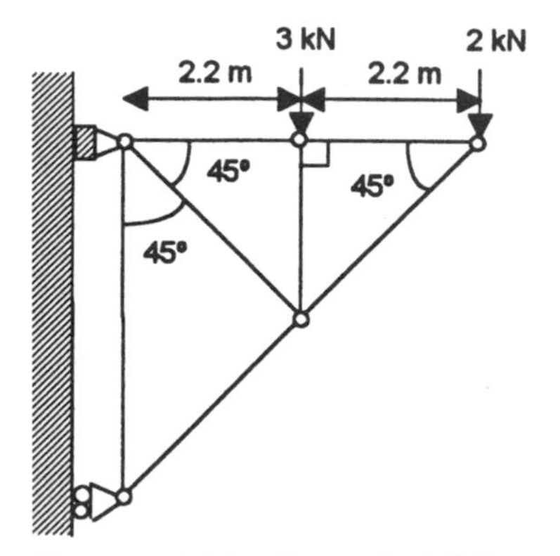

Figure 5.2.10 *Example 5.2.3*

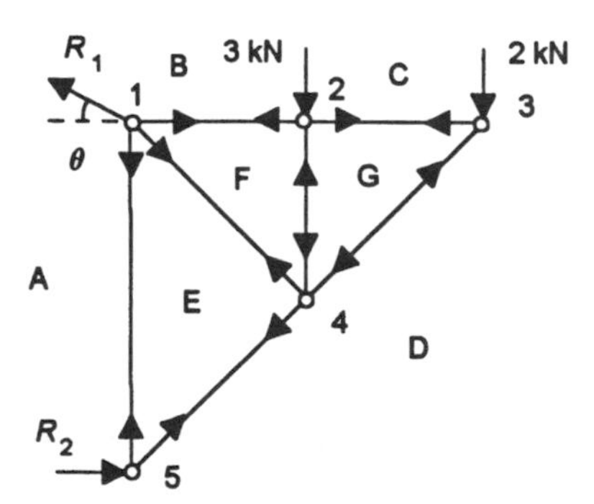

Figure 5.2.11 *Example 5.2.3*

Example 5.2.3

For the plane truss shown in Figure 5.2.10, determine the reactions at the supports and the forces in each member. The upper end of the truss is a pin connection to the wall while the lower end is held in contact by a roller.

Figure 5.2.11 shows the diagram labelled with Bow's notation. The reaction at the fixed end of the truss will be at some angle θ and thus can have both a horizontal and a vertical component; the reaction at the lower end will be purely a horizontal component.

Taking moments about junction 1 gives:

$$3 \times 2.2 + 2 \times 4.4 = 4.4R_2$$

Hence $R_2 = 3.5$ kN. The vertical components of the forces must be zero and so:

$$R_1 \sin\theta - 3 - 2 = 0$$

Hence $R_1 \sin\theta = 5$. The horizontal components must be zero and so:

$$R_1 \cos\theta = 3.5$$

Dividing these equations gives $\tan\theta = 5/3.5$ and so $\theta = 55.0°$.

Using the methods of joints, for joint 5 we have for the horizontal components:

$$F_{DE} \sin 45° - R_2 = 0$$

Thus $F_{DE} = 4.9$ kN. For the vertical components:

$$F_{DE} \cos 45° - F_{AE} = 0$$

and so $F_{AE} = 3.5$ kN.

For joint 3 we have for the vertical components:

$$2 - F_{DG} \sin 45° = 0$$

and so $F_{DG} = 2.8$ kN. For the horizontal components:

$$F_{CG} - F_{DG} \cos 45° = 0$$

and so $F_{CG} = 2.0$ kN.

For joint 2 we have for the vertical components:

$$3 - F_{FG} = 0$$

and so $F_{FG} = 3$ kN. For the horizontal components:

$$F_{CG} - F_{BF} = 0$$

and so $F_{BF} = 2.0$ kN.

For joint 1 we have for the vertical components:

$$F_{AE} + F_{EF} \cos 45° - R_1 \sin \theta = 0$$

and so $F_{EF} = 2.1$ kN. For the horizontal components:

$$F_{BF} + F_{EF} \cos 45° - R_1 \cos \theta = 0$$

This is correct to the accuracy with which these forces have already been calculated.

Thus the reactive forces are 6.1 kN at 55.0° to the horizontal and 3.5 kN horizontally. The forces in the members are $F_{AE} = +3.5$ kN, $F_{DE} = -4.9$ kN, $F_{DG} = -2.8$ kN, $F_{CG} = +2.0$ kN, $F_{BF} = +2.0$ kN, $F_{EF} = +2.1$ kN, $F_{FG} = -3.0$ kN.

Method of sections

This method is simpler to use than the method of joints when all that is required are the forces in just a few members of a truss. The method of sections considers the equilibrium of a part or section of a structure under the action of the external forces which act on it and the internal forces in the members joining the section to the rest of the truss. We thus imagine the truss to be cut at some particular place and the forces acting on one side of the cut are considered and the conditions for equilibrium applied. Since this can lead to only three equations, no more than three members of the truss should be cut by the section. The procedure is:

(1) Draw a line diagram of the structure.
(2) Label the diagram using Bow's notation or some other suitable notation.
(3) Put a straight line through the diagram to section it. No more than three members should be cut by the line and they should include those members for which the internal forces are to be determined.
(4) Consider one of the portions isolated by the section and then write equations for equilibrium of that portion. The equilibrium condition used can be chosen to minimize the arithmetic required. Thus to eliminate the forces in two cut parallel members, sum the forces perpendicular to them; to eliminate the forces in members whose lines of action intersect, take moments about their point of intersection. If necessary any unknown external forces or reactions at supports can be determined by considering the truss as an entity, ignoring all internal forces in members.

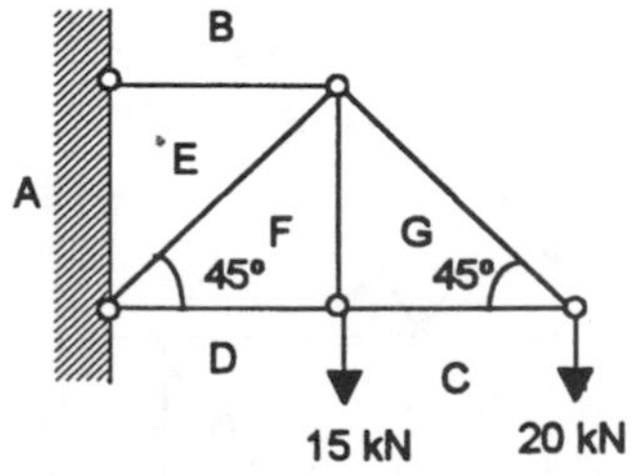

B
A
E
F
G
45°
45°
D
C
15 kN
20 kN

Figure 5.2.12 *Example 5.2.4*

Example 5.2.4

Determine, using the method of sections, the internal forces F_{BE} and F_{EF} for the plane pin-jointed truss shown in Figure 5.2.12.

Consider a vertical cut through the member concerned and then the equilibrium conditions for the section to the right of the cut (Figure 5.2.13). The forces on the cut members must be sufficient to give equilibrium.

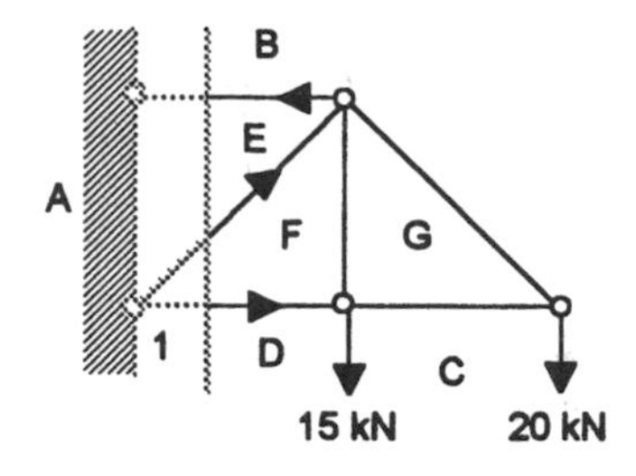

Figure 5.2.13 *Example 5.2.4*

In order to eliminate the unknowns at the cut of F_{EF} and F_{FD}, take moments about junction 1. We will assume the side of the square has members of length 1 unit. Thus, for equilibrium:

$$15 \times 1 + 20 \times 2 = F_{BE} \times 1$$

Hence $F_{BE} = 55$ kN and the member is in tension.

In order to eliminate the unknowns F_{BE} and F_{FD} at the cut, we can consider the equilibrium condition that there is no resultant force at the cut. Therefore, equilibrium of the vertical forces:

$$F_{EF} \sin 45° = 15 + 20$$

Hence $F_{EF} = 49.5$ kN and the member is in compression.

Example 5.2.5

Determine the force F_{AG} in the plane pin-jointed truss shown in Figure 5.2.14, the ends being supported on rollers. The members are all the same length.

Consider a diagonal cut through the member concerned, as in Figure 5.2.15, and then the equilibrium conditions for the section to the right of the cut.

In order to give an equilibrium condition which eliminates F_{GH} and F_{BH} from the calculation, take moments about junction 1 for the right-hand section. With each member taken to have a length of 1 unit:

$$F_{AG} \times 1 \sin 60° = R_2 \times 3$$

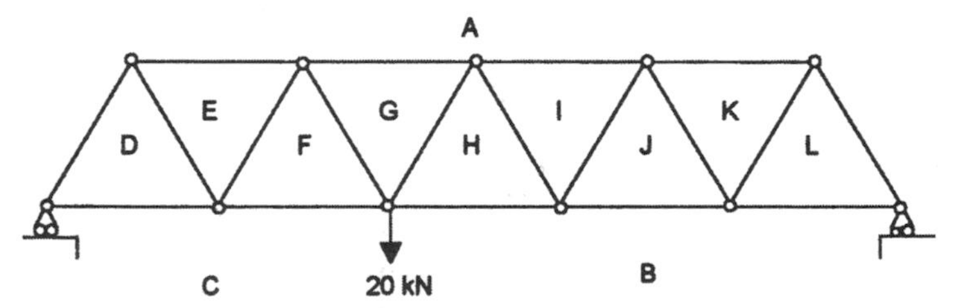

Figure 5.2.14 *Example 5.2.5*

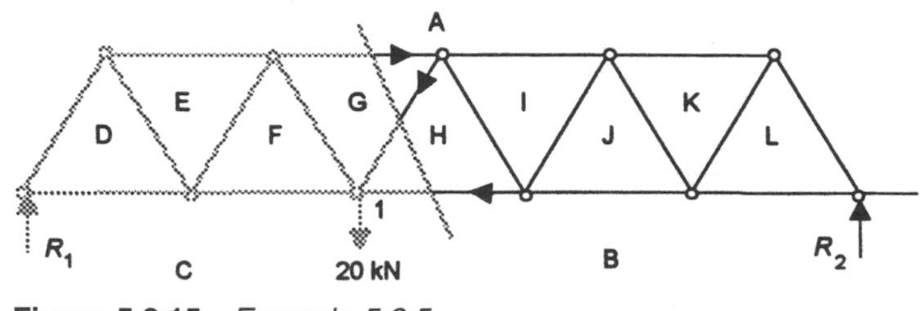

Figure 5.2.15 *Example 5.2.5*

We can determine R_2 by considering the equilibrium of the truss as an entity. Taking moments about the left-hand support:

$$R_2 \times 5 = 20 \times 2$$

Hence $R_2 = 8\,\text{kN}$ and so $F_{AG} = 27.7\,\text{kN}$ and is compressive.

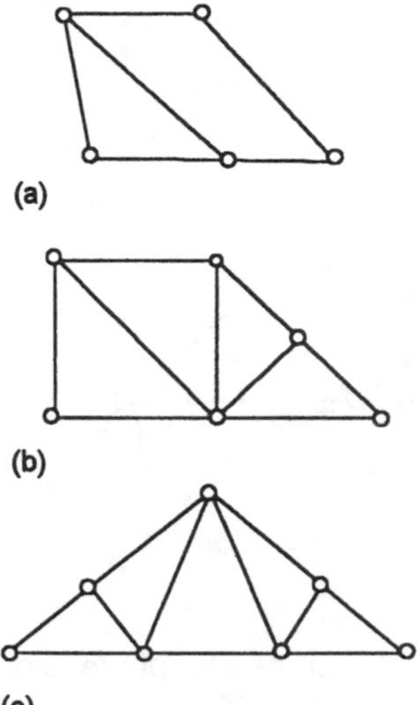

Figure 5.2.16 *Problem 1*

Problems 5.2.1

(1) Determine whether the frameworks shown in Figure 5.2.16 can be stable, are unstable or contain redundant members.

(2) Determine whether the pin-jointed trusses shown in Figure 5.2.17 can be stable, are unstable or contain redundant members.

(3) Figure 5.2.18 shows a framework of pin-jointed members resting on supports at each end and carrying loads of 60 kN and 90 kN. Determine the reactions at the supports and the forces in each member.

(4) Figure 5.2.19 shows a plane pin-jointed truss. Determine the reactions at the supports and the forces in the members.

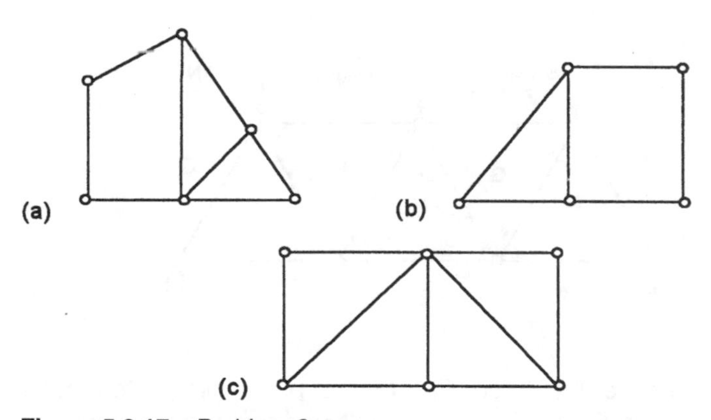

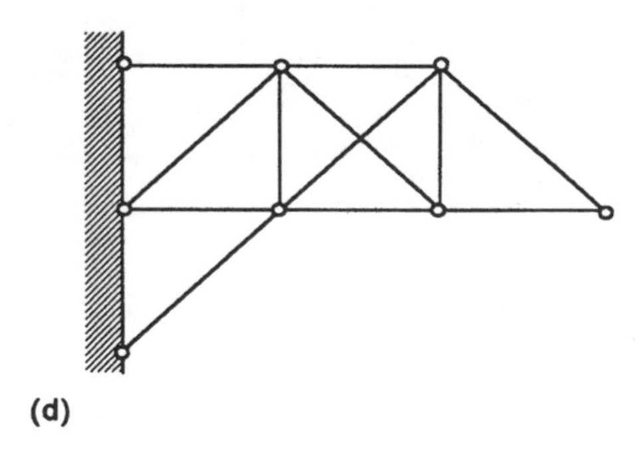

Figure 5.2.17 *Problem 2*

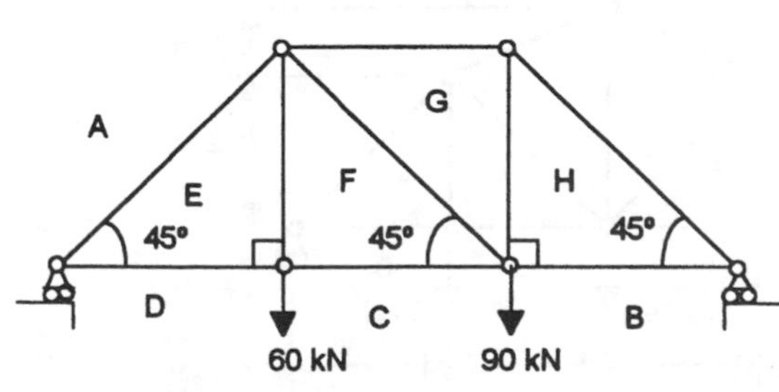

Figure 5.2.18 *Problem 3*

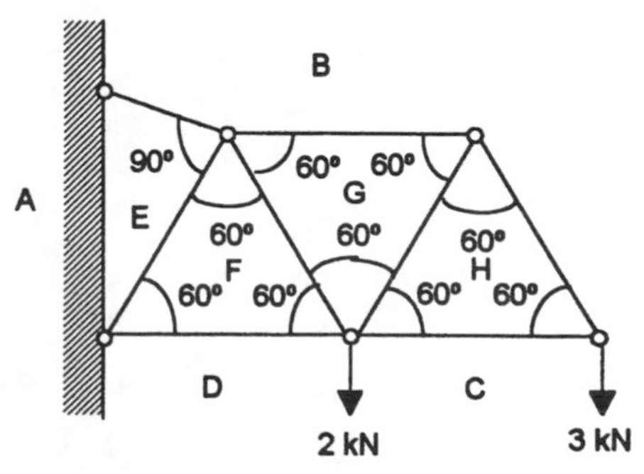

Figure 5.2.19 *Problem 4*

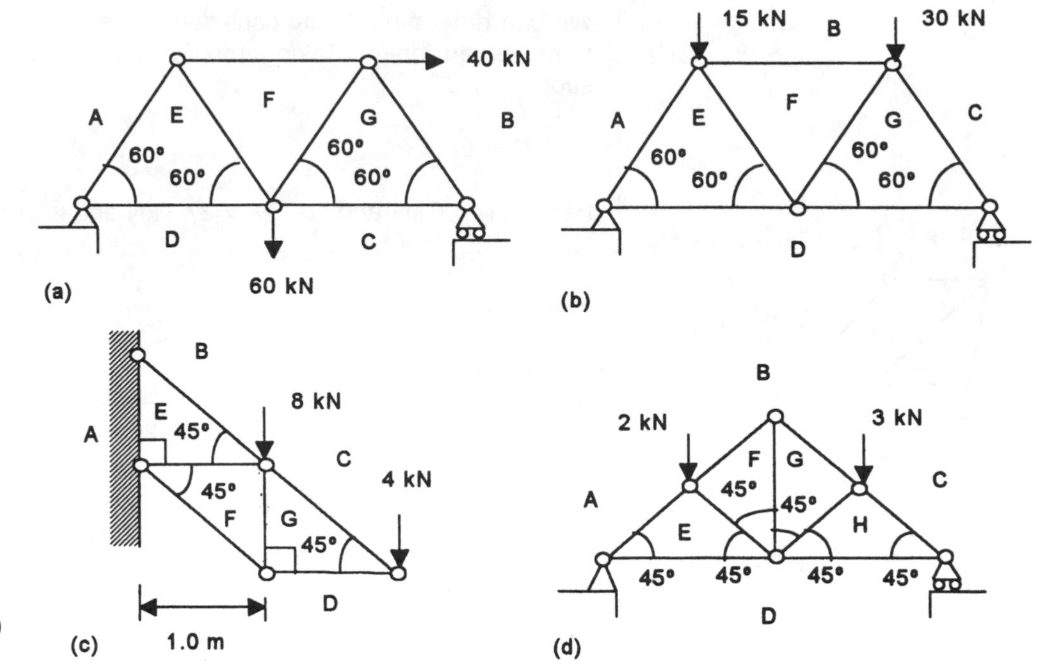

Figure 5.2.20 *Problem 5*

(5) Determine the forces in each member of the plane pin-jointed trusses shown in Figure 5.2.20.

(6) Determine the force F_{BG} in the plane pin-jointed truss shown in Figure 5.2.21. The ends of the truss are supported on rollers and each member is the same length.

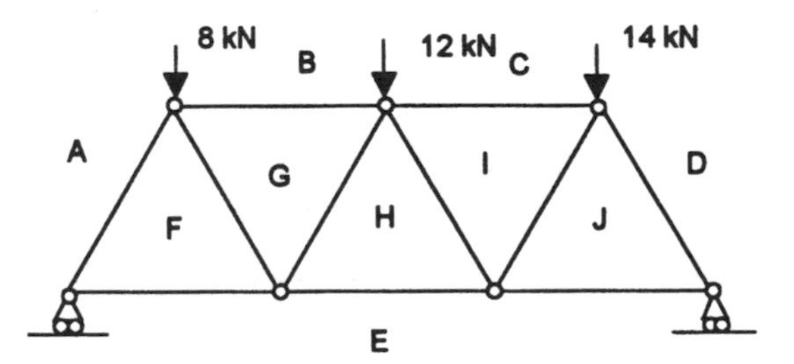

Figure 5.2.21 *Problem 6*

(7) Determine the force F_{AH} in the plane pin-jointed truss shown in Figure 5.2.22. The left-end support is pin jointed and the right-end support is a roller.

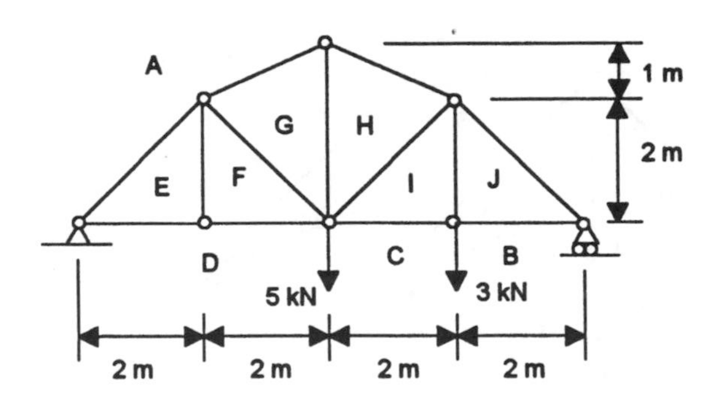

Figure 5.2.22 *Problem 7*

(8) Determine the forces F_{CI} and F_{HI} in the plane pin-jointed truss shown in Figure 5.2.23.

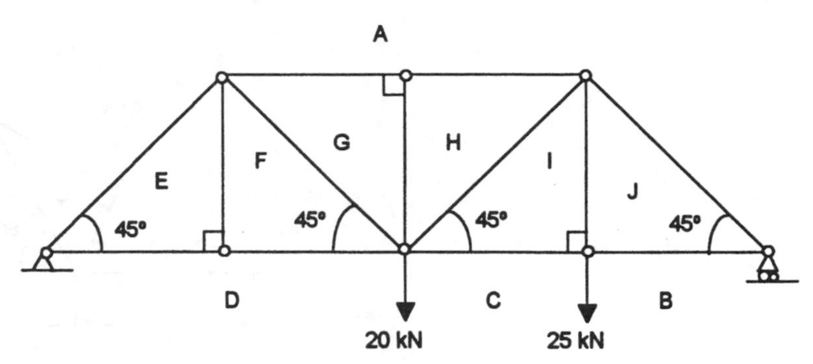

Figure 5.2.23 *Problem 8*

(9) Determine the forces F_{AE} and F_{DE} in the plane pin-jointed truss shown in Figure 5.2.24.

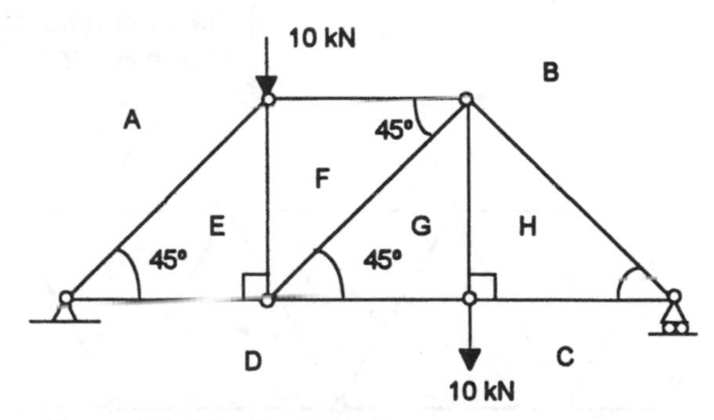

Figure 5.2.24 *Problem 9*

(10) Determine the force F_{FE} in the plane pin-jointed truss shown in Figure 5.2.25.

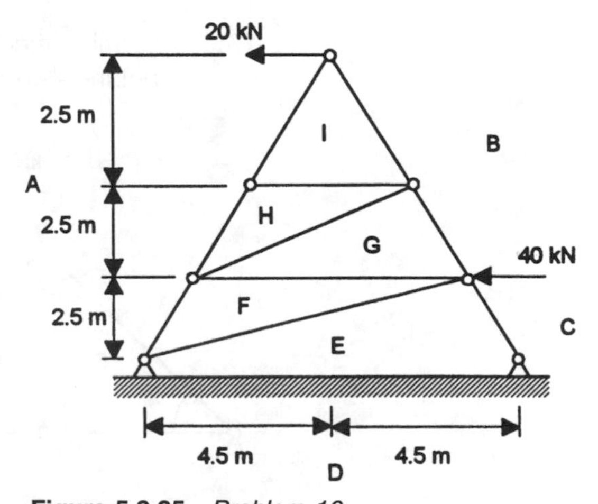

Figure 5.2.25 *Problem 10*

(11) Determine the forces F_{BG} and F_{GH} for the plane pin-jointed truss shown in Figure 5.2.26.

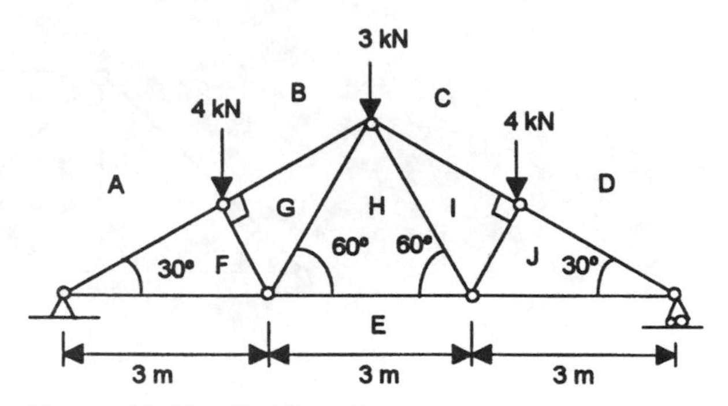

Figure 5.2.26 *Problem 11*

(12) Determine the reactions at the supports and the forces in the members of the plane pin-jointed structure shown in Figure 5.2.27.

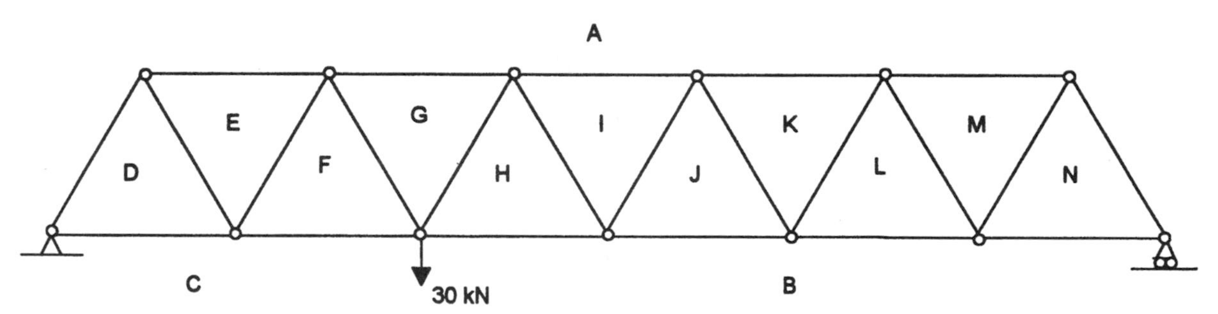

Figure 5.2.27 *Problem 12*

(13) Determine the forces F_{FG} and F_{DF} for the plane pin-jointed structure shown in Figure 5.2.28.

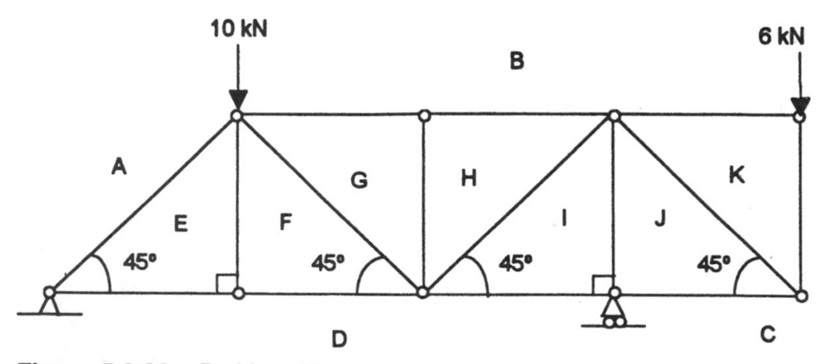

Figure 5.2.28 *Problem 13*

5.3 Stress and strain

In order to design safe structures it is necessary to know how members will deform under the action of forces and whether they might even break. For example, if we have forces of, say, 10 kN stretching a member of a structure, by how much will it extend and will the forces be enough to break it? In order to answer these questions, we need to take into account the length of the member and its cross-sectional area. To do this we introduce two terms, stress and strain.

Direct stress and strain

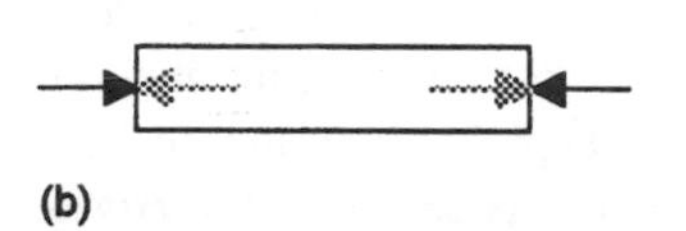

Figure 5.3.1 *(a) Tension, results in increase in length; (b) compression, results in decrease in length*

When external forces are applied to bodies, they deform and internal forces are set up which resist the deformation. Figure 5.3.1 shows such forces when bodies are in tension and compression; with tension the external forces result in an increase in length and with compression they result in a decrease. The term *direct stress* σ is used for the value of this internal force F per unit cross-sectional area when the force is at right angles to the area A (Figure 5.3.2). It is customary to denote tensile stresses as positive and compressive stresses as negative.

$$\sigma = \frac{F}{A} \tag{5.3.1}$$

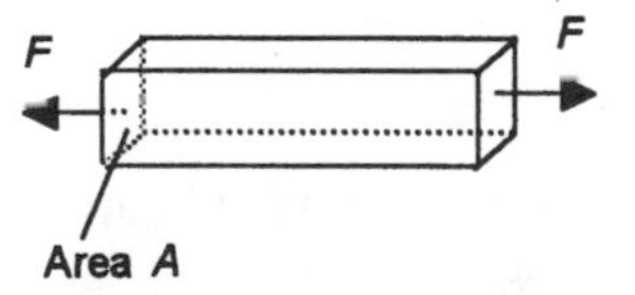

Figure 5.3.2 *Direct stress = F/A; direct because the force is at right angles to the area*

The SI unit of stress is N/m^2 and this is given the special name of pascal (Pa); 1 MPa = 10^6 Pa and 1 GPa = 10^9 Pa.

When a bar is subject to a direct stress it undergoes a change in length, the change in length per unit length stretched is termed the *direct strain* ε:

$$\varepsilon = \frac{\text{change in length}}{\text{length stretched}} \tag{5.3.2}$$

Since strain is a ratio of two lengths it is a dimensionless number, i.e. it has no units. It is sometimes expressed as a percentage; the strain being the percentage change in length. When the change in length is an increase in length then the strain is *tensile strain* and is positive; when the change in length is a decrease in length then the strain is *compressive strain* and is negative.

Example 5.3.1

A bar with a uniform rectangular cross-section of 20 mm by 25 mm is subjected to an axial force of 40 kN. Determine the tensile stress in the bar.

$$\sigma = \frac{F}{A} = \frac{40 \times 10^3}{0.020 \times 0.025} = 80 \times 10^6 \text{ Pa} = 80 \text{ MPa}$$

> **Example 5.3.2**
>
> Determine the strain experienced by a rod of length 100.0 cm when it is stretched by 0.2 cm.
>
> $$\text{Strain} = \frac{0.2}{100} = 0.0002$$
>
> As a percentage, the strain is 0.02%.

Stress–strain relationships

How is the strain experienced by a material related to the stress acting on it?

Consider first a simple situation. If gradually increasing tensile forces are applied to a strip of material then, for most engineering materials, the extension is initially proportional to the applied force (Figure 5.3.3). Within this proportionality region, a material is said to obey *Hooke's law*:

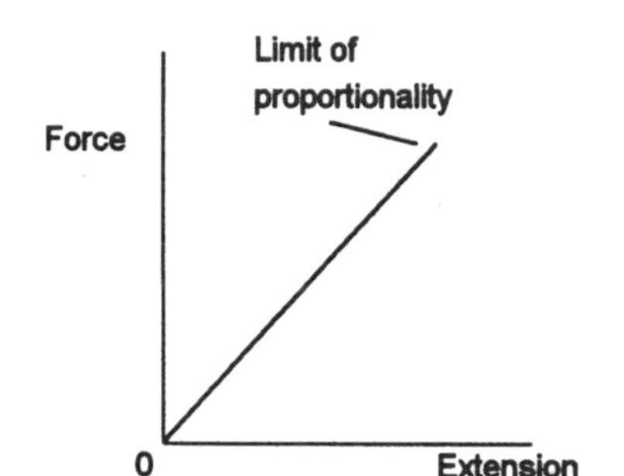

Figure 5.3.3 *Hooke's law is obeyed by a material when the extension is proportional to the applied force*

Extension $\propto$ applied force

Thus, provided the limit of proportionality is not exceeded, we can also state Hooke's law as strain is proportional to the stress producing it (Figure 5.3.4).

Strain $\propto$ stress

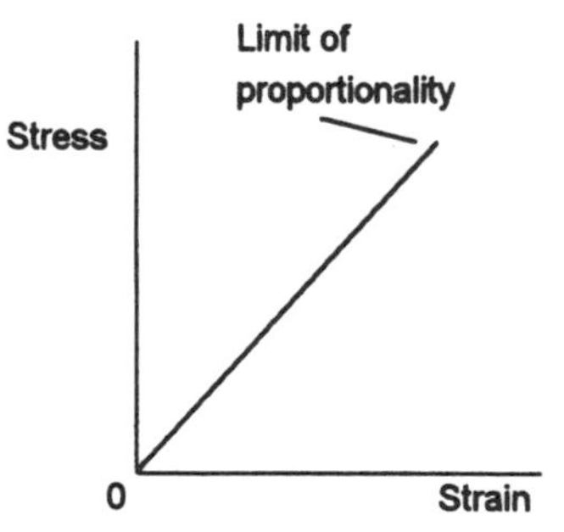

Figure 5.3.4 *Hooke's law is obeyed when the strain is proportional to the applied stress*

This law can generally be assumed to be obeyed within certain limits of stress by most of the metals used in engineering.

Within the limits to which Hooke's law is obeyed, the ratio of the direct stress σ to the strain ε is called the *modulus of elasticity E*:

$$E = \frac{\sigma}{\varepsilon} \tag{5.3.3}$$

Thus, for a bar of uniform cross-sectional area A and length L, subject to axial force F and extending by e:

$$E = \frac{FL}{Ae} \tag{5.3.4}$$

The unit of the modulus of elasticity is, since strain has no units, the unit of stress, i.e. Pa. Steel has a tensile modulus of about 210 GPa.

> **Example 5.3.3**
>
> A circular cross-section steel bar of uniform diameter 10 mm and length 1.000 m is subject to tensile forces of 12 kN. What will be the stress and strain in the bar? The steel has a modulus of elasticity of 200 GPa.

$$\text{Stress} = \frac{F}{A} = \frac{12 \times 10^3}{\frac{1}{4}\pi 0.010^2} = 88.4 \times 10^6\,\text{Pa} = 152.8\,\text{MP}$$

Assuming the limit of proportionality is not exceeded:

$$\text{Strain} = \frac{\sigma}{E} = \frac{152.8 \times 10^6}{200 \times 10^9} = 7.64 \times 10^{-4}$$

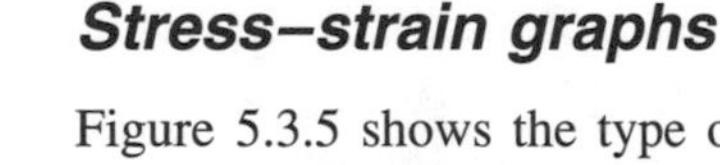

Stress–strain graphs

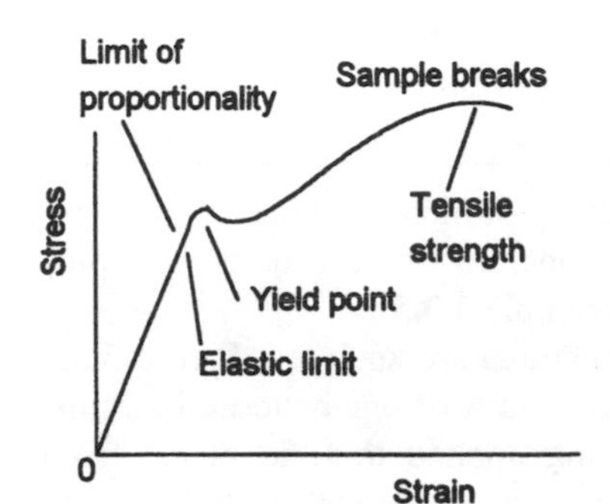

Figure 5.3.5 *Stress–strain graph for mild steel*

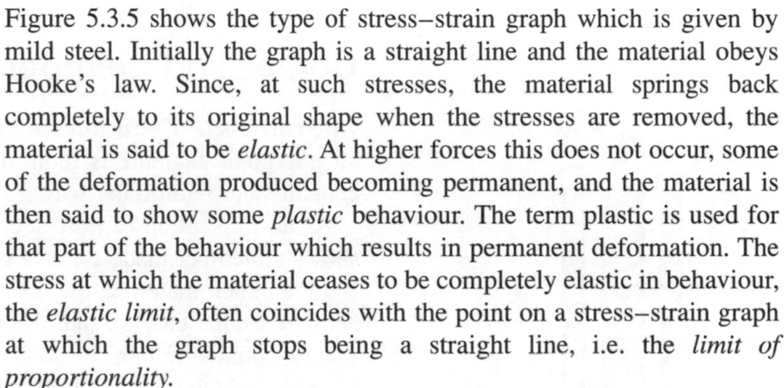

Figure 5.3.5 shows the type of stress–strain graph which is given by mild steel. Initially the graph is a straight line and the material obeys Hooke's law. Since, at such stresses, the material springs back completely to its original shape when the stresses are removed, the material is said to be *elastic*. At higher forces this does not occur, some of the deformation produced becoming permanent, and the material is then said to show some *plastic* behaviour. The term plastic is used for that part of the behaviour which results in permanent deformation. The stress at which the material ceases to be completely elastic in behaviour, the *elastic limit*, often coincides with the point on a stress–strain graph at which the graph stops being a straight line, i.e. the *limit of proportionality*.

The *strength* of a material is the ability of it to resist the application of forces without breaking. The term *tensile strength* is used for the maximum value of the tensile stress that a material can withstand without breaking; the *compressive strength* is the maximum compressive stress the material can withstand without becoming crushed.

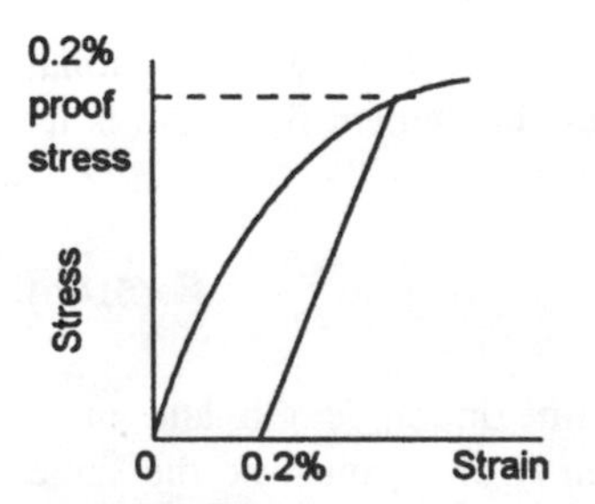

Figure 5.3.6 *0.2% proof stress obtained by drawing a line parallel to the straight line part of the graph but starting at a strain of 0.2%*

With some materials, e.g. mild steel, there is a noticeable dip in the stress–strain graph at some stress beyond the elastic limit and the strain increases without any increase in load. The material is said to have yielded and the point at which this occurs is the *yield point*. A carbon steel typically might have a tensile strength of 600 MPa and a yield stress of 300 MPa. Some materials, such as aluminium alloys (Figure 5.3.6), do not show a noticeable yield point and it is usual here to specify *proof stress*. The 0.2% proof stress is obtained by drawing a line parallel to the straight line part of the graph but starting at a strain of 0.2%. The point where this line cuts the stress–strain graph is termed the 0.2% yield stress. A similar line can be drawn for the 0.1% proof stress.

Poisson's ratio

When a material is longitudinally stretched it contracts in a transverse direction. The ratio of the transverse strain to the longitudinal strain is called *Poisson's ratio*.

$$\text{Poisson's ratio} = -\frac{\text{transverse strain}}{\text{longitudinal strain}} \qquad (5.3.5)$$

The minus sign is because when one of the strains is tensile the other is compressive. For most engineering metals, Poisson's ratio is about 0.3.

Example 5.3.4

A steel bar of length 1 m and square section 100 mm by 100 mm is extended by 0.1 mm. By how much will the width of the bar contract? Poisson's ratio is 0.3.

The longitudinal strain is 0.1/1000 = 0.0001. Thus, the transverse strain = $-0.3 \times 0.0001 = -3 \times 10^{-5}$ and so the change in width = original width × transverse strain = $100 \times (-3 \times 10^{-5}) = -3 \times 10^{-3}$ mm. The minus sign indicates that the width is reduced by this amount.

Compound members

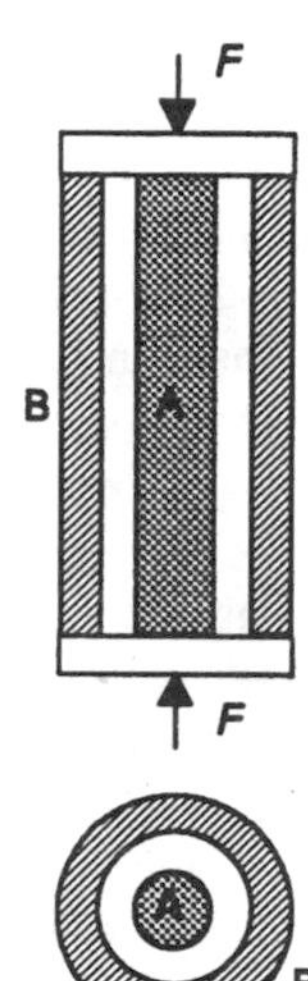

Figure 5.3.7 *A compound member with rod A inside tube B, both being rigidly fixed together at the ends*

Often members of structures are made up of more than one component; an important example is reinforced concrete where concrete columns contain steel reinforcing bars (see Example 5.3.5).

As an illustration of how such structures are analysed, Figure 5.3.7 shows a member made up of a central rod A of one material in a tube B of another material, the load being applied to rigid plates fixed across the tube ends so that the load is applied to both A and B. With such a compound bar, the load F applied is shared by the members. Thus if F_A is the force acting on member A and F_B is the force acting on member B:

$$F_A + F_B = F$$

If σ_A is the resulting stress in element A and A_A is its cross-sectional area then $\sigma_A = F_A/A_A$ and if σ_B is the stress in element B and A_B is its cross-sectional area then $\sigma_B = F_B/A_B$. Thus:

$$\sigma_A A_A + \sigma_B A_B = F \qquad (5.3.6)$$

Since the elements A and B are the same initial length and must remain together when loaded, the strain in A of ε_A must be the same as that in B of ε_B. Thus, assuming Hooke's law is obeyed, we must have:

$$\frac{\sigma_A}{E_A} = \frac{\sigma_B}{E_B} \qquad (5.3.7)$$

where E_A is the modulus of elasticity of the material of element A and E_B that of the material of element B.

Now consider another situation where we have a composite member consisting of two, or more, elements in series, e.g. as in Figure 5.3.8 where we have three rods connected end to end with the rods being of different cross-sections and perhaps materials. With just a single load we must have, for equilibrium, the same forces acting on each of the series members. Thus the forces stretching member A are the same as those stretching member B and the same as those stretching member C. The total extension of the composite bar will be the sum of the extensions arising for each series element.

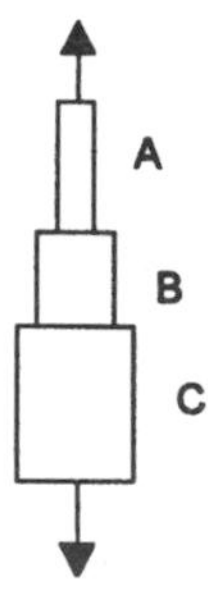

Figure 5.3.8 *Elements in series so that the forces applied to A, B and C are the same*

Example 5.3.5

A square section reinforced concrete column has a cross-section 450 mm by 450 mm and contains four steel reinforcing bars, each of diameter 25 mm. Determine the stresses in the steel and the concrete when the total load on the column is 1.5 MN. The steel has a modulus of elasticity of 200 GPa and the concrete a modulus of elasticity of 14 GPa.

The area of the column that is steel is $4 \times \frac{1}{4}\pi \times 25^2 = 1963\ \text{mm}^2$. The area of the column that is concrete is $450 \times 450 - 1963 = 200\,537\ \text{mm}^2$. The ratio of the stress on the steel σ_s to that on the concrete σ_c is:

$$\frac{\sigma_s}{E_s} = \frac{\sigma_c}{E_c}$$

$$\sigma_s = \frac{200}{14}\sigma_c = 14.3\sigma_c$$

The force F on the column is related to the stresses and areas of the components by:

$$F = \sigma_s A_s + \sigma_B A_B$$

$$1.5 \times 10^6 = 14.3\sigma_c \times 1963 \times 10^{-6} + \sigma_c \times 200\,537 \times 10^{-6}$$

Hence the stress on the concrete is 6.56 MPa and that on the steel is 93.8 MPa.

Example 5.3.6

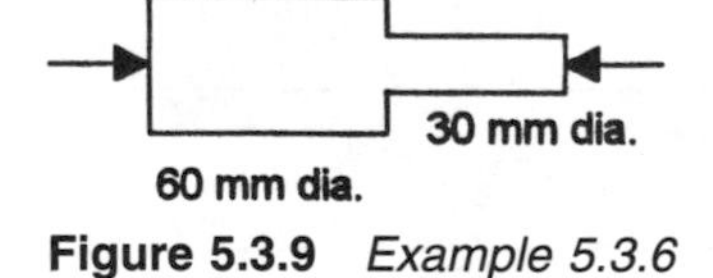

Figure 5.3.9 *Example 5.3.6*

A rod is formed with one part of it having a diameter of 60 mm and the other part a diameter of 30 mm (Figure 5.3.9) and is subject to an axial force of 20 kN. What will be the stresses in the two parts of the rod?

Each part will experience the same force and thus the stress on the larger diameter part is $20 \times 10^3/(\frac{1}{4}\pi \times 0.060^2) = 7.1$ MPa and the stress on the smaller diameter part will be $20 \times 10^3/(\frac{1}{4}\pi \times 0.030^2) = 28.3$ MPa.

Temperature stresses

Telephone engineers, when suspending cables between telegraph poles, always allow some slack in the cables. This is because if the temperature drops then the cable will decrease in length and if it becomes taut then the stresses produced could be high enough to break cables. There are many situations in engineering where we have to consider the stresses that can be produced as a result of temperature changes.

When the temperature of a body is changed it changes in length and if this expansion or contraction is wholly or partially resisted, stresses are set up in the body. Consider a bar of initial length L_0. If the temperature is now raised by θ and the bar is free to expand, the length increases to $L_\theta = L_0(1 + \alpha\theta)$, where α is the coefficient of linear expansion of the bar material. The change in length of the bar is thus $L_\theta - L_0 = L_0(1 + \alpha\theta) - L_0 = L_0\alpha\theta$. If this expansion is prevented, it is as if a bar of length $L_0(1 + \alpha\theta)$ has been compressed to a length L_0 and so the resulting compressive strain ε is:

$$\varepsilon = \frac{L_0\alpha\theta}{L_0(1 + \alpha\theta)}$$

Since $\alpha\theta$ is small compared with 1, then:

$$\varepsilon = \alpha\theta$$

If the material has a modulus of elasticity E and Hooke's law is obeyed, the stress σ produced is:

$$\sigma = \alpha\theta E \qquad (5.3.8)$$

The stress is thus proportional to the coefficient of linear expansion, the change in temperature and the modulus of elasticity.

Example 5.3.7

Determine the stress produced per degree change in temperature for a fully restrained steel member if the coefficient of linear expansion for steel is 12×10^{-6} per °C and the modulus of elasticity of the steel is 200 GPa.

$$\sigma = \alpha\theta E = 12 \times 10^{-6} \times 1 \times 200 \times 10^9$$

$$= 2.4 \times 10^6 \text{ Pa} = 2.4 \text{ MPa}$$

Example 5.3.8

A steel wire is stretched taut between two rigid supports at 20°C and is under a stress of 20 MPa. What will be the stress in the wire when the temperature drops to 10°C? The coefficient of linear expansion for steel is 12×10^{-6} per °C and the modulus of elasticity is 200 GPa.

The effect of the drop in temperature is to produce a tensile stress of:

$$\sigma = \alpha\theta E = 12 \times 10^{-6} \times 10 \times 200 \times 10^9 = 24 \text{ MPa}$$

The total stress acting on the wire will be the sum of the thermal stress and the initial stress 24 + 20 = 44 MPa.

Composite bars

Consider a composite bar with materials in parallel, such as a circular bar A inside a circular tube B, with the two materials having different coefficients of expansion, α_A and α_B, and different modulus of elasticity values, E_A and E_B, but the same initial length L and attached rigidly together (Figure 5.3.10). The two materials are considered to be initially unstressed. The temperature is then changed by θ.

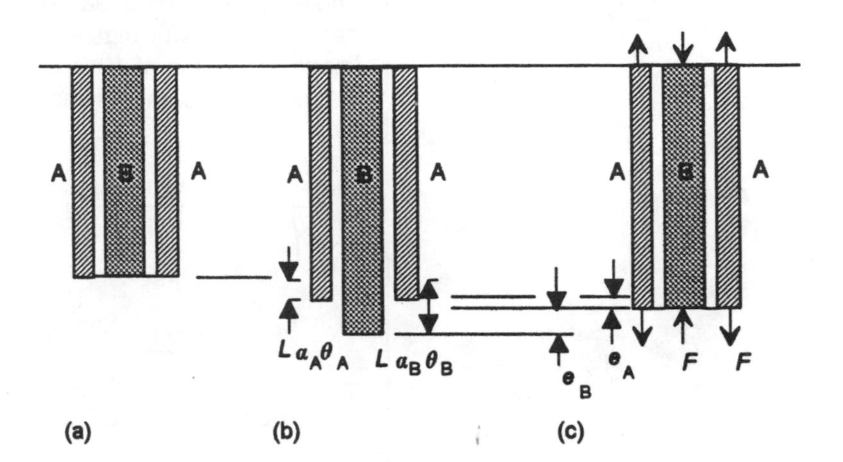

Figure 5.3.10 *Composite bar: (a) initially; (b) when free to expand; (c) when the expansion is restrained by being fixed together*

If the two members had not been fixed to each other, when the temperature changed they would have expanded; A would have changed its length by $L\alpha_A\theta_A$ and B its length by $L\alpha_B\theta_B$ and there would be a difference in length between the two members at temperature θ of $(\alpha_A - \alpha_B)\theta L$ (Figure 5.3.10(b)). However, the two members are rigidly fixed together and so this difference in length is eliminated by compressing member B with a force F and extending A with a force F (Figure 5.3.10(c)).

The extension e_A of A due to the force F is:

$$e_A = \frac{FL}{E_A A_A}$$

where A_A is its cross-sectional area. The contraction e_B of B due to the force F is:

$$e_B = \frac{FL}{E_B A_B}$$

where A_B is its cross-sectional area. But $e_A + e_B = (\alpha_A - \alpha_B)\theta L$ and so:

$$(\alpha_A - \alpha_B)\theta L = FL\left(\frac{1}{E_A A_A} + \frac{1}{E_B A_B}\right)$$

$$F = \frac{(\alpha_A - \alpha_B)\theta}{\left(\dfrac{1}{E_A A_A} + \dfrac{1}{E_B A_B}\right)} \quad (5.3.9)$$

Example 5.3.9

A steel rod has a diameter of 30 mm and is fitted centrally inside copper tubing of internal diameter 35 mm and external diameter 60 mm. The rod and tube are rigidly fixed together at each end but are initially unstressed. What will be the stresses produced in each by a temperature increase of 100°C? The copper has a modulus of elasticity of 100 GPa and a coefficient of linear expansion of 20×10^{-6}/°C; the steel has a modulus of elasticity of 200 GPa and a coefficient of linear expansion of 12×10^{-6}/°C.

The force compressing the copper and extending the steel is:

$$F = \frac{(\alpha_A - \alpha_B)\theta}{\left(\frac{1}{E_A A_A} + \frac{1}{E_B A_B}\right)}$$

$$= \frac{(20 - 12) \times 10^{-6} \times 100}{\frac{1}{100 \times 10^9 \times \frac{1}{4}\pi(0.060^2 - 0.035^2)} + \frac{1}{200 \times 10^9 \times \frac{1}{4}\pi 0.03}}$$

$$= 64.3\ \text{kN}$$

The compressive stress acting on the copper is thus:

$$\sigma_A = \frac{64.3 \times 10^3}{\frac{1}{4}\pi(0.060^2 - 0.035^2)} = 34.5\ \text{MPa}$$

The tensile stress acting on the steel is:

$$\sigma_B = \frac{64.3 \times 10^3}{\frac{1}{4}\pi 0.030^2} = 91.0\ \text{MPa}$$

Shear stress

When forces are applied in such a way as to tend to slide one layer of a material over an adjacent layer (Figure 5.3.11), the material is said to be subject to shear. The areas over which forces act are in the same plane as the line of action of the forces. The force per unit area is called the *shear stress*:

$$\text{Shear stress} = \frac{\text{force}}{\text{area}} \qquad (5.3.10)$$

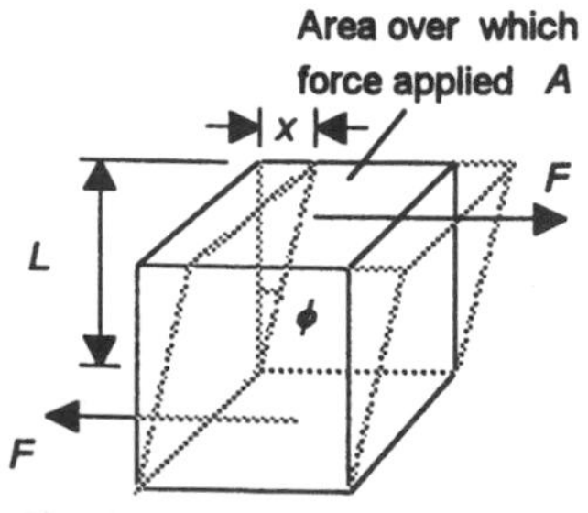

Figure 5.3.11 *Shear with the forces tending to slide one face over another*

The unit of shear stress is the pascal (Pa).

The deformation of a material subject to shear is an angular change ϕ and shear strain is defined as being the angular deformation:

$$\text{Shear strain} = \phi \qquad (5.3.11)$$

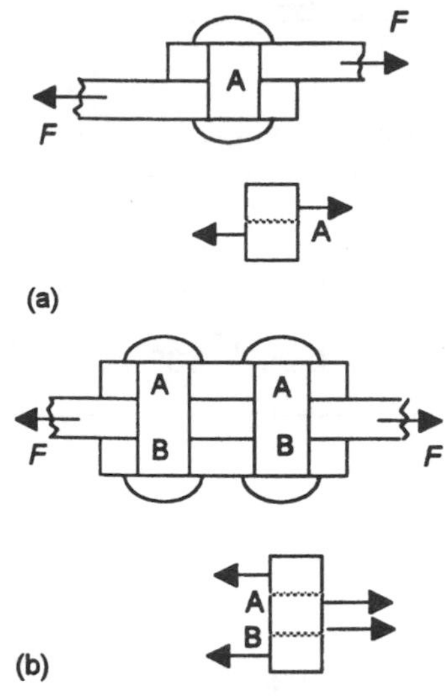

Figure 5.3.12 *(a) Lap joint giving single shear; (b) double cover butt joint giving double shear*

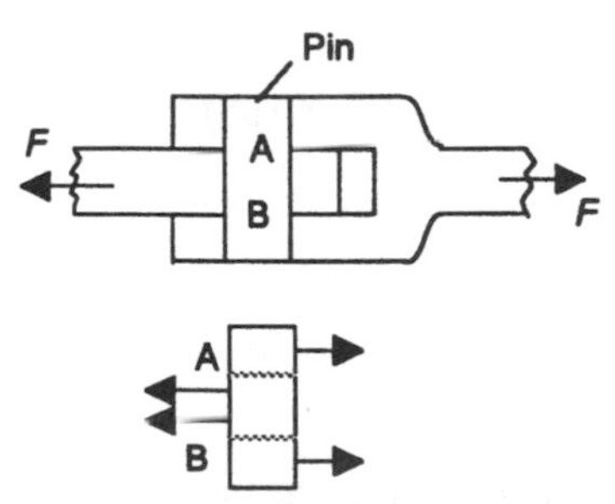

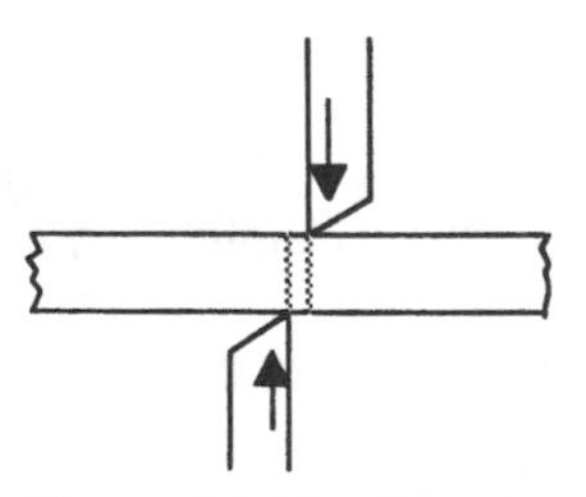

Figure 5.3.13 *Example 5.3.11*

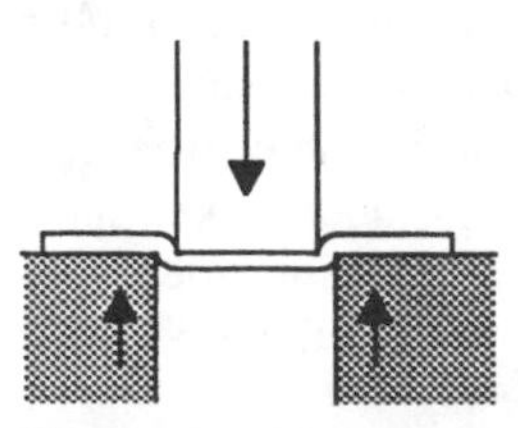

Figure 5.3.15 *Punching a hole involves shear forces*

Figure 5.3.14 *Cropping a plate involves shear forces*

The unit used is the radian and, since the radian is a ratio, shear strain can be either expressed in units of radians or without units. For the shear shown in Figure 5.3.11, $\tan \phi = x/L$ and, since for small angles $\tan \phi$ is virtually the same as ϕ expressed in radians:

$$\text{Shear strain} = \frac{x}{L} \qquad (5.3.12)$$

Figure 5.3.12 shows an example of shear occurring in a fastening, in this case riveted joints. In Figure 5.3.12(a), a simple lap joint, the rivet is in shear as a result of the forces applied to the plates joined by the rivet. The rivet is said to be in *single shear*, since the bonding surface between the members is subject to just a single pair of shear forces; there is just one surface A subject to the shear forces. In Figure 5.3.12(b) the rivets are used to produce a double cover butt joint; the rivets are then said to be in *double shear* since there are two shear surfaces A and B subject to shear forces.

Example 5.3.10

What forces are required to shear a lap joint made using a 25 mm diameter rivet if the maximum shear stress it can withstand is 250 MPa?

The rivet is in single shear and thus force = shear stress × area = $250 \times 10^6 \times \frac{1}{4}\pi \times 0.025^2 = 1.2 \times 10^5$ N = 120 kN.

Example 5.3.11

Calculate the maximum load that can be applied to the coupling shown in Figure 5.3.13 if the pin has a diameter of 8 mm and the maximum shear stress it can withstand is 240 MPa.

The pin is in double shear, at A and B, and thus the forces applied to each shear surface are $\frac{1}{2}F$. Thus, force = shear stress × area = $2 \times 240 \times 10^6 \times \frac{1}{4}\pi \times 0.008^2 = 20 \times 10^3$ N = 24.1 kN.

Shear strength

The *shear strength* of a material is the maximum shear stress that the material can withstand before failure occurs. An example of where such shear stresses are applied are when a guillotine is used to crop a material (Figure 5.3.14). The area over which the shear forces are being applied is the cross-sectional area of the plate being cropped. Another example is when a punch is used to punch holes in a material (Figure 5.3.15). In this case the area over which the punch force is applied is the plate thickness multiplied by the perimeter of the hole being punched.

Example 5.3.12

What is the force which the guillotine has to apply to crop a plate of mild steel 0.50 m wide and 1 mm thick if the shear strength of the steel is 200 MPa?

$$\text{Force} = \text{shear strength} \times \text{area} = 200 \times 10^6 \times 0.50 \times 0.001$$
$$= 160 \times 10^3\,\text{N} = 160\,\text{kN}$$

Example 5.3.13

What is the maximum diameter hole that can be punched in an aluminium plate of thickness 10 mm if the punching force is limited to 20 kN? The shear strength of the aluminium is 90 MPa.

Shear strength = (punch force)/(area being sheared) and so:

$$\text{Area} = \pi d \times 10 \times 10^{-3}$$
$$= \text{force/shear strength} = (20 \times 10^3)/(90 \times 10^6)$$

and thus d, the diameter of the hole, is 7.1 mm.

Shear modulus

Within some limiting stress, metals usually have the shear stress proportional to the shear strain. The *shear modulus* (or *modulus of rigidity*) is then:

$$\text{Shear modulus } G = \frac{\text{shear stress}}{\text{shear strain}} \qquad (5.3.13)$$

The SI unit for the shear modulus is the pascal (Pa). For mild steel a typical value of the shear modulus is 75 GPa.

Strain energy

If you stretch a rubber band and then let go, it is fairly obvious that there is energy stored in the band when stretched and that it is released when you let go. Think of a catapult; energy is stored in the stretched rubber when it is stretched and released when it is let go. As a consequence the catapult can be used to propel objects over considerable distances. Likewise, ropes which are stretched when used to moor a boat have stored energy which can be released if they break – quite often with disastrous consequences to anyone who is in the way of the rope when it springs back; the energy released can be quite high.

Consider a spring, or a length of material, being stretched by tensile forces and for which Hooke's law is obeyed. If a force F produces an

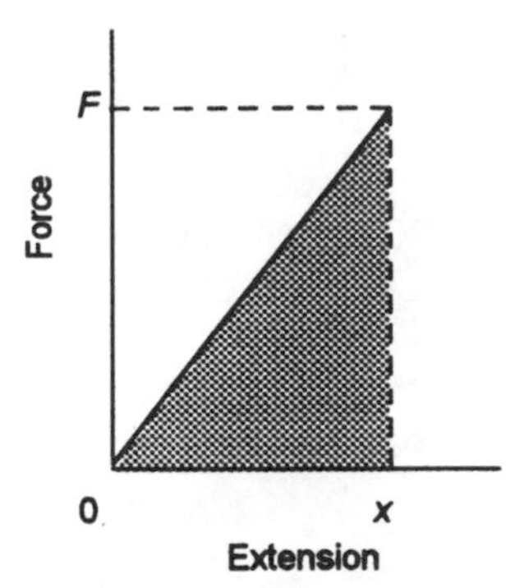

Figure 5.3.16 *Work done in extending to extension x is area under the graph*

extension x, as in Figure 5.3.16, the average force applied is $\frac{1}{2}F$ and so the work done in stretching the material is $\frac{1}{2}Fx$. This is the area under the graph. Since the material obeys Hooke's law we can write $F = kx$, where k is a constant which is often termed the spring constant or force constant, and so:

$$\text{Work done} = \tfrac{1}{2}kx^2 \qquad (5.3.14)$$

The energy is in joules (J) when x is in m and k in N/m.

This work results in energy being stored in the stretched material, this being generally referred to as *strain energy* or *elastic potential energy*. Think of stretching a spring or a length of rubber. Work has to be done to stretch it and results in energy being stored in the stretched material which is released when the spring or length of rubber is released.

The increase in stored energy in a spring when it is initially at extension x_1 and is then stretched to x_2 is:

$$\text{Increase in stored energy} = \tfrac{1}{2}k(x_2{}^2 - x_1{}^2) \qquad (5.3.15)$$

Example 5.3.14

Determine the energy stored in a spring when it is stretched from an initially unstretched state to give an extension of 10 mm if the spring constant is 20 N/m.

$$\text{Stored energy} = \tfrac{1}{2}kx^2 = \tfrac{1}{2} \times 20 \times 0.010 = 0.1\text{ J}$$

Strain energy in terms of stress and strain

The work done in stretching the material is $\frac{1}{2}Fx$ and so if the volume of the material is AL then the work done per unit volume is $\frac{1}{2}Fx/AL$ and, since F/A is the stress and x/L is the strain:

$$\text{Work done per unit volume} = \tfrac{1}{2}\text{ stress} \times \text{strain} \qquad (5.3.16)$$

and thus, for a material obeying Hooke's law with stress σ = modulus of elasticity $E \times$ strain:

$$\text{Work done per unit volume} = \tfrac{1}{2}\sigma \times \frac{\sigma}{E} = \frac{\sigma^2}{2E} \qquad (5.3.17)$$

The work done, and hence energy stored in the material, is in joules (J) when the stress is in Pa, the modulus in Pa and volume in m^3. Thus, for a given material, the energy stored is proportional to the square of the stress.

The strain energy per unit volume when a material is stretched to the limit of proportionality is called the *modulus of resilience* of the material and represents the ability of the material to absorb energy within the elastic range.

Example 5.3.15

A steel cable of length 10 m and cross-sectional area 1200 mm^2 is being used to lift a load of 3000 kg. What will be the strain energy stored in the cable? The modulus of elasticity of the steel is 200 GPa.

The strain energy per unit volume = $\frac{1}{2}$ stress × strain and thus if the modulus of elasticity is E, the cross-sectional area is A and the length L:

$$\text{Strain energy} = \tfrac{1}{2}\frac{F}{A} \times \frac{F/A}{E} \times AL = \frac{F^2L}{2AE}$$

$$= \frac{(3000 \times 9.8)^2 \times 10}{2 \times 1200 \times 10^{-6} \times 200 \times 10^9} = 18\ \text{J}$$

Problems 5.3.1

(1) A steel bar with a uniform cross-sectional area of 500 mm^2 is stretched by forces of 10 kN. What is the tensile stress in the bar?

(2) What is the strain experienced by a rod of length 2.000 m when it is compressed and its length reduces by 0.6 mm?

(3) A rod has a length of 2.000 m and a cross-sectional area of 200 mm^2. Determine the elongation of the rod when it is subject to tensile forces of 20 kN. The material has a modulus of elasticity of 200 GPa.

(4) A material has a tensile strength of 800 MPa. What force will be needed to break a bar of that material if it has a cross-sectional area of 200 mm^2?

(5) A steel bar has a rectangular cross-section 70 mm by 20 mm and is subject to a tensile longitudinal load of 200 kN. Determine the decrease in the lengths of the sides of the resulting cross-section if the material has an elastic modulus of 200 GPa and Poisson's ratio of 0.3.

(6) By how much will a steel tie rod of length 3 m and diameter 30 mm increase in length when subject to a tensile load of 100 kN. The material has a modulus of elasticity of 200 GPa.

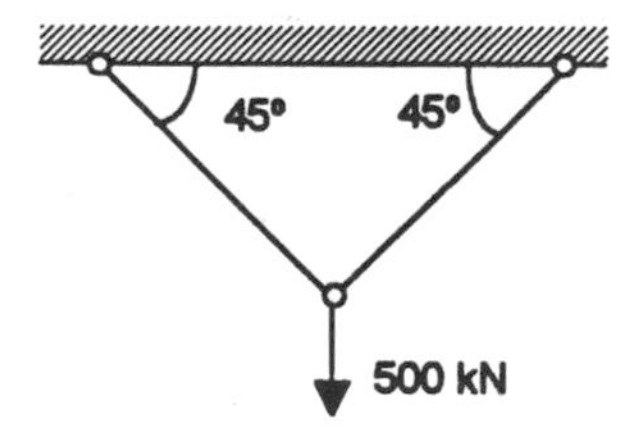

Figure 5.3.17 *Problem 7*

(7) Figure 5.3.17 shows a pin-jointed structure supporting a load of 500 kN. What should be the cross-sectional area of the supporting bars if the stress in them should not exceed 200 MPa?

(8) A steel rod of length 4 m and square cross-section 25 mm by 25 mm is stretched by a load of 20 kN. Determine the elongation of the rod. The material has a modulus of elasticity of 200 GPa.

(9) Figure 5.3.18 shows a plane pin-jointed truss supporting a single point load of 480 kN. Determine the tensile forces F_{FG} and F_{DC} and the cross-sectional areas required for those members if the tensile stress in them is not to exceed

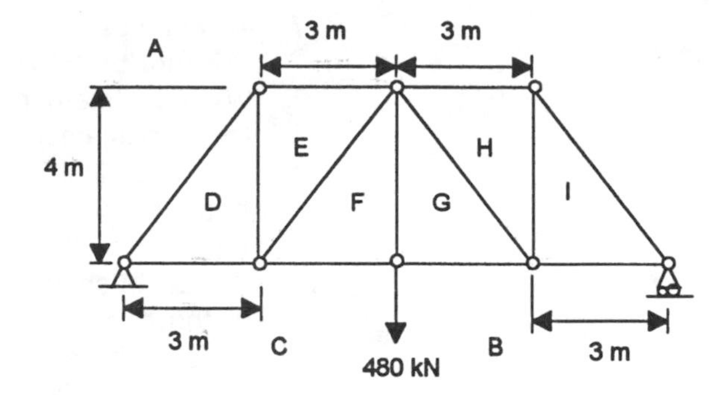

Figure 5.3.18 *Problem 9*

200 MPa. The modulus of elasticity should be taken as 200 GPa.

(10) A steel bolt is not to be exposed to a tensile stress of more than 200 MPa. What should be the minimum diameter of the bolt if the load is 700 kN?

(11) The following results were obtained from a tensile test of a steel. The test piece had a diameter of 10 mm and a gauge length of 50 mm. Plot the stress–strain graph and determine (a) the tensile strength, (b) the yield stress, (c) the tensile modulus.

Load/kN	0	5	10	15	20	25	30	32.5	35.8
Ext./mm	0	0.016	0.033	0.049	0.065	0.081	0.097	0.106	0.250

(12) A reinforced concrete column is uniformly 500 mm square and has a reinforcement of four steel rods, each of diameter 25 mm, embedded in the concrete. Determine the compressive stresses in the concrete and the steel when the column is subject to a compressive load of 1000 kN, the modulus of elasticity of the steel being 200 GPa and that of the concrete 14 GPa.

(13) A steel bolt (Figure 5.3.19) has a diameter of 25 mm and carries an axial tensile load of 50 kN. Determine the average tensile stress at the shaft section aa and the screwed section bb if the diameter at the root of the thread is 21 mm.

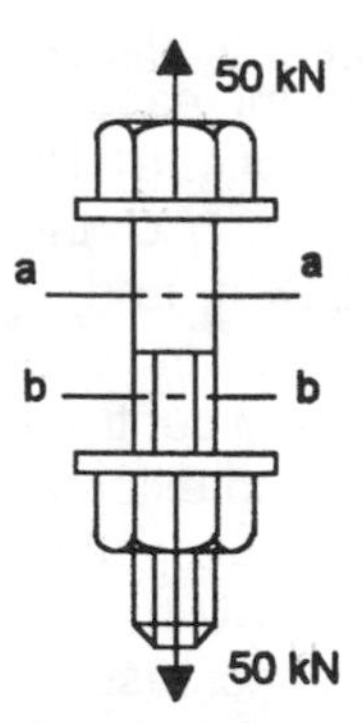

Figure 5.3.19 *Problem 13*

(14) A reinforced concrete column has a rectangular cross-section 220 mm by 200 mm and is reinforced by four steel rods. What diameter rods will be required if the stress in the concrete must not exceed 7 MPa when the axial load is 500 kN? The steel has a modulus of elasticity 15 times that of the concrete.

(15) A composite bar consists of a steel rod 400 mm long and 40 mm diameter fixed to the end of a copper rod having a length of 800 mm. Determine the diameter of the copper rod if each element is to extend by the same amount when the composite is subject to an axial load. The modulus of elasticity of the steel is twice that of the copper.

(16) An electrical distribution conductor consists of a steel wire of diameter 5 mm which has been covered with a 2 mm thickness of copper. What will be the stresses in each material when it is subject to an axial force of 2.6 kN? The modulus of elasticity of the steel is 200 GPa and that of the copper 120 GPa.

(17) A reinforced concrete column is to have a square section 250 mm by 250 mm and is required to support a load of 700 kN. Determine the minimum number of steel reinforcement rods, each of diameter 6 mm, which will be needed if the stress in the concrete is not to exceed 8 MPa. The steel has a modulus of elasticity of 200 GPa and the concrete a modulus of 12 GPa.

(18) A steel bush at 20°C is unstressed. What will be the stress in the bush when its temperature is raised to 60°C if its expansion is completely restricted? The steel has a modulus of elasticity of 200 GPa and a coefficient of linear expansion of 12×10^{-6}/°C.

(19) Determine the stress produced per degree change in temperature for a fully restrained aluminum member if the coefficient of linear expansion for aluminium is 22×10^{-6} per °C and the modulus of elasticity is 74 GPa.

(20) A steel rod is clamped at both ends. What will be the stresses produced in the rod if the temperature rises by 60°C? The steel has a modulus of elasticity of 200 GPa and a coefficient of linear expansion of 12×10^{-6}/°C.

(21) A brass rod has a diameter of 40 mm and is fitted centrally inside steel tubing of internal diameter 40 mm and external diameter 60 mm. The rod and tube are rigidly fixed together at each end but are initially unstressed. What will be the stresses produced in each by a temperature increase of 80°C? The brass has a modulus of elasticity of 90 GPa and a coefficient of linear expansion of 20×10^{-6}/°C, the steel a modulus of elasticity of 200 GPa and a coefficient of linear expansion of 11×10^{-6}/°C.

(22) A copper rod has a diameter of 45 mm and is fitted centrally inside steel tubing of internal diameter 50 mm and external diameter 80 mm. The rod and tube are rigidly fixed together at each end but are initially unstressed. What will be the stresses produced in each by a temperature increase of 100°C? The copper has a modulus of elasticity of 120 GPa and a coefficient of linear expansion of 16.5×10^{-6}/°C; the steel has a modulus of elasticity of 200 GPa and a coefficient of linear expansion of 11.5×10^{-6}/°C.

(23) A brass rod has a diameter of 25 mm and is fitted centrally inside steel tubing of internal diameter 30 mm and external diameter 50 mm. The rod and tube are rigidly fixed together at each end but are initially unstressed. What will be the stresses produced in each by a temperature increase of 100°C? The brass has a modulus of elasticity of 100 GPa and a coefficient of linear expansion of 19×10^{-6}/°C; the steel has a modulus of elasticity of 200 GPa and a coefficient of linear expansion of 12×10^{-6}/°C.

(24) What is the minimum diameter required for the bolt in the coupling shown in Figure 5.3.20 if the shear stress in the bolt is not to exceed 90 MPa when the forces applied to the coupling are 30 kN?

(25) Determine the force necessary to punch a hole of 25 mm diameter through a 10 mm thick plate of steel if the material has a shear strength of 300 MPa.

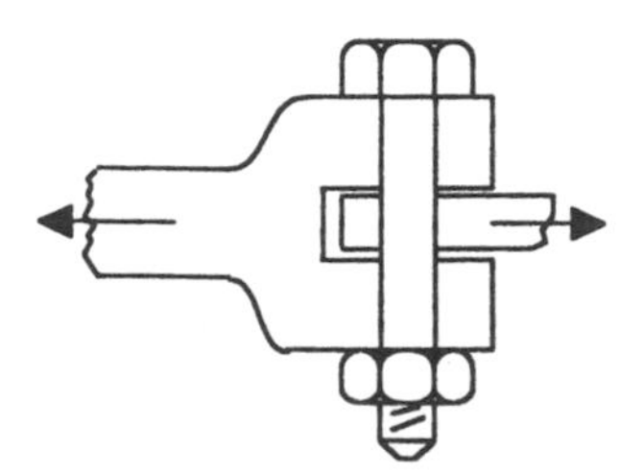

Figure 5.3.20 *Problems 24 and 28*

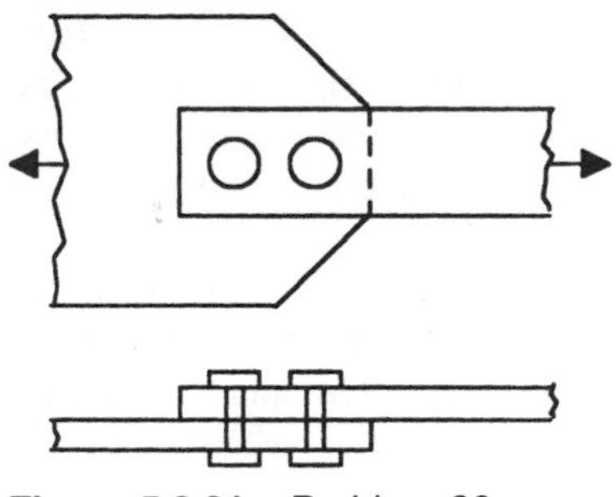
Figure 5.3.21 *Problem 30*

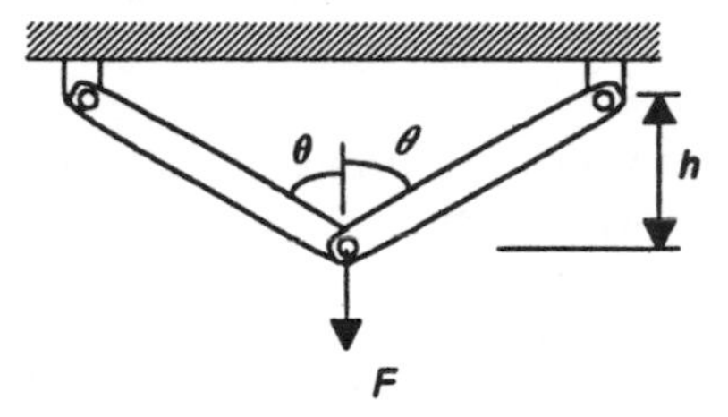

Figure 5.3.22 *Problem 33*

(26) What will be the shear strain produced by a shear stress of 150 MPa if the shear modulus is 85 GPa.

(27) Two steel plates are joined by a lap joint employing four rivets. Determine the rivet diameter required if the shear stress in the rivets is not to exceed 75 MPa when the plates are subject to tensile forces of 60 kN.

(28) For the bolted joint shown in Figure 5.3.20, determine the maximum tensile load that can be applied to it if the bolt has a diameter of 10 mm and the maximum shear stress it can withstand is 200 MPa.

(29) What force is required to punch a hole of 20 mm diameter through a 10 mm thick steel plate if the shear strength of the plate is 400 MPa?

(30) A bar is attached to a gusset plate by two 20 mm diameter bolts, as in Figure 5.3.21. What will be the shear stress acting on each bolt when tensile forces of 70 kN are applied?

(31) What is the strain energy stored in a steel bar of rectangular cross-section 50 mm × 30 mm and length 600 mm when it is subject to an axial load of 200 kN? The steel has a modulus of elasticity of 200 GPa.

(32) A cable of cross-sectional area 1250 mm^2 and length 12 m has a load of 30 kN suspended from it. What will be the strain energy stored in the cable? The cable material has a modulus of elasticity of 200 GPa.

(33) Determine the strain energy stored in the members of the pin-jointed structure shown in Figure 5.3.22 when there is a load F and each member has the same cross-sectional area A and modulus of elasticity E.

5.4 Beams

A *beam* can be defined as a structural member which is subject to forces causing it to bend and is a very common structural member. This section is about such structural members and the analysis of the forces, moments and stresses concerned.

Bending occurs when forces are applied which have components at right angles to the member axis and some distance from a point of support; as a consequence beams become curved. Generally beams are horizontal and the loads acting on the beam act vertically downwards.

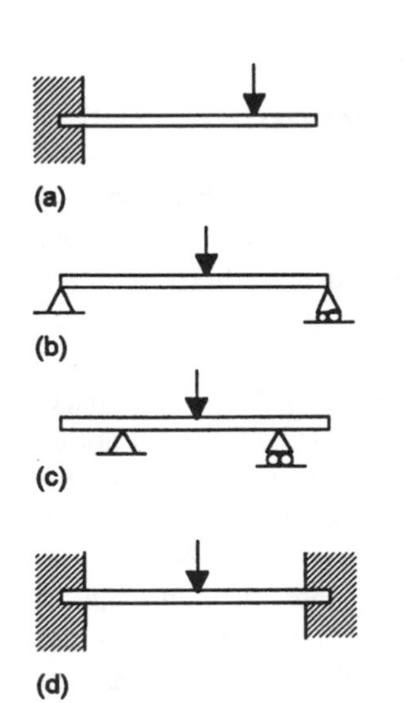

Figure 5.4.1 *Examples of beams: (a) cantilever; (b) simply supported; (c) simply supported with overhanging ends; (d) built-in*

Beams

The following are common types of beams:

- *Cantilever* (Figure 5.4.1(a)) This is rigidly fixed at just one end, the rigid fixing preventing rotation of the beam when a load is applied to the cantilever. Thus there will be a moment due to a load and this, for equilibrium, has to be balanced by a resisting moment at the fixed end. At a free end there are no reactions and no resisting moments.
- *Simply supported beam* (Figure 5.4.1(b)) This is a beam which is supported at its ends on rollers or smooth surfaces or one of these combined with a pin at the other end. At a supported end or point there are reactions but no resisting moments.

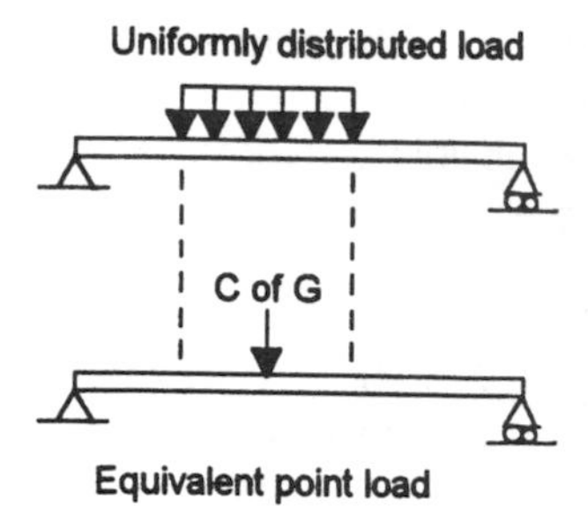

Figure 5.4.2 *Distributed load can be replaced by a single point load at the centre of gravity*

- *Simply supported beam with overhanging ends* (Figure 5.4.1(c)) This is a simple supported beam with the supports set in some distance from the ends. At a support there are reactions but no resisting moments; at a free end there are no reactions and no resisting moments.
- *Built-in beam (encastre)* (Figure 5.4.1(d)) This is a beam which is built-in at both ends and so both ends are rigidly fixed. Where an end is rigidly fixed there is a reaction force and a resisting moment.

The loads that can be carried by beams may be concentrated or distributed. A concentrated load is one which can be considered to be applied at a point while a distributed load is one that is applied over a significant length of the beam. With a distributed load acting over a length of beam, the distributed load may be replaced by an equivalent concentrated load at the centre of gravity of the distributed length. For a uniformly distributed load, the centre of gravity is at the midpoint of the length (Figure 5.4.2). Loading on a beam may be a combination of fixed loads and distributed loads.

Beams can have a range of different forms of section and the following (Figure 5.4.3) are some commonly encountered forms:

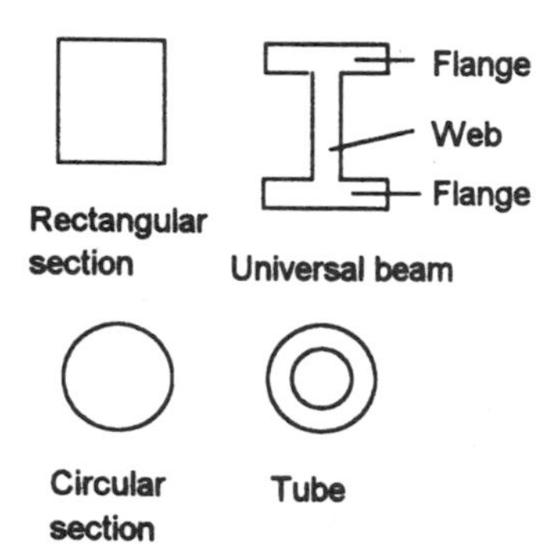

Figure 5.4.3 *Forms of cross-sections of beams*

- Simple rectangular sections, e.g. as with the timber joists used in the floor construction of houses. Another example is the slab which is used for floors and roofs in buildings; this is just a rectangular section beam which is wide and shallow.
- An I-section, this being termed the *universal beam*. The universal beam is widely used and available from stockists in a range of sizes and weights. The I-section is an efficient form of section in that the material is concentrated in the flanges at the top and bottom of the beam where the highest stresses will be found.
- Circular sections or tubes, e.g. tubes carrying liquids and supported at a number of points.

Shear force and bending moment

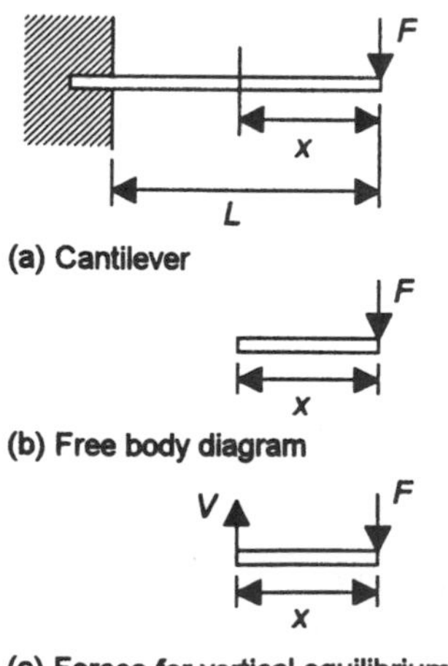

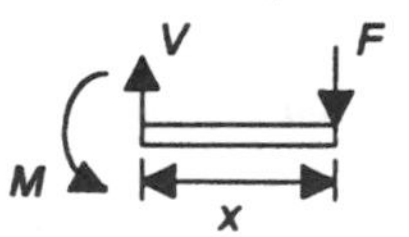

Figure 5.4.4 *Loaded cantilever and equilibrium*

There are two important terms used in describing the behaviour of beams: shear force and bending moment. Consider a cantilever (Figure 5.4.4(a)) which has a concentrated load F applied at the free end. If we make an imaginary cut through the beam at a distance x from the free end, we will think of the cut section of beam (Figure 5.4.4(b)) as a *free body*, isolated from the rest of the beam and effectively floating in space. With just the force indicated in Figure 5.4.4(b)), the body would not be in equilibrium. However, since it is in equilibrium, we must have for vertical equilibrium a vertical force V acting on it such that $V = F$ (Figure 5.4.4(c)). This force V is called the *shear force* because the combined action of V and F on the section is to shear it (Figure 5.4.5). In general, the shear force at a transverse section of a beam is the algebraic sum of the external forces acting at right angles to the axis of the beam on one side of the section concerned.

In addition to vertical equilibrium we must also have the section of beam in rotational equilibrium. For this we must have a moment M applied (Figure 5.4.4(d)) at the cut so that $M = Fx$. This moment is

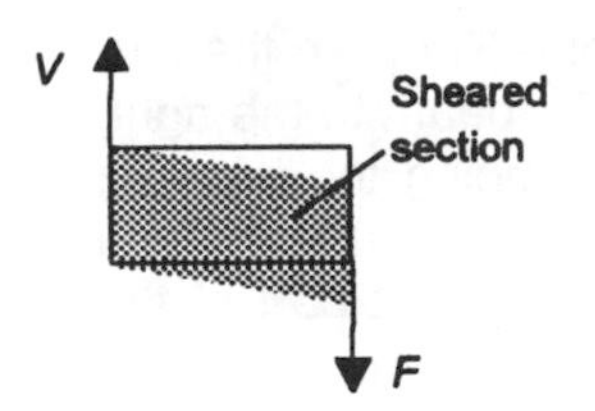

Figure 5.4.5 *Shear on a beam section*

termed the *bending moment*. In general, the bending moment at a transverse section of a beam is the algebraic sum of the moments about the section of all the forces acting on one (either) side of the section concerned.

Sign conventions

The conventions most often used for the signs of shear forces and bending moments are:

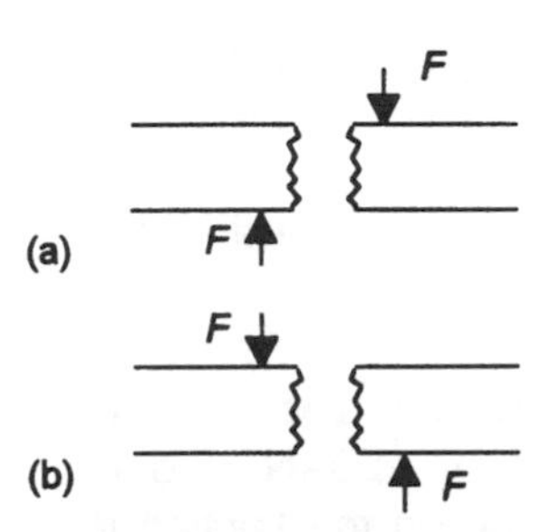

Figure 5.4.6 *Shear force: (a) positive; (b) negative*

- *Shear force*
  When the left-hand side of the beam section is being pushed upwards and the right-hand side downwards, i.e. the shear forces on either side of the section are in a clockwise direction, then the shear force is taken as being positive (Figure 5.4.6(a)). When the left-hand side of the beam is being pushed downwards and the right-hand side upwards, i.e. when the shear forces on either side of a section are in an anticlockwise direction, the shear force is taken as being negative (Figure 5.4.6(b)).
- *Bending moment*
  Bending moments are positive if they give rise to sagging (Figure 5.4.7(a)) and negative if they give rise to hogging (Figure 5.4.7(b)).

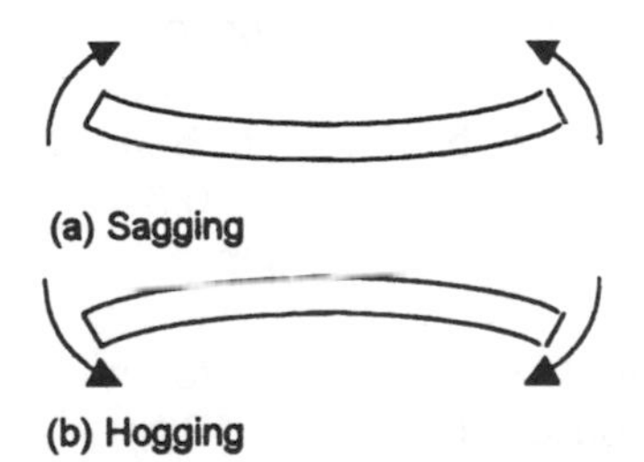

Figure 5.4.7 *Bending moment: (a) positive; (b) negative*

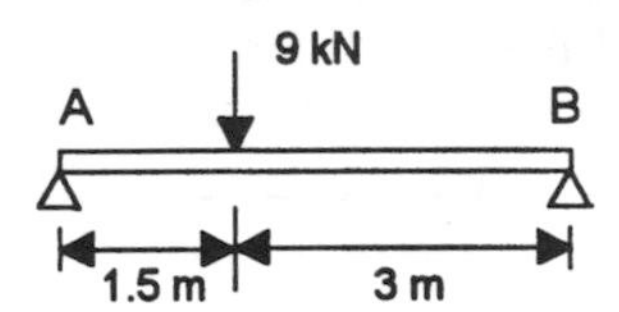

Figure 5.4.8 *Example 5.4.1*

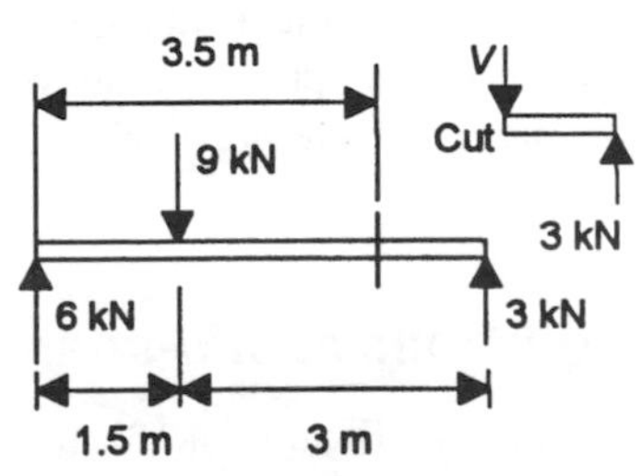

Figure 5.4.9 *Example 5.4.1*

Example 5.4.1

Determine the shear force and bending moment at points 0.5 m and 1.5 m from the left-hand end of a beam of length 4.5 m which is supported at its ends and subject to a point load of 9 kN a distance of 1.5 m from the left-hand end (Figure 5.4.8). Neglect the weight of the beam.

The reactions at the supports can be found by taking moments about the left-hand end support A:

$$R_B \times 4.5 = 9 \times 1.5$$

to give $R_B = 3$ kN. Since, for vertical equilibrium we have:

$$R_A + R_B = 9$$

then $R_A = 6$ kN. The forces acting on the beam are thus as shown in Figure 5.4.9.

If we make an imaginary cut in the beam at 3.5 m from the left-hand end (Figure 5.4.9), then the force on the beam to the right of the cut is 3 kN upwards and that to the left is 9 – 6 = 3 kN downwards. The shear force V is thus negative and –3 kN.

If we make an imaginary cut in the beam at 0.5 m from the left-hand end (Figure 5.4.10), then the force on the beam to the right of the cut is 9 – 3 = 6 kN downwards and that to the left is 6 kN upwards. The shear force V is thus positive and +6 kN.

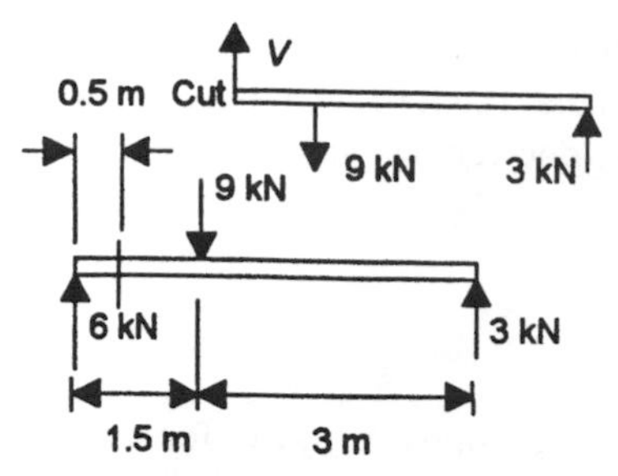

Figure 5.4.10 *Example 5.4.1*

The bending moment at a distance of 3.5 m from the left-hand end of the beam, for that part of the beam to the right, is $3 \times 1 = 3$ kN m. Since the beam is sagging the bending moment is +3 kN m. At a distance of 0.5 m from the left-hand end of the beam, the bending moment, for that part of the beam to the right, is $3 \times 4 - 9 \times 0.5 = +7.5$ kN m.

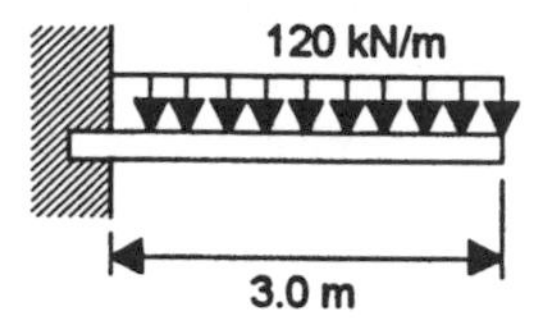

Figure 5.4.11 *Example 5.4.2*

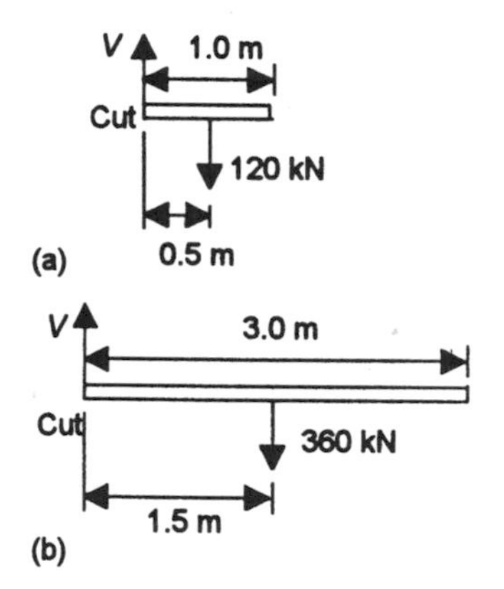

Figure 5.4.12 *Example 5.4.2*

Example 5.4.2

A uniform cantilever of length 3.0 m (Figure 5.4.11) has a uniform weight per metre of 120 kN. Determine the shear force and bending moment at distances of (a) 1.0 m and (b) 3.0 m from the free end if no other loads are carried by the beam.

(a) With a cut 1.0 m from the free end, there is 1.0 m of beam to the right with a weight of 120 kN (Figure 5.4.12(a)). Thus, since the total force on the section to the right of the cut is 120 kN the shear force V is +120 kN; it is positive because the forces are clockwise.

The weight of this section can be considered to act at its centre of gravity which, because the beam is uniform, is at its midpoint. Thus the 120 kN weight force can be considered to be 0.5 m from the cut end of the right-hand section and so the bending moment is $-120 \times 0.5 = -60$ kN m; it is negative because there is hogging.

(b) At 3.0 m from the free end, there is 3.0 m of beam to the right and it has a weight of 360 kN (Figure 5.4.12(b)). Thus, since the total force on the section to the right of the cut is 360 N the shear force V is +360 kN; it is positive because the forces are clockwise.

The weight of this section can be considered to act at its midpoint, a distance of 1.5 m from the free end. Thus the bending moment is $-360 \times 1.5 = -540$ kN m; it is negative because there is hogging.

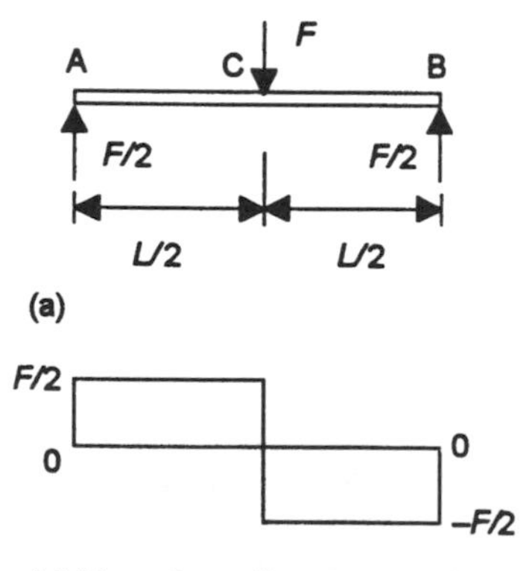

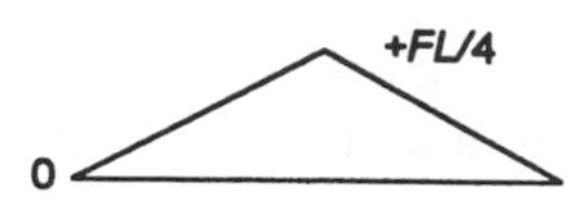

Figure 5.4.13 *(a) Simply supported beam with point load; (b) shear force diagram; (c) bending moment diagram*

Shear force and bending moment diagrams

Shear force diagrams and *bending moment diagrams* are graphs used to show the variations of the shear forces and bending moments along the length of a beam. The convention that is generally adopted is to show positive shear forces and bending moments plotted above the axial line of the beam and below it if negative.

Simply supported beam with point load at mid-span

For a simply supported beam with a central load F (Figure 5.4.13), the reactions at each end will be $F/2$.

At point A, the forces to the right are $F - F/2$ and so the shear force at A is $+F/2$. This shear force value will not change as we move along the beam from A until point C is reached. To the right of C we have just a force of $-F/2$ and this gives a shear force of $-F/2$. To the left of C we have just a force of $+F/2$ and this gives a shear force of $+F/2$. Thus at point C, the shear force takes on two values. Figure 5.4.13(b) shows the shear force diagram.

At point A, the moments to the right are $F \times L/2 - F/2 \times L = 0$. The bending moment is thus 0. At point C the moment to the right is $F/2 \times L/2$ and so the bending moment is $+ FL/4$; it is positive because sagging is occurring. At point B the moment to the left is $F \times L/2 - F/2 \times L = 0$. Between A and C the bending moment will vary, e.g. at one-quarter the way along the beam it is $FL/8$. In general, between A and C the bending moment a distance x from A is $Fx/2$ and between C and B is $Fx/2 - F(x - L/2) = F/2(L - x)$. Figure 5.4.13(c) shows the bending moment diagram. The maximum bending moment occurs under the load and is $\frac{1}{4}FL$.

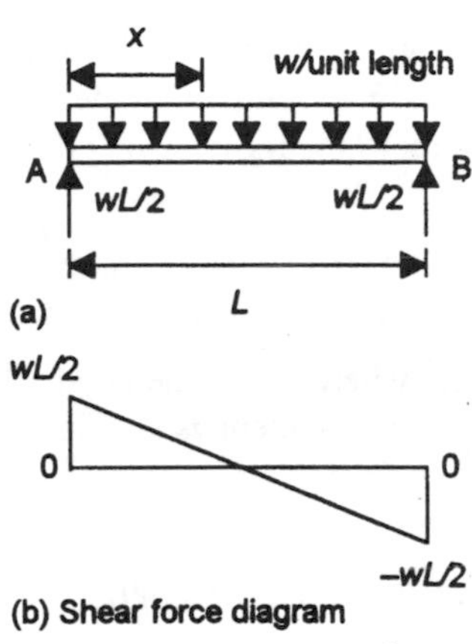

Figure 5.4.14 *(a) Simply supported beam with distributed load; (b) shear force diagram; (c) bending moment diagram*

Simply supported beam with uniformly distributed load

A simply supported beam which carries just a uniformly distributed load of w/unit length (Figure 5.4.14) will have the reactions at each end as $wL/2$.

For the shear force a distance x from the left-hand end of the beam, the load acting on the left-hand section of beam is wx and thus the shear force is:

$$V = wL/2 - wx = w(\tfrac{1}{2}L - x)$$

When $x = \frac{1}{2}L$, the shear force is zero. When $x < \frac{1}{2}L$ the shear force is positive and when $x > \frac{1}{2}L$ it is negative. Figure 5.4.14(b) shows the shear force diagram.

At A the moment due to the beam to the right is $-wL \times L/2 + wL/2 \times L = 0$. At the midpoint of the beam the moment is $-wL/2 \times L/4 + wL/2 \times L/2 = wL^2/8$ and so the bending moment is $+wL^2/8$. At the quarter-point along the beam, the moment due to the beam to the right is $-3L/4 \times 3L/8 + wL/2 \times 3L/4 = 3wL^2/32$. In general, the bending moment due to the beam at distance x from A is:

$$M = -wx \times x/2 + wL/2 \times x = -wx^2/2 + wLx/2$$

Differentiating the equation gives $\mathrm{d}M/\mathrm{d}x = -wx + wL/2$. Thus $\mathrm{d}M/\mathrm{d}x = 0$ at $x = L/2$. The bending moment is thus a maximum at $x = L/2$ and so the bending moment at this point is $wL^2/8$. Figure 5.4.14(c) shows the bending moment diagram.

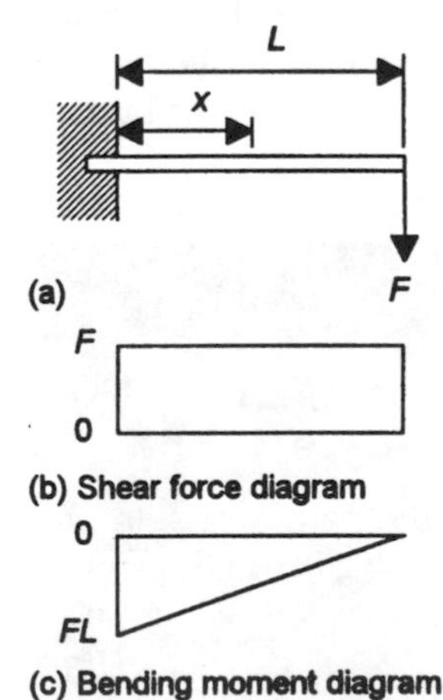

Figure 5.4.15 *(a) Cantilever with point load at free end; (b) shear force diagram; (c) bending moment diagram*

Cantilever with point load at free end

For a cantilever which carries a point load F at its free end (Figure 5.4.15(a)) and for which the weight of the beam is neglected, the shear force at any section will be $+F$, the shear force diagram thus being as shown in Figure 5.4.15(b). The bending moment at a distance x from the fixed end is:

$$M = -F(L - x)$$

The minus sign is because the beam shows hogging. We have $dM/dx = F$ and thus the bending moment diagram is a line of constant slope F. At the fixed end, when $x = 0$, the bending moment is FL. At the free end, when $x = L$, the bending moment is 0.

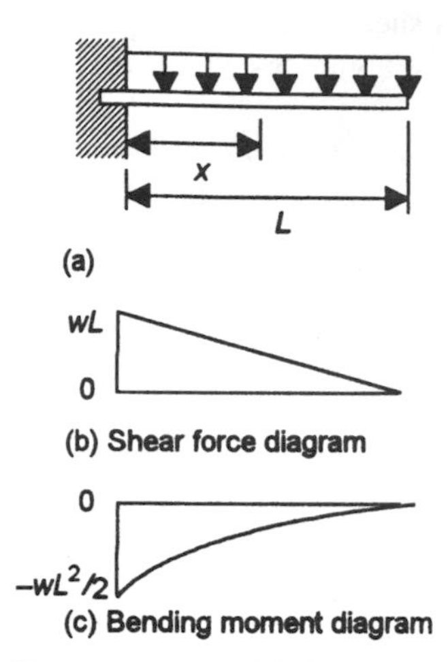

Figure 5.4.16 *(a) Cantilever with uniformly distributed load; (b) shear force diagram; (c) bending moment diagram*

Cantilever with uniformly distributed load

Consider a cantilever which has just a uniformly distributed load of w per unit length (Figure 5.4.16(a)). The shear force a distance x from the fixed end is:

$$V = +w(L - x)$$

Thus at the fixed end the shear force is $+wL$ and at the free end it is 0. Figure 5.4.16(b) shows the shear force diagram. The bending moment at a distance x from the fixed end is, for the beam to the right of the point, given by:

$$M = -w(L - x) \times (L - x)/2 = -\tfrac{1}{2}w(L - x)^2$$

This is a parabolic function. At the fixed end, where $x = 0$, the bending moment is $-\tfrac{1}{2}wL^2$. At the free end the bending moment is 0. Figure 5.4.16(c) shows the bending moment diagram.

General points about shear force and bending moment diagrams

From the above examples of beams, we can conclude that:

- Between point loads, the shear force is constant and the bending moment variation with distance is a straight line.
- Throughout a length of beam which has a uniformly distributed load, the shear force varies linearly with distance and the bending moment varies with distance as a parabola.
- The bending moment is a maximum when the shear force is zero.
- The shear force is a maximum when the slope of the bending moment diagram is a maximum and zero when the slope is zero.
- For point loads, the shear force changes abruptly at the point of application of the load by an amount equal to the size of the load.

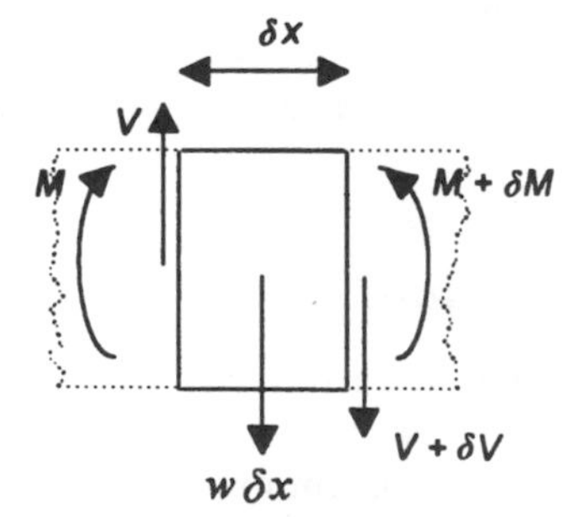

Figure 5.4.17 *Small segment of a beam supporting a uniformly distributed load w per unit length*

Mathematics in action

Maximum value of the bending moment occurs at a point of zero shear force

The following is a proof of the above statement. Consider a very small segment of beam (Figure 5.4.17) of length δx and which is supporting a uniformly distributed load of w per unit length. The load on the segment is $w\ \delta x$ and can be considered to act through its centre. The values of the shear force V and bending moment M increase by δV and δM from one end of the segment to the other.

If we take moments about the left-hand edge of the segment then:

$$M + V\,\delta x = w\,\delta x \times \delta x/2 + M + \delta M$$

Neglecting multiples of small quantities gives $V\,\delta x = \delta M$ and hence, as δx tends to infinitesimally small values, we can write:

$$V = \frac{dM}{dx} \qquad (5.4.1)$$

Thus, the shear force is the gradient of the bending moment–distance graph and so $V = 0$ when this gradient $dM/dx = 0$. The gradient of a curve is zero when it is a maximum or a minimum. Thus a local maximum or minimum bending moment occurs when the shear force is zero.

Integrating the above equation gives:

$$M = \int V\,dx \qquad (5.4.2)$$

Hence the bending moment at any point is represented by the area under the shear force diagram up to that point.

Example 5.4.3

A cantilever has a length of 3 m and carries a point load of 3 kN at a distance of 2 m from the fixed end and a uniformly distributed load of 2 kN/m over a 2 m length from the fixed end. Draw the shear force and bending moment diagrams and determine the position and size of the maximum bending moment and maximum shear force.

Figure 5.4.18(a) shows the beam. To the right of C there are no forces and so the shear force for that section of the beam is zero. To the right of A the total force is 4 + 3 = 7 kN and so this is the value of the shear force at A. At a point 1 m from A the total force to the right is 2 + 3 = 5 kN and so this is the shear force at that point. The shear force a distance x from A, with x less than 2 m, is $2(2 - x) + 3$ kN. At C, when $x = 2$, the shear force is 3 kN. Figure 5.4.18(b) shows the shear force diagram.

The bending moment at C for the beam to the right is 0. At A the bending moment for the beam to the right is $-(4 \times 1 + 2 \times 3) = -10$ kN m; it is negative because there is hogging. At a distance x from the fixed end, the bending moment is $-[2(2 - x) \times \frac{1}{2}(2 - x) + 3 \times (2 - x)]$ kN m. Figure 5.4.18(c) shows the resulting bending moment diagram.

The maximum bending moment will be at a point on the shear force diagram where the shear force is zero and is thus at the fixed end and has the value 10 kN m.

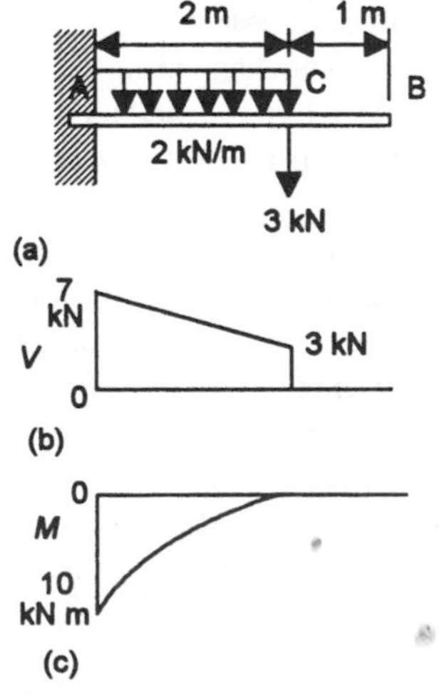

Figure 5.4.18 *Example 5.4.3*

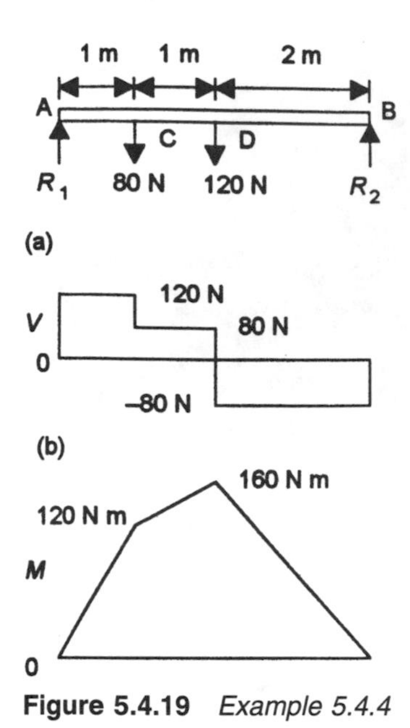

Figure 5.4.19 *Example 5.4.4*

Example 5.4.4

A beam of length 4 m is supported at its ends and carries loads of 80 N and 120 N at 1 m and 2 m from one end (Figure 5.4.19(a)). Draw the shear force and bending moment diagrams and determine the position and size of the maximum bending moment. Neglect the weight of the beam.

Taking moments about A gives:

$$4R_2 = 80 \times 1 + 120 \times 2$$

Hence $R_2 = 80$ N. Since $R_1 + R_2 = 80 + 120$, then $R_1 = 120$ N.

At any point in AC the shear force is 80 + 120 − 80 = 120 N. In CD the shear force is 120 − 80 = 40 N. In DB the shear force is −80 N. Figure 5.4.19(b) shows the resulting shear force diagram.

At any point a distance x from A in AC the bending moment is $120x$, reaching 120 N m at C. At any point a distance x from A in CD the bending moment is $120x - 80(x - 1) = 40x + 80$, reaching 160 N m at D. At any point a distance x from A in DB the bending moment is $120x - 80(x - 1) - 120(x - 2) = -80x + 320$, reaching 0 at B. Figure 5.4.19(c) shows the resulting bending moment diagram.

Localized maximum bending moments will be at points on the shear force diagram where the shear force is zero. The maximum is at 2 m from A and has the value 160 N m.

Bending stress

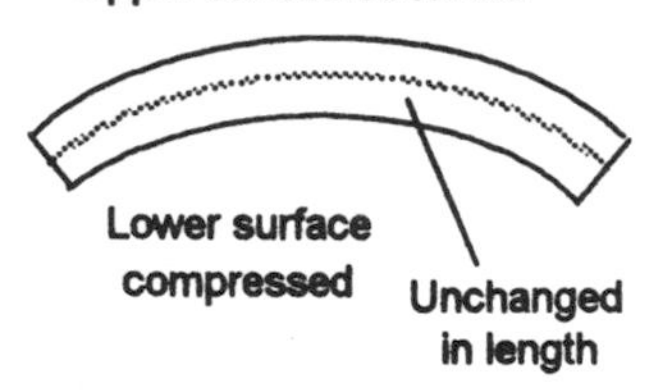

Figure 5.4.20 *Bending stretches the upper surface and contracts the lower surface, in between there is an unchanged in length surface*

The following is a derivation of equations which enable us to determine the stresses produced in beams by bending. When a bending moment is applied to a beam it bends, as in Figure 5.4.20, and the upper surface becomes extended and in tension and the lower surface becomes compressed and in compression. The upper surface increasing in length and the lower surface decreasing in length implies that between the upper and lower surface there is a plane which is unchanged in length when the beam is bent. This plane is called the *neutral plane* and the line where the plane cuts the cross-section of the beam is called the *neutral axis*.

Consider a beam, or part of a beam, and assume that it is bent into a circular arc. A section through the beam aa which is a distance y from the neutral axis (Figure 5.4.21) has increased in length as a consequence of the beam being bent and the strain it experiences is the change in length ΔL divided by its initial unstrained length L. For circular arcs, the arc length is the radius of the arc multiplied by the angle it subtends, and thus, $L + \Delta L = (R + y)\theta$. The neutral axis NA will, by definition, be unstrained and so for it we have $L = R\theta$. Hence, the strain on aa is:

$$\text{Strain} = \frac{\Delta L}{L} = \frac{(R + y)\theta - R\theta}{R\theta} = \frac{y}{R} \qquad (5.4.3)$$

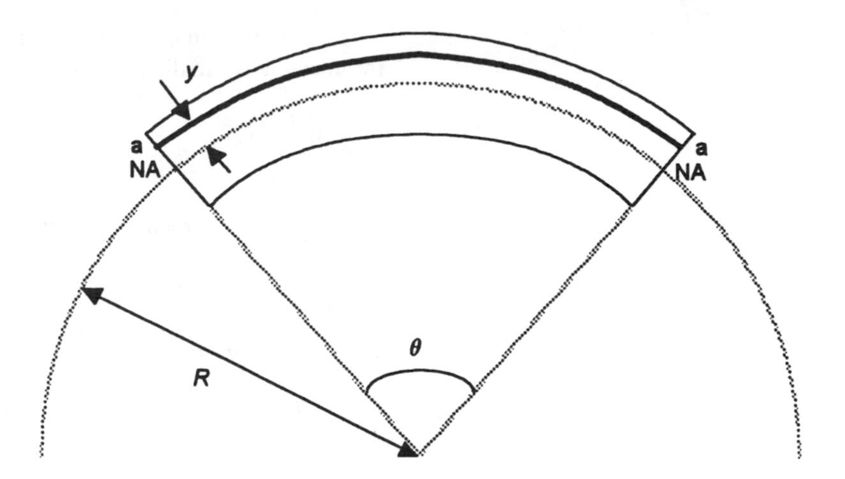

Figure 5.4.21 *Bending into a circular arc*

The strain thus varies linearly through the thickness of the beam being larger the greater the distance y from the neutral axis. The maximum strains thus occur at the outer surfaces of a beam, these being the most distant from the neutral axis.

Provided we can use Hooke's law, the stress due to bending which is acting on aa is:

$$\text{Stress } \sigma = E \times \text{strain} = \frac{Ey}{R} \qquad (5.4.4)$$

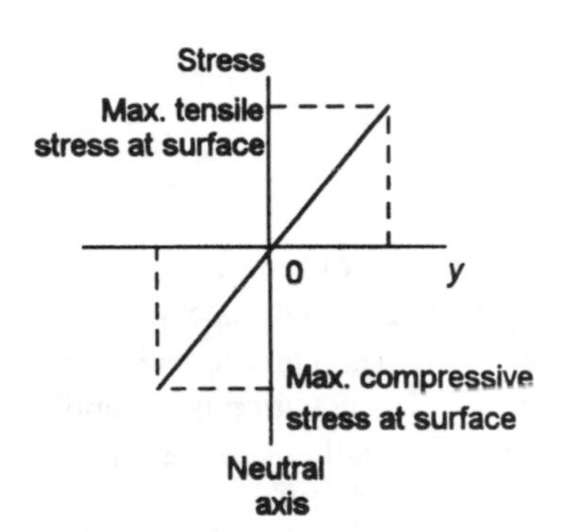

Figure 5.4.22 *Stress variation across beam section*

The maximum stresses thus occur at the outer surfaces of a beam, these being the most distant from the neutral axis. Figure 5.4.22 shows how the stress will vary across the section of rectangular cross-section beam, such a beam having a central neutral axis. It is because the highest stresses will be at the maximum distances from the neutral axis that the I-section girder makes most effective use of material, placing most material where the stresses are the highest.

The general bending equations

For the beam, shown in Figure 5.4.21, which has been bent into the arc of a circle, if the element aa has a cross-sectional area δA in the cross-section of the beam at a distance y from the neutral axis of radius R, the element will be stretched as a result of the bending with stress $\sigma = Ey/R$, where E is the modulus of elasticity of the material. The forces stretching this element aa are:

$$\text{Force} = \text{stress} \times \text{area} = \sigma\, \delta A = \frac{Ey}{R}\, \delta A$$

The moment of the force acting on this element about the neutral axis is:

$$\text{Moment} = \text{force} \times y = \frac{Ey}{R}\, \delta A \times y = \frac{E}{R}\, y^2\, \delta A$$

The total moment M produced over the entire cross-section is the sum of all the moments produced by all the elements of area in the cross-section. Thus, if we consider each such element of area to be infinitesimally small, we can write:

$$M = \int \frac{E}{R} y^2 \, dA = \frac{E}{R} \int y^2 \, dA$$

The integral is termed the *second moment of area I* of the section:

$$I = \int y^2 \, dA$$

Thus we can write:

$$M = \frac{EI}{R} \tag{5.4.5}$$

Since the stress σ on a layer a distance y from the neutral axis is yE/R:

$$M = \frac{\sigma I}{y} \tag{5.4.6}$$

The above two equations are generally combined and written as the *general bending formula*:

$$\frac{M}{I} = \frac{\sigma}{y} = \frac{E}{R} \tag{5.4.7}$$

This formula is only an exact solution for the case of pure bending, i.e. where the beam is bent into the arc of a circle. This only occurs where the bending moment is constant; however, many beam problems involve bending moments which vary along the beam. The equation is generally still used since it provides answers which are usually accurate enough for engineering design purposes.

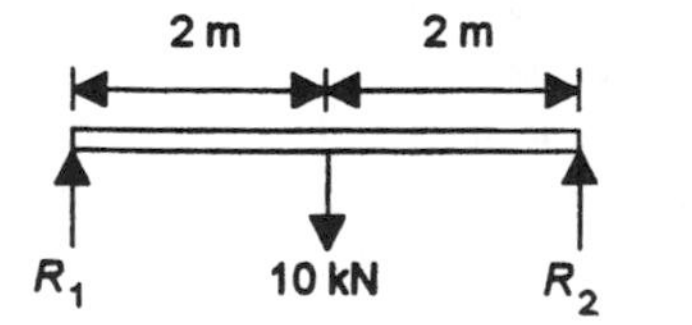

Figure 5.4.23 *Example 5.4.5*

Example 5.4.5

A uniform rectangular cross-section horizontal beam of length 4 m and depth 150 mm rests on supports at its ends (Figure 5.4.23) and supports a concentrated load of 10 kN at its midpoint. Determine the maximum tensile and compressive stresses in the beam if it has a second moment of area of 2.8×10^{-5} m^4 and the weight of the beam can be neglected.

Taking moments about the left-hand end gives $4R_2 = 2 \times 10$ and so $R_2 = 5$ kN. Since $R_1 + R_2 = 10$ then $R_1 = 5$ kN. The maximum bending moment will occur at the midpoint and is 10 kN m. The maximum bending stress will occur at the cross-section where the bending moment is a maximum and on the outer surfaces of the beam, i.e. $y = \pm 75$ mm. Hence:

$$\sigma = \frac{My}{I} = \pm \frac{10 \times 10^3 \times 0.075}{2.8 \times 10^{-5}} = \pm 26.8 \text{ MPa}$$

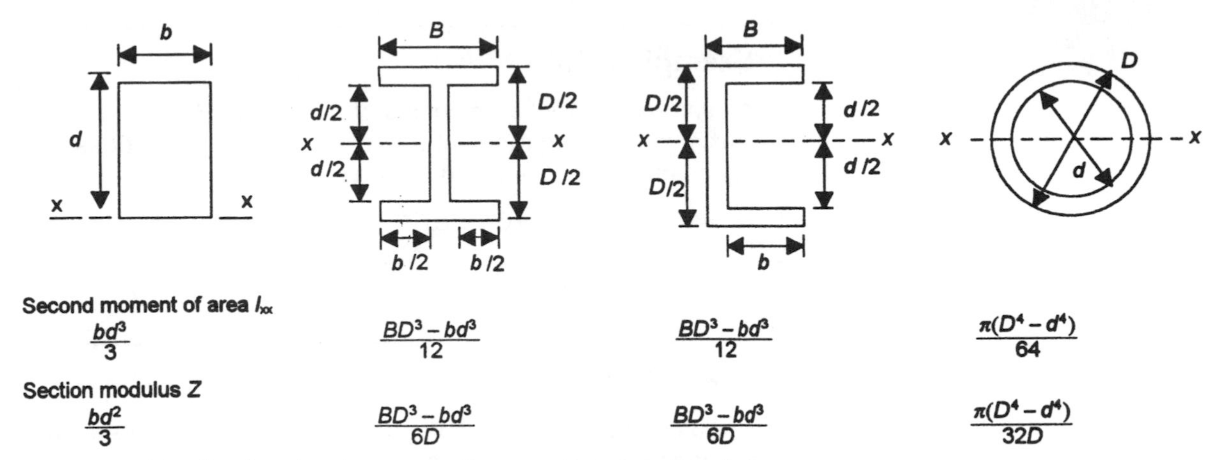

Figure 5.4.24 *Section data for commonly encountered sectional shapes*

Section modulus

For a beam which has been bent, the maximum stress σ_{max} will occur at the maximum distance y_{max} from the neutral axis. Thus:

$$M = \frac{I}{y_{max}} \sigma_{max}$$

The quantity I/y_{max} is a purely geometric function and is termed the *section modulus Z*. Thus we can write:

$$M = Z\sigma_{max} \tag{5.4.8}$$

with:

$$Z = \frac{I}{y_{max}} \tag{5.4.9}$$

Figure 5.4.24 gives values for typical sections. Standard section handbooks give values of the second moment of area and section modulus for different standard section beams, Table 5.4.1 being an illustration of the information provided.

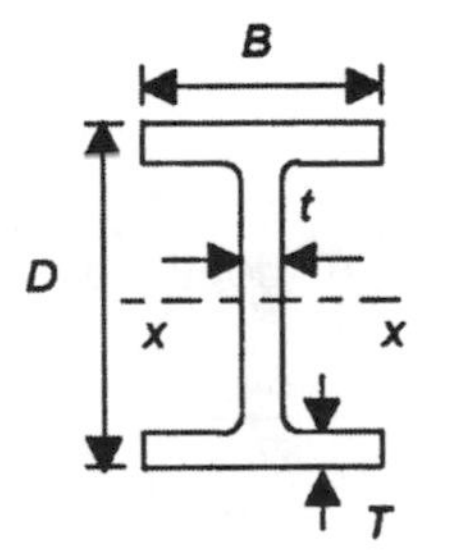

Figure 5.4.25 *Table 5.4.1*

Table 5.4.1 Universal beams (Figure 5.4.25)

Size mm	*Mass per metre* kg	*Depth of section D* mm	*Width of section B* mm	*Web thickness t* mm	*Flange thickness T* mm	*Second moment of area I_{xx}* cm^4	*Section modulus Z* cm^3	*Area of section* cm^2
838 × 292	226	850.9	293.8	16.1	26.8	340 000	7990	289.0
	194	840.7	292.4	14.7	21.7	279 000	6550	247.0
	176	834.9	291.6	14.0	18.8	246 000	5890	224.0
762 × 267	197	769.6	268.0	15.6	25.4	240 000	6230	251.0
	147	753.9	265.3	12.9	17.5	169 000	4480	188.0
686 × 254	170	692.9	255.8	14.5	23.7	170 000	4910	217.0
	140	683.5	253.7	12.4	19.0	136 000	3990	179.0
	125	677.9	253.0	11.7	16.2	118 000	3480	160.0

Example 5.4.6

A beam has a section modulus of $3 \times 10^6\,\text{mm}^3$, what is the maximum bending moment that can be used if the stress in the beam must not exceed 6 MPa?

$$M = Z\sigma_{max} = 3 \times 10^6 \times 10^{-9} \times 6 \times 10^6 = 18\,\text{kN m}$$

Example 5.4.7

A rectangular cross-section beam of length 4 m rests on supports at each end and carries a uniformly distributed load of 10 kN/m. If the stress must not exceed 8 MPa, what will be a suitable depth for the beam if its width is 100 mm?

For a simply supported beam with a uniform distributed load over its full length, the maximum bending moment is $wL^2/8$ and thus the maximum bending moment for this beam is $10 \times 4^2/8 = 20\,\text{kN m}$. Hence, the required section modulus is:

$$Z = \frac{M}{\sigma_{max}} = \frac{20 \times 10^3}{8 \times 10^6} = 2.5 \times 10^{-3}\,\text{m}^3$$

For a rectangular cross-section $Z = bd^2/6$ and thus:

$$d = \sqrt{\frac{6Z}{b}} = \sqrt{\frac{6 \times 2.5 \times 10^{-3}}{0.100}} = 0.387$$

A suitable beam might thus be one with a depth of 400 mm.

Moments of area

The product of an area and its distance from some axis is called the first moment of area; the product of an area and the square of its distance from an axis is the second moment of area. By considering the first moment of area we can determine the location of the neutral axis; the second moment of area enables us, as indicated above, to determine the stresses present in bent beams.

First moment of area and the neutral axis

In the above discussion of the bending of beams it was stated that the position of the neutral axis for a rectangular cross-section beam was its central axis. Here we derive the relationship which enables the position of the neutral axis to be determined for any cross-section. Consider a beam bent into the arc of a circle. The forces acting on a segment a distance y from the neutral axis are:

$$\text{Force} = \text{stress} \times \text{area} = \sigma \delta A = \frac{Ey}{R}\,\delta A$$

The total longitudinal force will be the sum of all the forces acting on such segments and thus, when we consider infinitesimally small areas, is given by:

$$\text{Total force} = \int \frac{Ey}{R}\,\mathrm{d}A = \frac{E}{R}\int y\,\mathrm{d}A$$

But since the beam is only bent, it is only acted on by a bending moment and there is no longitudinal force stretching the beam. Thus, since E and R are not zero, we must have:

$$\int y\,\mathrm{d}A = 0 \qquad (5.4.10)$$

The integral $\int y\,\mathrm{d}A$ is called the *first moment of area* of the section.

The only axis about which we can take such a moment and obtain 0 is an axis through the centre of the area of the cross-section, i.e. the centroid of the beam. Thus the neutral axis must pass through the centroid of the section when the beam is subject to just bending. Hence we can determine the position of the neutral axis by finding the position of the centroid.

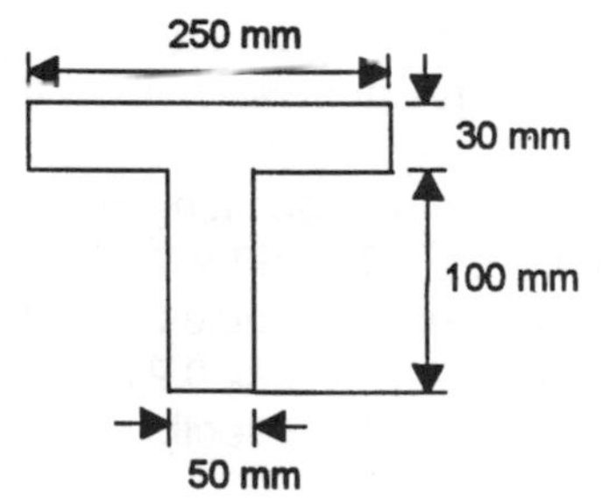

Figure 5.4.26 *Example 5.4.8*

Example 5.4.8

Determine the position of the neutral axis for the T-section beam shown in Figure 5.4.26

The neutral axis will pass through the centroid. We can consider the T-section to be composed of two rectangular sections with the centroid of each at its centre. Hence, taking moments about the base of the T-section:

$$\text{Total moment} = 250 \times 30 \times 115 + 100 \times 50 \times 50$$
$$= 1.11 \times 10^6\,\text{mm}^4$$

Hence the distance of the centroid from the base is (total moment)/(total area):

$$\text{Distance from base} = \frac{1.11 \times 10^6}{250 \times 30 + 100 \times 50} = 89\,\text{mm}$$

Second moment of area

The integral $\int y^2\,\mathrm{d}A$ defines the *second moment of area I* about an axis and can be obtained by considering a segment of area δA some distance y from the neutral axis, writing down an expression for its second moment of area and then summing all such strips that make up the section concerned, i.e. integrating. As indicated in the discussion of the general bending equation, the second moment of area is needed if we are to relate the stress produced in a beam to the applied bending moment.

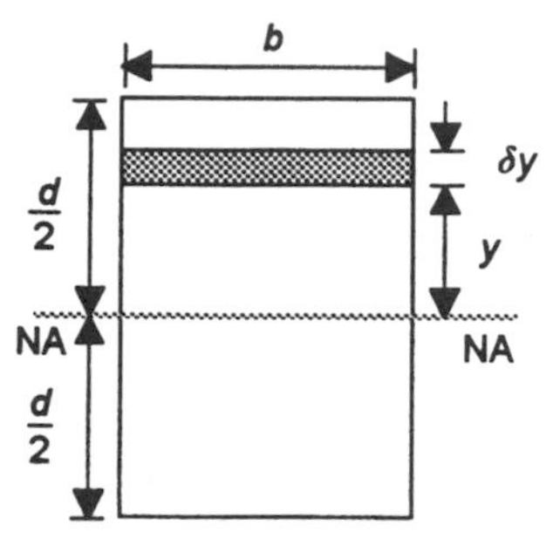

Figure 5.4.27 *Second moment of area*

Mathematics in action

Second moment of area

As an illustration of the derivation of a second moment of area from first principles, consider a rectangular cross-section of breadth b and depth d (Figure 5.4.27). For a layer of thickness δy a distance y from the neutral axis, which passes through the centroid, the second moment of area for the layer is:

Second moment of area of strip $= y^2 \delta A = y^2 b\, \delta y$

The total second moment of area for the section is thus:

$$\text{Second moment of area} = \int_{-d/2}^{d/2} y^2 b\, \mathrm{d}y = \frac{bd^3}{12} \qquad (5.4.11)$$

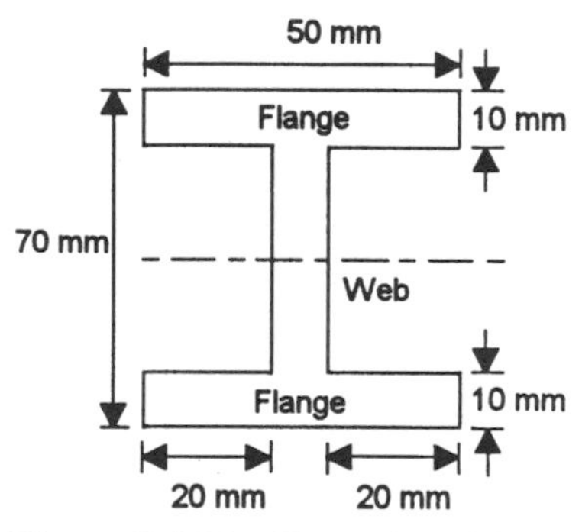

Figure 5.4.28 *Example 5.4.9*

Example 5.4.9

Determine the second moment of area about the neutral axis of the I-section shown in Figure 5.4.28.

We can determine the second moment of area for such a section by determining the second moment of area for the entire rectangle containing the section and then subtracting the second moments of area for the rectangular pieces 'missing' (Figure 5.4.29).

Thus for the rectangle containing the entire section, the second moment of area is given by $I = bd^3/12 = (50 \times 70^3)/12 = 1.43 \times 10^6\,\text{mm}^4$. Each of the 'missing' rectangles will have a second moment of area of $(20 \times 50^3)/12 = 0.21 \times 10^6\,\text{mm}^4$. Thus the second moment of area of the I-section is $1.43 \times 10^6 - 2 \times 0.21 \times 10^6 = 1.01 \times 10^6\,\text{mm}^4$.

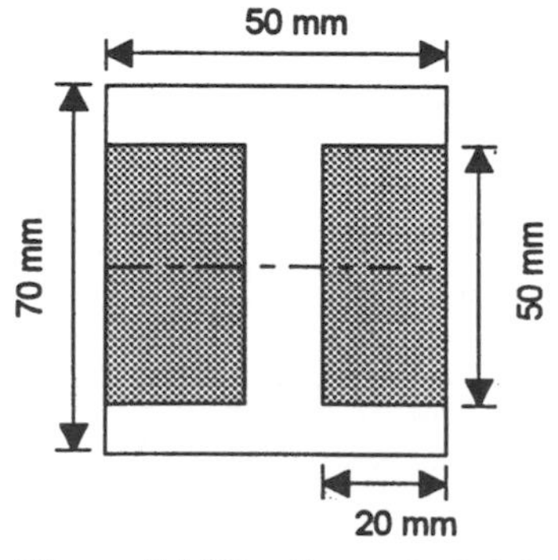

Figure 5.4.29 *Example 5.4.9*

Parallel axis theorem

The parallel axis theorem is useful when we want to determine the second moment of area about an axis which is parallel to the one for which we already know the second moment of area. If we had a second moment of area $I = \int y^2\, \mathrm{d}A$ of an area about an axis and then considered a situation where the area was moved by a distance h from the axis, the new second moment of area I_h, would be:

$$I_h = \int (y + h)^2\, \mathrm{d}A = \int y^2\, \mathrm{d}A + 2\,h \int y\, \mathrm{d}A + h^2 \int \mathrm{d}A$$

But $\int y\, \mathrm{d}A = 0$ and $\int \mathrm{d}A = A$. Hence:

$$I_h = I + Ah^2 \qquad (5.4.12)$$

Thus the second moment of area of a section about an axis parallel with an axis through the centroid is equal to the second moment of area about the axis through the centroid plus the area of the section multiplied by the square of the distance between the parallel axes. This is called the *theorem of parallel axes.*

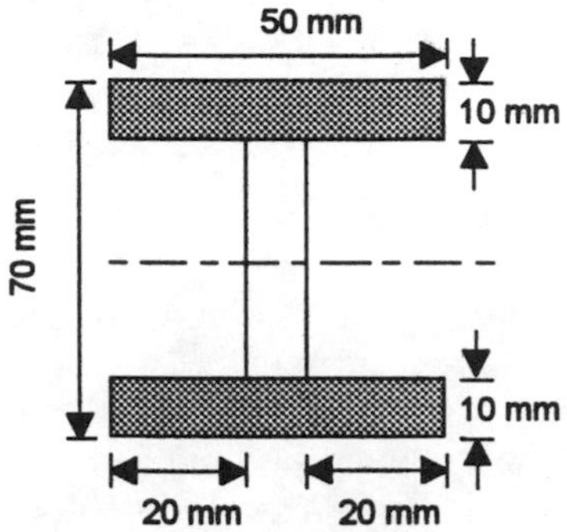

Figure 5.4.30 *Example 5.4.10*

Example 5.4.10

Determine, using the theorem of parallel axes, the second moment of area about the neutral axis of the I-section shown in Figure 5.4.28.

This is a repeat of Example 5.4.9. Consider the section as three rectangular sections with the web being the central rectangular section and the flanges a pair of rectangular sections with their neutral axes displaced from the neutral axis of the I-section by 30 mm (Figure 5.4.30). The central rectangular section has a second moment of area of $I = bd^3/12 = (10 \times 50^3)/12 = 0.104 \times 10^6\,\text{mm}^4$. Each of the outer rectangular areas will have a second moment of area given by the theorem of parallel axes as $I_h = I + Ah^2 = (50 \times 10^3)/12 + 50 \times 10 \times 30^2 = 0.454 \times 10^6\,\text{mm}^4$. Thus the second moment of area of the I-section is $1.01 \times 10^6\,\text{mm}^4$.

Perpendicular axis theorem

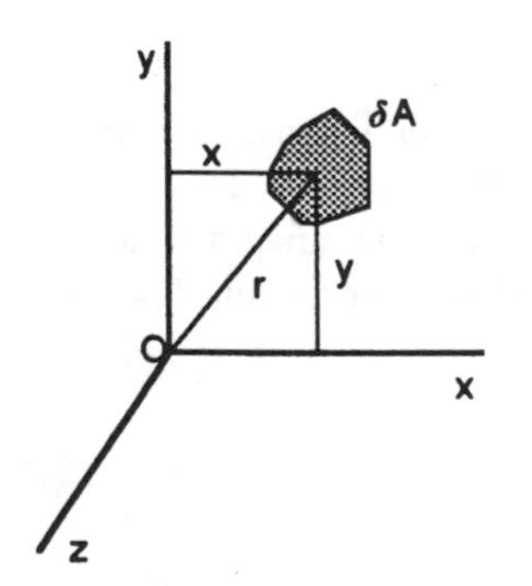

Figure 5.4.31 *Perpendicular axes and an elemental section*

The perpendicular axis theorem is useful when we want the second moment of area about an axis which is perpendicular to one about which we already know the second moment of area. Consider the second moment of an area with respect to an axis which is perpendicular to the plane of a section. Such an axis is termed the polar axis. Thus for the elemental section δA shown in Figure 5.4.31, this is the z-axis. The second moment of area of this section about the x-axis is:

$$I_x = \int y^2\,dA$$

The second moment of area of this section about the y-axis is:

$$I_y = \int x^2\,dA$$

Thus:

$$I_x + I_y = \int y^2\,dA + \int x^2\,dA = \int (y^2 + x^2)\,dA$$

But $r^2 = y^2 + x^2$ and thus

$$I_x + I_y = \int r^2\,dA$$

Since r is the distance of the elemental area from the z-axis, then the integral of $r^2\,dA$ is the second moment of area I_z about that axis, often termed the *polar second moment*. Thus:

$$I_z = I_x + I_y \tag{5.4.13}$$

Thus the second moment of area with respect to an axis through a point perpendicular to a section is equal to the sum of the second moments of area with respect to any two mutually perpendicular axes in the plane of the area through the same point. This is known as the *perpendicular axis theorem.*

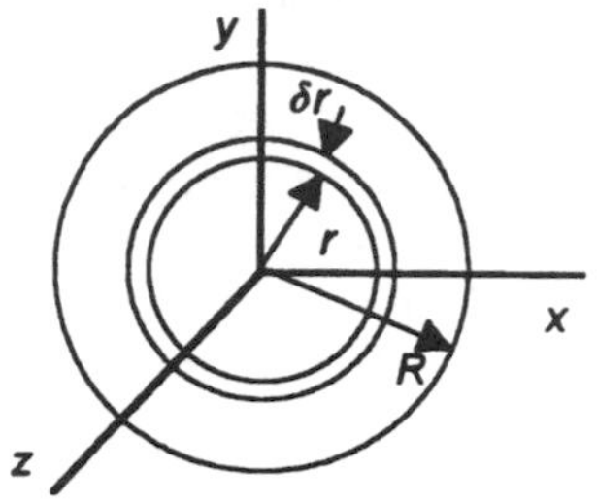

Figure 5.4.32 *Circular section*

Mathematics in action

Second moments of area of a circular section

To obtain the second moment of area I_z about an axis perpendicular to the plane of the circular section, consider an annular element of radius r and thickness δr (Figure 5.4.32). The area of the element is $2\pi r\,dr$. Thus:

$$I_z = \int_0^R 2\pi r \times r^2\,dx = \frac{\pi R^4}{2}$$

This is the polar second moment of area. To obtain the second moments of area about two mutually perpendicular axes x and y we can use the perpendicular axes theorem. Since, as a result of symmetry, we have $I_x = I_y$, then:

$$I_x = I_y = \frac{I_z}{2} = \frac{\pi R^4}{4} \qquad (5.4.14)$$

Radius of gyration

A geometric property which is widely quoted for beams, of various cross-sections, is the *radius of gyration k* of a section. This is the distance from the axis at which the area may be imagined to be concentrated to give the same second moment of area I. Thus $I = Ak^2$ and so:

$$k = \sqrt{\frac{I}{A}} \qquad (5.4.15)$$

Example 5.4.11

Determine the radius of gyration of a rectangular area about an axis passing through its centroid.

Since $I = bd^3/12$ and $A = bd$, then $k^2 = I/A = d^2/12$ and thus $k = d/2\sqrt{3}$.

Beam deflections

By how much does a beam deflect when it is bent as a result of the application of a bending moment? The following is a derivation of the basic differential equation which can be used to derive deflections. We then look at how we can solve the differential equation to obtain the required deflections.

The differential equation

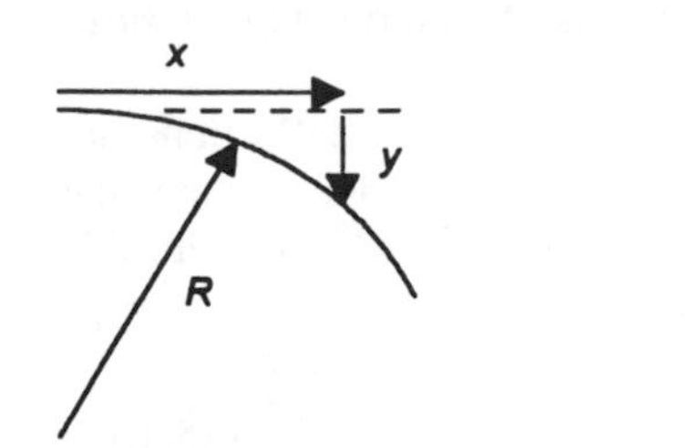

Figure 5.4.33 *Beam deflection of y at distance x*

When a beam is bent (Figure 5.4.33) as a result of the application of a bending moment M it curves with a radius R given by the general bending equation in 4.3.3 as:

$$\frac{1}{R} = \frac{M}{EI}$$

This radius can be expressed in terms of the vertical deflection y as $1/R = \mathrm{d}^2y/\mathrm{d}x^2$, where x is the distance along the beam from the chosen origin at which the deflection is measured, and so:

$$M = EI\frac{\mathrm{d}^2y}{\mathrm{d}x^2}$$

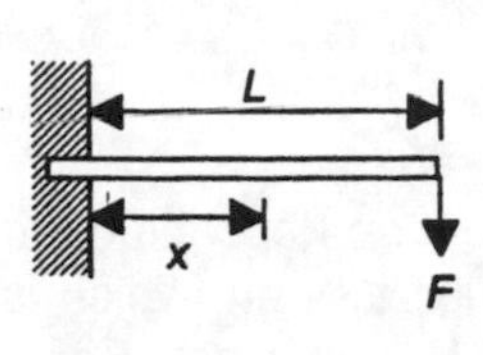

$$\frac{\mathrm{d}y}{\mathrm{d}x} = \frac{F}{EI}\left(Lx - \frac{x^2}{2}\right)$$

$$y = \frac{Fx^2}{6EI}(3L - x)$$

$$\left(\frac{\mathrm{d}y}{\mathrm{d}x}\right)_{max} = \frac{FL^3}{3EI}$$

at free end

$$y_{max} = \frac{FL^2}{2EI}$$

at free end

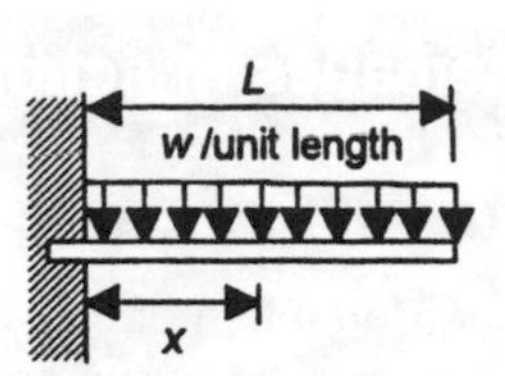

$$\frac{\mathrm{d}y}{\mathrm{d}x} = \frac{w}{6EI}(3L^2x - 3Lx^2 + x^3$$

$$y = \frac{wx^2}{24EI}(6L^2 - 4Lx + x^2)$$

$$\left(\frac{\mathrm{d}y}{\mathrm{d}x}\right)_{max} = \frac{wL^3}{6EI}$$

at free end

$$y = \frac{wL^4}{8EI}$$

at free end

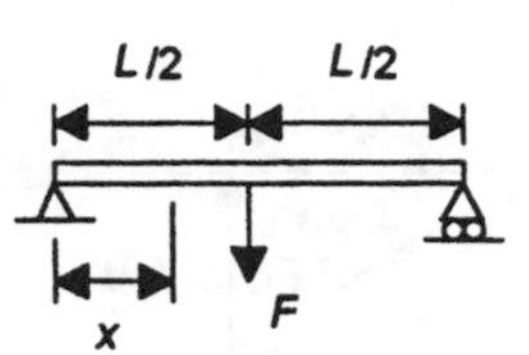

$$\frac{\mathrm{d}y}{\mathrm{d}x} = -\frac{F}{16EI}(4x^2 - L^2)$$

$$y = \frac{Fx}{48EI}(3L^2 - 4x^2)$$

$$\left(\frac{\mathrm{d}y}{\mathrm{d}x}\right)_{max} - \frac{FL^2}{16EI}$$

at ends

$$y_{max} = \frac{FL^3}{48EI}$$

at mid span

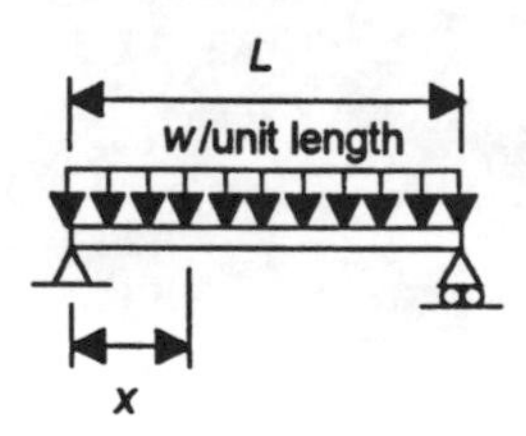

$$\frac{\mathrm{d}y}{\mathrm{d}x} = -\frac{w}{24EI}(6Lx^2 + 4wx^3 + L^3)$$

$$y = \frac{wx}{24EI}(x^3 - 2Lx^2 + L^3)$$

$$\left(\frac{\mathrm{d}y}{\mathrm{d}x}\right)_{max} = \frac{wL^3}{24EI}$$

at ends

$$y_{max} = \frac{5wL^4}{384EI}$$

at midpoint

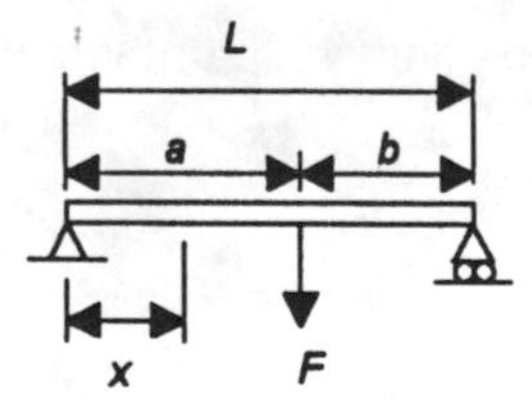

For $0 \le x \le a$

$$y = \frac{Fbx}{6EIL}(L^2 - b^2 - x^2)$$

For $a \le x \le b$

$$y = \frac{Fbx}{6EIL}(L^2 - b^2 - x^2) + \frac{F}{6EI}(x - a)^3$$

$$y_{max} = \frac{Fb}{9\sqrt{3}\,EIL}(L^2 - b^2)^{3/2}$$

at

$$x = \sqrt{\frac{L^2 - b^2}{3}}$$

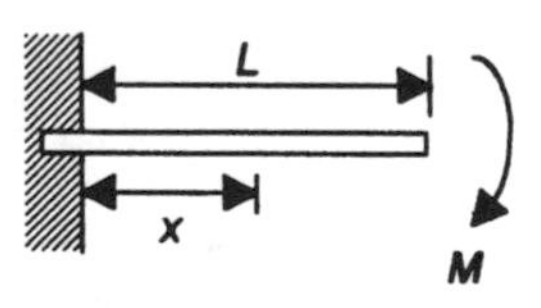

$$\frac{\mathrm{d}y}{\mathrm{d}x} = \frac{Mx}{EI}$$

$$y = \frac{Mx^2}{2EI}$$

$$\left(\frac{\mathrm{d}y}{\mathrm{d}x}\right)_{max} = \frac{ML}{EI}$$

at free end

$$y_{max} = \frac{ML^2}{2EI}$$

at free end

Figure 5.4.34 *Slopes and deflections of beams*

This differential equation provides the means by which the deflections of beams can be determined. The product EI is often termed the *flexural rigidity* of a beam.

The sign convention that is generally used is that deflections are measured in a downward direction and defined as positive and thus the bending moment is negative, since there is hogging. Hence, with this convention, the above equation is written as:

$$M = -EI\frac{d^2y}{dx^2} \tag{5.4.16}$$

Figure 5.4.34 lists solutions of the above equation for some common beam loading situations.

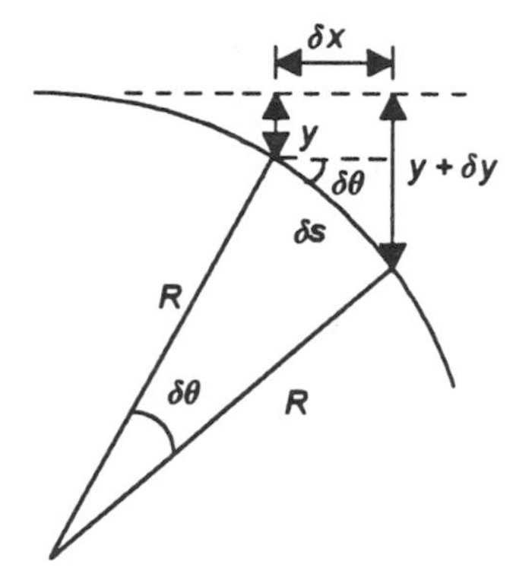

Figure 5.4.35 *The deflection curve of radius R*

Mathematics in action

The deflection curve

This is a derivation of the differential equation describing the curvature of a beam when bent. Consider a beam which is bent into a circular arc and the radius R of the arc. For a segment of circular arc (Figure 5.4.35), the angle $\delta\theta$ subtended at the centre is related to the arc length δs by $\delta s = R\,\delta\theta$. Because the deflections obtained with beams are small δx is a reasonable approximation to δs and so we can write $\delta x = R\,\delta\theta$ and $1/R = \delta\theta/\delta x$. The slope of the straight line joining the two end points of the arc is $\delta y/\delta x$ and thus $\tan\delta\theta = \delta y/\delta x$. Since the angle will be small we can make the approximation that $\delta\theta = \delta y/\delta x$. Hence:

$$\frac{1}{R} = \frac{\delta\theta}{\delta x} = \frac{\delta}{\delta x}\left(\frac{\delta y}{\delta x}\right)$$

In the limit we can thus write:

$$\frac{1}{R} = \frac{d^2y}{dx^2} \tag{5.4.17}$$

Deflections by double integration

The deflection y of a beam can be obtained by integrating the differential equation $d^2y/dx^2 = -M/EI$ with respect to x to give:

$$\frac{dy}{dx} = -\frac{1}{EI}\int M\,dx + A \tag{5.4.18}$$

with A being the constant of integration and then carrying out a further integration with respect to x to give:

$$y = -\frac{1}{EI}\int\left[\int M\,dx + A\right] + B = -\frac{1}{EI}\int\int M\,dx + Ax + B \tag{5.4.19}$$

with B being a constant of integration.

At the point where maximum deflection occurs, the slope of the deflection curve will be zero and thus the point of maximum deflection can be determined by equating dy/dx to zero.

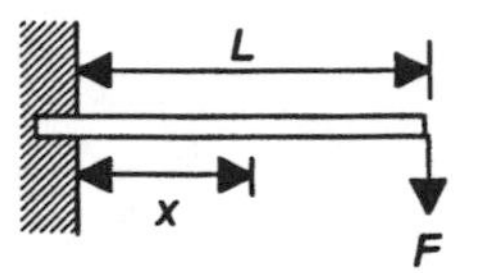

Figure 5.4.36 *Example 5.4.12*

Example 5.4.12

Show that the deflection y at the free end of a cantilever (Figure 5.4.36) of length L when subject to just a point load F located at the free end is given by $y = FL^3/3EI$.

The bending moment a distance x from the fixed end is $M = -F(L - x)$ and so the differential equation becomes:

$$EI\frac{d^2y}{dx^2} = -M = F(L - x)$$

Integrating with respect to x gives:

$$EI\frac{dy}{dx} = FLx - \frac{Fx^2}{2} + A$$

Since we have the slope $dy/dx = 0$ at the fixed end where $x = 0$, then we must have $A = 0$ and so:

$$EI\frac{dy}{dx} = FLx - \frac{Fx^2}{2}$$

Integrating again gives:

$$EIy = \frac{FI\,x^2}{2} - \frac{Fx^3}{6} + B$$

Since we have the deflection $y = 0$ at $x = 0$, then we must have $B = 0$ and so:

$$EIy = \frac{FLx^2}{2} - \frac{Fx^3}{6} = \frac{Fx^2}{6}(3L - x)$$

When $x = L$:

$$y = \frac{FL^3}{3EI}$$

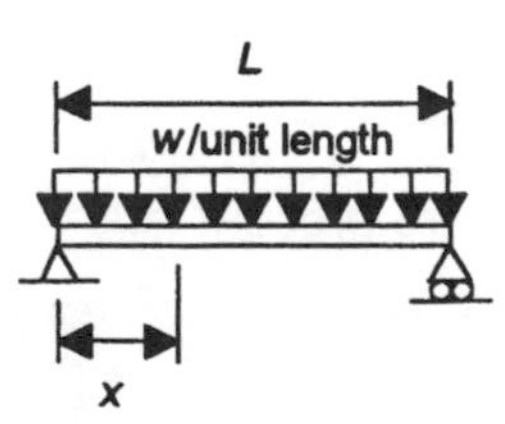

Figure 5.4.37 *Example 5.4.13*

Example 5.4.13

Show that the deflection y at the mid-span of a beam of length L supported at its ends (Figure 5.4.37) and subject to just a uniformly distributed load of w/unit length is given by $y = 5wL^4/384EI$.

The bending moment a distance x from the left-hand end is $M = \frac{1}{2}wLx - wx\frac{1}{2}x$ and so the differential equation becomes:

$$EI\frac{d^2y}{dx^2} = -M = -\frac{wLx}{2} + \frac{wx^2}{2}$$

Integrating with respect to x gives:

$$EI\frac{dy}{dx} = -\frac{wLx^2}{4} + \frac{wx^3}{6} + A$$

Since the slope $dy/dx = 0$ at mid-span where $x = \frac{1}{2}L$, then $A = wL^3/24$ and so:

$$EI\frac{dy}{dx} = -\frac{wLx^2}{4} + \frac{wx^3}{6} + \frac{wL^3}{24}$$

Integrating with respect to x gives:

$$EIy = -\frac{wLx^3}{12} + \frac{wx^4}{24} + \frac{wL^3x}{24} + B$$

Since $y = 0$ at $x = 0$, then $B = 0$ and so:

$$EIy = -\frac{wLx^3}{12} + \frac{wx^4}{24} + \frac{wL^3x}{24} = \frac{wx}{24}(x^3 - 2Lx^2 + L^3)$$

When $x = \frac{1}{2}L$:

$$y = \frac{5wL^4}{384EI}$$

Superposition

When a beam is subject to more than one load, while it is possible to write an equation for the bending moment and solve the differential equation, a simpler method is to consider the deflection curve and the deflection for each load separately and then take the algebraic sum of the separate deflection curves and deflections to give the result when all the loads are acting. The following example illustrates this.

Example 5.4.14

Determine the deflection curve for a cantilever of length L which is loaded by a uniformly distributed load of w/unit length over its entire length and a concentrated load of F at the free end.

For just a uniformly distributed load:

$$y = \frac{wx^2}{24EI}(6L^2 - 4Lx + x^2)$$

For just the concentrated load:

$$y = \frac{Fx^2}{6EI}(3L - x)$$

Thus the deflection curve when both loads are present is:

$$y = \frac{wx^2}{24EI}(6L^2 - 4Lx + x^2) + \frac{Fx^2}{6EI}(3L - x)$$

Macaulay's method

For beams with a number of concentrated loads, i.e. where there are discontinuities in the bending moment diagram, we cannot write a single bending moment equation to cover the entire beam but have to write separate equations for each part of the beam between adjacent loads. Integration of each such expression then gives the slope and deflection relationships for each part of the beam. An alternative is to write a single equation using *Macaulay's brackets*. The procedure is:

(1) Take the point from which distances are measured as the left-hand end of the beam. Denote the distance to a section as x.
(2) Write the bending moment expression at a section at the right-hand end of the beam in terms of the loads to the left of the section.
(3) Do not simplify the bending moment expression by expanding any of the terms involving the distance x which are contained within brackets. Indeed it is customary to use { } for such brackets since we will treat them differently from other forms of bracket.
(4) Integrate the bending moment equation, keeping all the bracketed terms within the { } brackets.
(5) Apply the resulting slope and deflection equations to any part of the beam but all terms within brackets which are negative are taken as being zero.

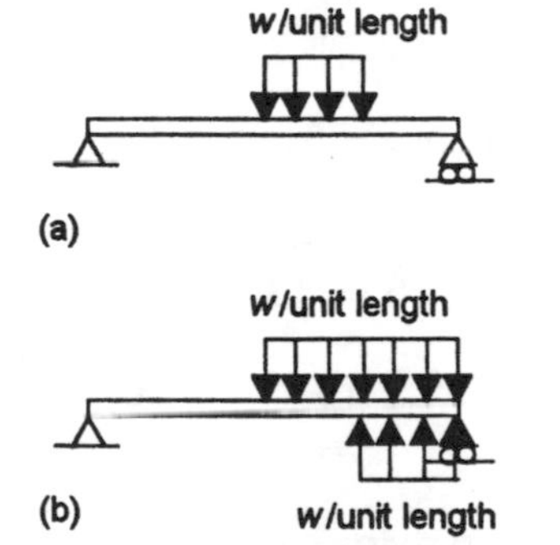

Figure 5.4.38 *Distributed load*

Where a distributed load does not extend to the right-hand end of the beam, as, for example, shown in Figure 5.4.38(a), then introduce two equal but opposite dummy loads, as in Figure 5.4.38(b). These two loads will have no net effect on the behaviour of the beam.

The following examples illustrate the use of this method to determine the deflections of beams.

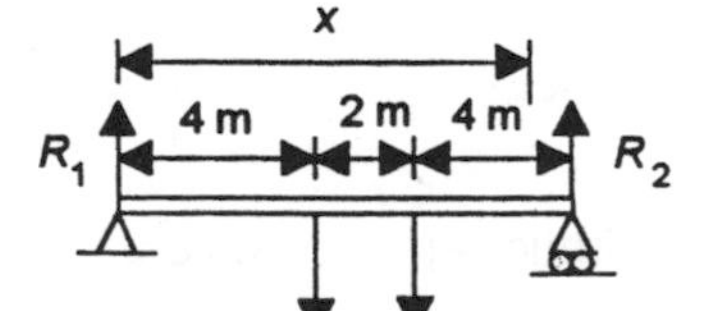

Figure 5.4.39 *Example 5.4.15*

Example 5.4.15

A beam of length 10 m is supported at its ends and carries a point load of 2 kN a distance of 4 m from the left-hand end and another point load of 5 kN a distance of 6 m from the left-hand end (Figure 5.4.39). Determine (a) the deflection under the 2 kN load, (b) the deflection under the 5 kN load and (c) the maximum deflection. The beam has a modulus of elasticity of 200 GPa and a second moment of area about the neutral axis of 10^8 mm^4.

Taking moments about the left-hand end gives $10R_2 = 4 \times 2 + 6 \times 5$ and hence $R_2 = 3.8$ kN. Since $R_1 + R_2 = 7$ then $R_1 = 3.2$ kN.

For a section near the right-hand end a distance x from the left-hand end, the bending moment for the section to the left $3.2x - 2\{x - 4\} - 5\{x - 6\}$. Hence:

$$EI\frac{d^2y}{dx^2} = -(3.2x - 2\{x-4\} - 5\{x-6\})$$

Integrating with respect to x gives:

$$EI\frac{dy}{dx} = -(1.6x^2 - \{x - 4\}^2 - 2.5\{x - 6\}^2 + A)$$

Integrating again with respect to x gives:

$$EIy = -(0.53x^3 - 0.33\{x - 4\}^3 - 0.83\{x - 6)^3 + Ax + B)$$

When $x = 0$, $y = 0$ and so $B = 0$. When $x = 10$, $y = 0$ and thus $0 = 0.53 \times 1000 - 0.33 \times 6^3 - 0.83 \times 4^3 + 10A$ and so $A = -40.6$.

$$EIy = -(0.53x^3 - 0.33\{x - 4\}^3 - 0.83\{x - 6)^3 - 40.6x)$$

(a) The deflection under the 2 kN load is at $x = 4$ m and thus we have:

$$EIy = -0.53 \times 4^3 + 40.6 \times 4 = 128.5$$

Hence $y = 128.5 \times 10^3/(200 \times 10^9 \times 10^8 \times 10^{-12}) = 8.5 \times 10^{-3}$ m = 8.5 mm.

(b) The deflection under the 5 kN load is at $x = 6$ m and thus we have:

$$EIy = -0.53 \times 6^3 + 0.33 \times 2^2 + 40.6 \times 6 = 131.8$$

Hence $y = 131.8 \times 10^3/(200 \times 10^9 \times 10^8 \times 10^{-12}) = 6.6 \times 10^{-3}$ m = 6.6 mm.

(c) The maximum deflection occurs when $dy/dx = 0$ and so:

$$1.6x^2 - \{x - 4\}^2 - 2.5\{x - 6\}^2 - 40.6 = 0$$

The maximum deflection can be judged to be between the two loads and thus we omit the $\{x - 6\}$ term to give:

$$1.6x^2 - (x^2 - 4x + 16) - 40.6 = 0$$

$$0.6x^2 + 8x - 56.6 = 0$$

Using the equation for roots of a quadratic equation gives:

$$x = \frac{-8 \pm \sqrt{64 + 4 \times 0.6 \times 56.6}}{1.2} = 5.1$$

Hence the maximum deflection is given by:

$$EIy = -0.53 \times 5.1^3 + 0.33 \times 1.1^3 + 40.6 \times 5.1 = 137.2$$

Note that in the above equation the forces have been in kN. Hence $y = 137.2 \times 10^3/(200 \times 10^9 \times 10^8 \times 10^{-12}) = 6.9 \times 10^{-3}$ m = 6.9 mm.

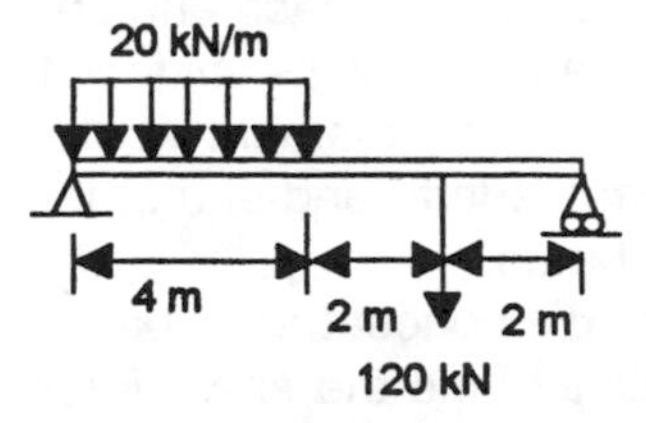

Figure 5.4.40 *Example 5.4.16*

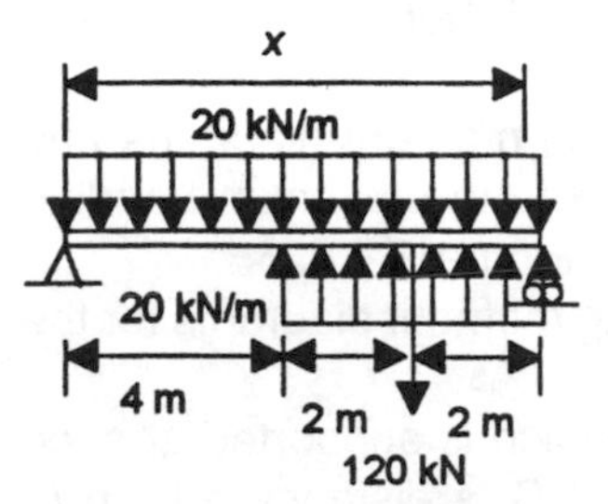

Figure 5.4.41 *Example 5.4.16*

Example 5.4.16

A beam of length 8 m is supported at both ends and subject to a distributed load of 20 kN/m over the 4 m length of the beam from the left-hand support and a point load of 120 kN at 6 m from the left-hand support (Figure 5.4.40). Determine the mid-span deflection. The beam has a modulus of elasticity of 200 GPa and a second moment of area about the neutral axis of $10^8\,\text{mm}^4$.

Taking moments about the left-hand end gives, for the right-hand end support reaction R_2, $8R_2 = 80 \times 2 + 120 \times 6$ and so $R_2 = 110\,\text{kN}$. Since $R + R_2 = 80 + 120$, then $R_1 = 90\,\text{kN}$.

To apply the Macaulay method we need to introduce two dummy distributed loads so that distributed loads are continuous from the right-hand end. Thus the loads are as in Figure 5.4.41. The bending moment a distance x from the left-hand end is $90x - \frac{1}{2}x \times 20x + \frac{1}{2}\{x-4\} \times 20\{x-4\} - 120\{x-6\}$. Hence we can write:

$$EI\frac{d^2y}{dx^2} = -(90x - 10x^2 + 10\,\{x-4\}^2 - 120\,\{x-6\})$$

Integrating with respect to x gives:

$$EI\frac{dy}{dx} = -\left(45x^2 - \frac{10}{3}x^3 + \frac{10}{3}\{x-4\}^3 - 60\{x-6\}^2 + A\right)$$

Integrating again with respect to x gives:

$$EIy = -\left(\frac{45}{3}x^3 - \frac{10}{12}x^4 + \frac{10}{12}\{x-4\}^4 - 20\{x-6\}^3 + Ax + B\right)$$

When $x = 0$ we have $y = 0$ and so $B = 0$. When $x = 8$ then $y = 0$ and so the above equation gives $0 = -(15 \times 8^3 - 10 \times 8^4/12 + 10 \times 4^4/12 + 8A)$. Hence $A = -540$ and so we have:

$$EIy = -\left(\frac{45}{3}x^3 - \frac{10}{12}x^4 + \frac{10}{12}\{x-4\}^4 - 20\{x-6\}^3 + 540x\right)$$

The deflection at mid-span, i.e. $x = 4$, is thus:

$$EIy = -15 \times 4^3 + 10 \times 4^4/12 + 540 \times 4 = 1413.3$$

Note that in the above equation the forces have been in kN. With $EI = 200 \times 10^9 \times 10^8 \times 10^{-12}$ then $y = 1413.3 \times 10^3/(200 \times 10^5) = 0.071\,\text{m}$.

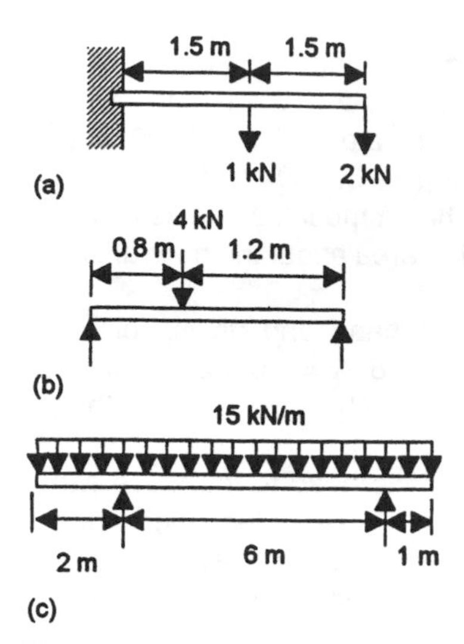

Figure 5.4.42 *Problem 4*

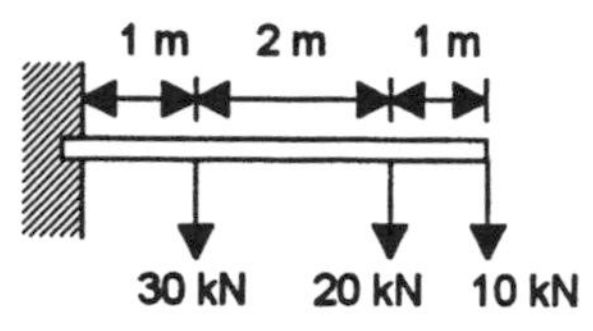

Figure 5.4.43 *Problem 5*

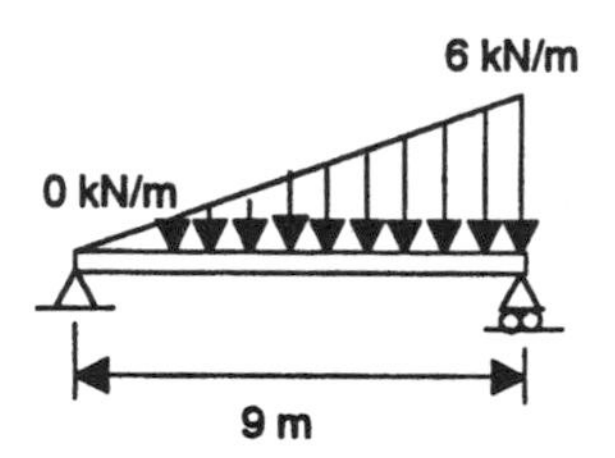

Figure 5.4.44 *Problem 10*

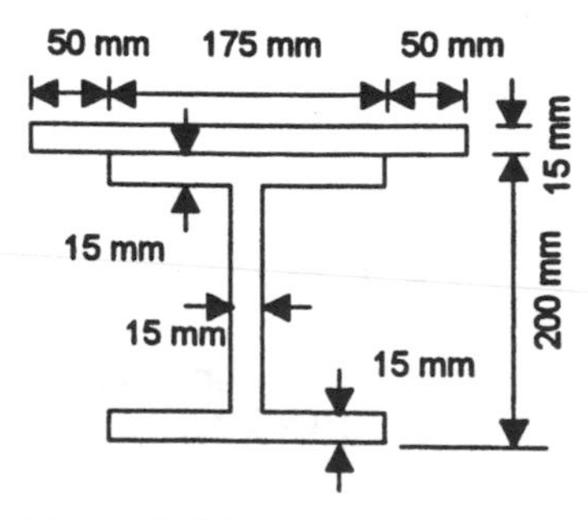

Figure 5.4.45 *Problem 19*

Problems 5.4.1

(1) A beam of length 4.0 m rests on supports at each end and a concentrated load of 100 N is applied at its midpoint. Determine the shear force and bending moment at distances of (a) 3.0 m, (b) 1.5 m from the left-hand end of the beam. Neglect the weight of the beam.

(2) A cantilever has a length of 2 m and a concentrated load of 1 kN is applied to its free end. Determine the shear force and bending moment at distances of (a) 1.5 m, (b) 1.0 m from the free end. Neglect the weight of the beam.

(3) A cantilever of length 4.0 m has a uniform weight per metre of 1 kN. Determine the shear force and bending moment at (a) 2.0 m, (b) 1.0 m from the free end. Neglect the weight of the beam.

(4) Draw the shear force and bending moment diagrams and determine the position and size of the maximum bending moment for the beams shown in Figure 5.4.42.

(5) Draw the shear force and bending moment diagrams for the beam shown in Figure 5.4.43.

(6) For a beam AB of length $a + b$ which is supported at each end and carries a point load F at C, a distance a from A, show that (a) the shear force a distance x from A between A and C is $Fb/(a + b)$ and the bending moment $Fbx/(a + b)$, (b) between C and B the shear force is $-Fa/(a + b)$ and the bending moment $Fa(a + b - x)/(a + b)$, (c) the maximum bending moment is $Fab/(a + b)$.

(7) A beam AB has a length of 9 m and is supported at each end. If it carries a uniformly distributed load of 6 kN/m over the 6 m length starting from A, by sketching the shear force and bending moment diagrams determine the maximum bending moment and its position.

(8) A beam AB has a length of 4 m and is supported at 0.5 m from A and 1.0 m from B. If the beam carries a uniformly distributed load of 6 kN/m over its entire length, by sketching the shear force and bending moment diagrams determine the maximum bending moment and its position.

(9) A beam AB has a length of 8 m and is supported at end A and a distance of 6 m from A. The beam carries a uniformly distributed load of 10 kN/m over its entire length and a point load of 80 kN a distance of 3 m from A and another point load of 65 kN a distance of 8 m from A. By sketching the shear force and bending moment diagrams determine the maximum bending moment and its position.

(10) By sketching the shear force and bending moment diagrams, determine the maximum bending moment and its position for the beam in Figure 5.4.44 where there is a non-uniform distributed load, the load rising from 0 kN/m at one end to 6 kN/m at the other end.

(11) A uniform rectangular cross-section horizontal beam of length 6 m and depth 400 mm rests on supports at its ends and supports a uniformly distributed load of 20 kN/m. Determine the maximum tensile and compressive stresses in the beam if it has a second moment of area of 140×10^6 mm^4.

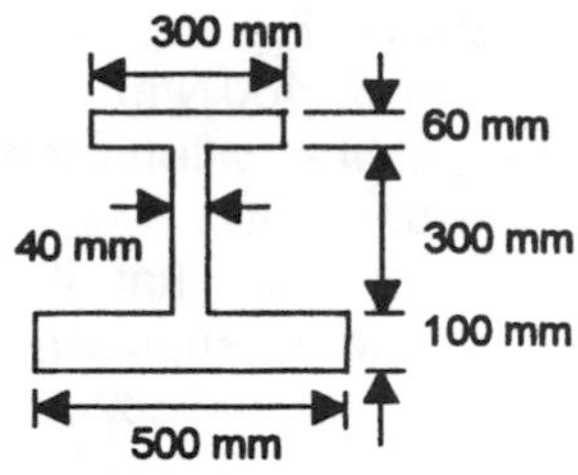

Figure 5.4.46 *Problem 20*

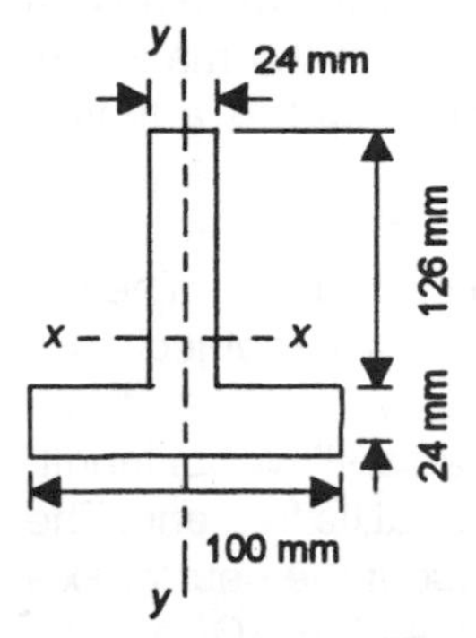

Figure 5.4.47 *Problem 21*

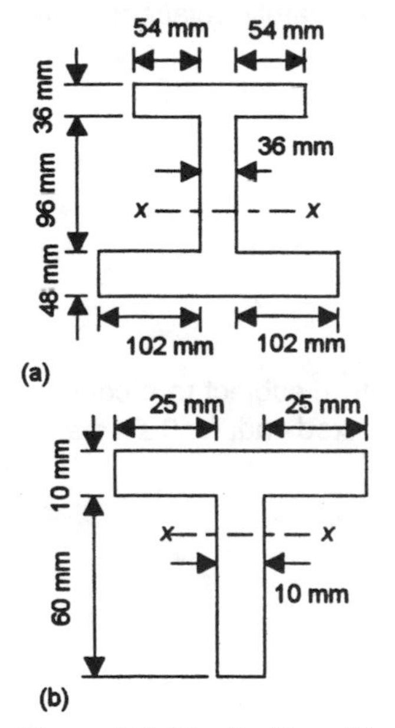

Figure 5.4.48 *Problem 24*

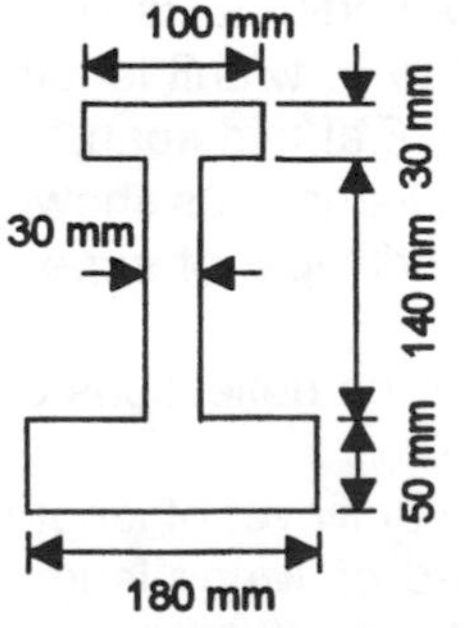

Figure 5.4.49 *Problem 26*

(12) A steel tube has a section modulus of 5×10^{-6} m^3. What will be the maximum allowable bending moment on the tube if the bending stresses must not exceed 120 MPa?

(13) An I-section beam has a section modulus of 3990 cm^3. What will be the maximum allowable bending moment on the beam if the bending stresses must not exceed 120 MPa?

(14) Determine the section modulus required of a steel beam which is to span a gap of 6 m between two supports and support a uniformly distributed load over its entire length, the total distributed load being 65 kN. The maximum bending stress permissible is 165 MPa.

(15) An I-section girder has a length of 6 m, a depth of 250 mm and a weight of 30 kN/m. If it has a second moment of area of 120×10^{-6} m^4 and is supported at both ends, what will be the maximum stress?

(16) A rectangular section beam has a length of 4 m and carries a uniformly distributed load of 10 kN/m. If the beam is to have a width of 120 mm, what will be the maximum permissible beam depth if the maximum stress in the beam is to be 80 MPa?

(17) Determine the second moment of area of an I-section, about its horizontal neutral axis when the web is vertical, if it has rectangular flanges each 144 mm by 18 mm, a web of thickness 12 mm and an overall depth of 240 mm.

(18) Determine the second moment of area of a rectangular section 100 mm × 150 mm about an axis parallel to the 100 mm side if the section has a central hole of diameter 50 mm.

(19) A compound girder is made up of an I-section with a plate on the top flange, as shown in Figure 5.4.45. Determine the position of the neutral axis above the base of the section.

(20) Determine the position of the neutral axis above the base of the I-section shown in Figure 5.4.46.

(21) Determine the second moment of areas about the *xx* and *yy* axes through the centroid for Figure 5.4.47.

(22) Determine the radius of gyration of a section which has a second moment of area of 2.5×10^6 mm^4 and an area of 100 mm^2.

(23) Determine the radius of gyration about an axis passing through the centroid for a circular area of diameter d and in the plane of that area.

(24) Determine the position of the neutral axis *xx* and the second moment of area I_{xx} about the neutral axis for the sections shown in Figure 5.4.48.

(25) Determine the second moment of area of a rectangular section tube of external dimensions 200 mm × 300 mm and wall thickness 24 mm about the axes parallel to the sides and through the centroid.

(26) A steel beam of length 12 m has the uniform section shown in Figure 5.4.49 and rests on supports at each end. Determine the maximum stress in the beam under its own weight if the steel has a density of 7900 kg/m^3.

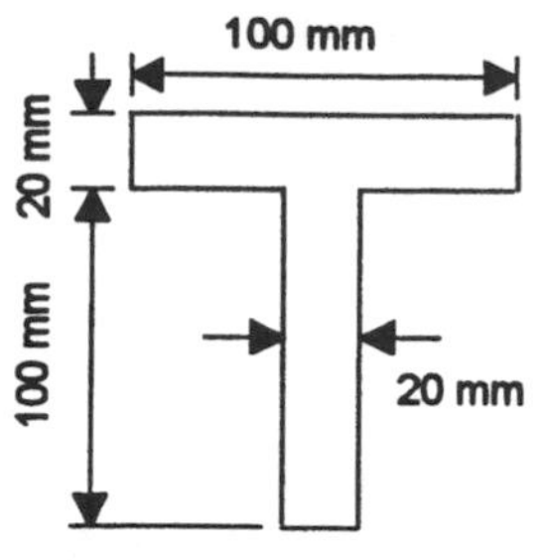

Figure 5.4.50 *Problem 27*

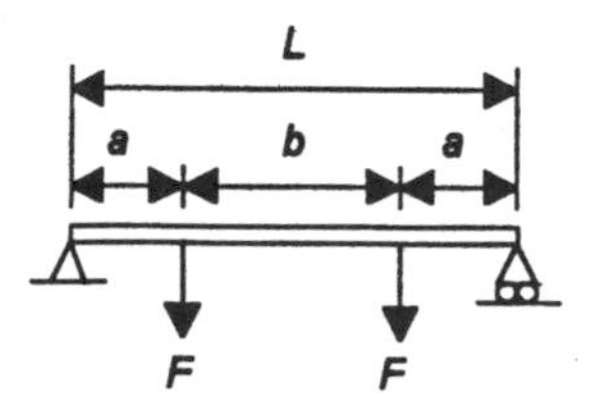

Figure 5.4.51 *Problem 35*

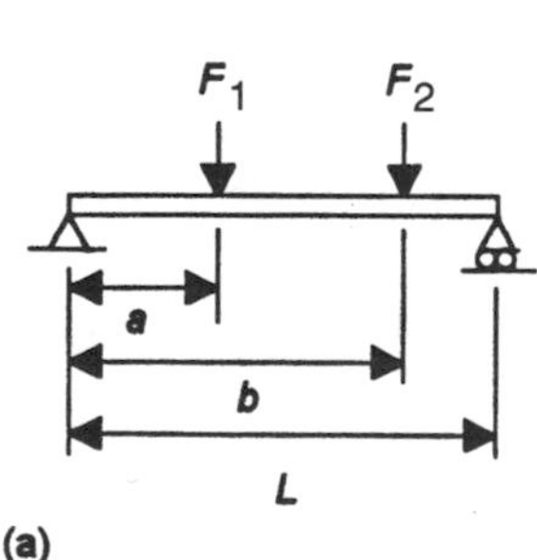

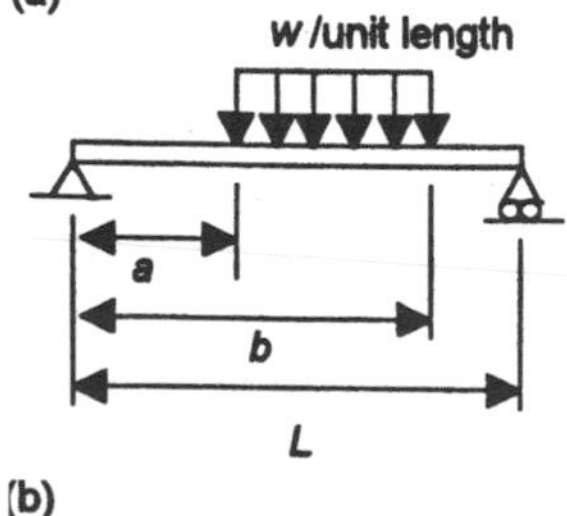

Figure 5.4.52 *Problem 36*

(27) A cantilever has the uniform section shown in Figure 5.4.50 and a length of 1.8 m. If it has a weight of 400 N/m and supports a load of 160 N at its free end, what will be the maximum tensile and compressive stresses in it?

(28) Show, by the successive integration method, that the deflection y at the mid-span of a beam of length L supported at its ends and subject to just a point load F at mid-span is given by $y = FL^3/48EI$.

(29) A circular beam of length 5 m and diameter 100 mm is supported at both ends. Determine the maximum value of a central span point load that can be used if the maximum deflection is to be 6 mm. The beam material has a tensile modulus of 200 GPa.

(30) A rectangular cross-section beam 25 mm × 75 mm has a length of 3 m and is supported at both ends. Determine the maximum deflection of the beam when it is subject to a distributed load of 2500 N/m. $E = 200$ GPa.

(31) Determine the maximum deflection of a cantilever of length 3 m when subject to a point load of 50 kN at its free end. The beam has a second moment of area about the neutral axis of 300×10^6 mm^4 and a tensile modulus of 200 GPa.

(32) Show that for a cantilever of length L subject to a uniformly distributed load of a length a immediately adjacent to the fixed end, for $0 \le x \le a$:

$$y = \frac{wx^2}{24EI}(6a^2 - 4ax + x^2)$$

and for $a \le x \le L$:

$$y = \frac{wa^3}{24EI}(4x - a)$$

(33) Show that for a cantilever of length L subject to a concentrated load F a distance a from the fixed end, for $0 \le x \le a$:

$$y = \frac{Fx^2}{6EI}(3a - x)$$

and for $a \le x \le L$:

$$y = \frac{Fa^2}{6EI}(3x - a)$$

(34) Determine, using the principle of superposition, the mid-span deflection of a beam of length L supported at its ends and subject to a uniformly distributed load of w/unit length over its entire length and a point load of F at mid-span.

(35) A beam of length L is subject to two point loads F, as shown in Figure 5.4.51. Determine, using the principle of superposition, the deflection curve.

(36) Determine the Macaulay expressions for the deflections of the beams shown in Figure 5.4.52.

(37) Determine the maximum deflection of a cantilever of length L which has a uniformly distributed load of w/unit length extending from the fixed end to the midpoint of the beam.

(38) Determine the mid-span deflection of a beam of length 8 m which is supported at each end and carries a point load of 20 kN a distance of 2 m from the left-hand end, another point load of 40 kN a distance of 6 m from the left-hand end and a distributed load of 15 kN/m over the length of the beam between the point loads. $EI = 36 \times 10^6$ N m^2.

(39) Determine the mid-span deflection of a beam of length L which is supported at each end and carries a uniformly distributed load of w/unit length from one-quarter span to three-quarter span.

5.5 Cables

Flexible cables are structural members that are used in such applications as suspension bridges, electrical transmission lines and telephone cables. Such cables may support a number of concentrated loads and/or a distributed load. The following is an analysis of such situations. It should be noted that cables are flexible and so can only resist tensile forces.

A cable with a point load

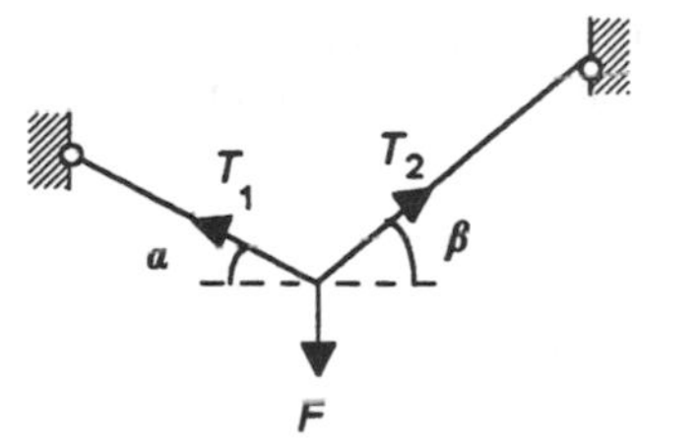

Figure 5.5.1 *Cable supporting a point load*

Consider a cable supporting a single point load (Figure 5.5.1) and assume the cable itself is so light that its weight is negligible. The points at which the cable are tethered are not assumed to be on the same level. The tensile forces in the two sections of the cable can be determined by resolving the forces at the load point. Thus:

$$T_1 \cos \alpha = T_2 \cos \beta \tag{5.5.1}$$

$$F = T_1 \sin \alpha + T_2 \sin \beta \tag{5.5.2}$$

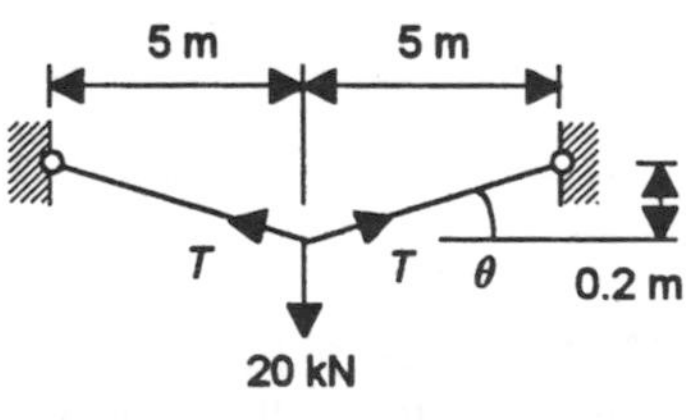

Figure 5.5.2 *Example 5.5.1*

> **Example 5.5.1**
>
> A cable of negligible mass is tethered between two points at the same height and 10 m apart (Figure 5.5.2). When a load of 20 kN is suspended from its midpoint, it sags by 0.2 m. Determine the tension in the cable.
>
> Because there is symmetry about the midpoint, the tension will be the same in each half of the cable. Thus $2T \sin \theta = 2T[0.2/\sqrt{(5^2 + 0.2^2)}] = 20$ and so $T = 250$ kN.

Cable with distributed load

In the above discussion there was just a single point load; the cable was thus assumed to form straight lines between the supports and the point load. Suppose, however, we have a distributed load over the length of the cable. This type of loading is one that can be assumed to occur with suspension bridges where a uniform weight per length bridge desk is

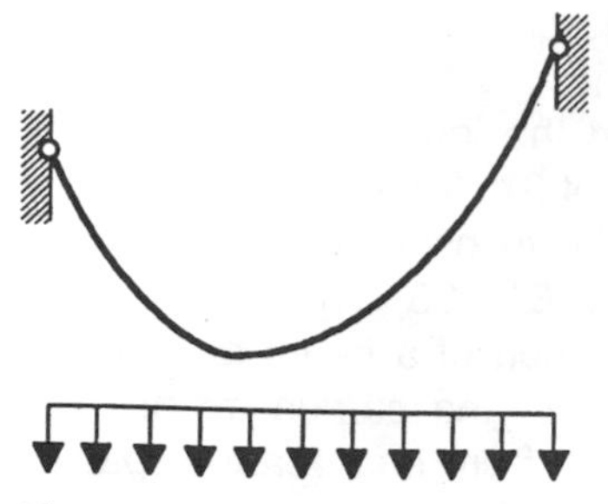

Figure 5.5.3 *Cable supporting a distributed load*

suspended from numerous points along the length of the cable and gives a uniformly distributed load per unit horizontal length (Figure 5.5.3); the weight of the cable is not distributed uniformly with horizontal distance but is neglected. A cable which sags under its own weight will, however, have a uniform load per unit length of cable and not per unit horizontal distance. These two types of distributed loading will result in the cables sagging into different shapes. The curve in which a flexible cable sags under its own weight is called a *catenary*; the curve into which it will sag under a uniform distributed load per unit horizontal distance is a *parabola*.

Uniform distributed load per unit horizontal distance

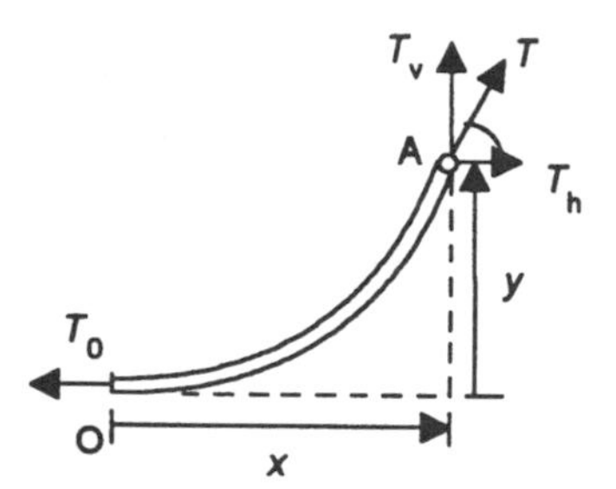

Figure 5.5.4 *Segment of cable*

The following is a derivation of an equation to give the tension in a cable subject to a uniform distributed load per unit horizontal distance w and suspended from two fixed points at the same height. The system is symmetrical and Figure 5.5.4 shows the forces acting on a section of the cable between a fixture point A and the midpoint O. The vertical component T_v of the tension T at the point of suspension B must be half the distributed load acting on the entire cable and so wx. For equilibrium we must have the sum of the clockwise moments on the segment of cable equal to the anticlockwise moments and thus, taking moments about O, $wx \times x/2 + T_h y = T_v x = (wx)x$. Hence $T_h = wx^2/2y$. Hence the tension in the cable is:

$$T = \sqrt{(wx)^2 + \left(\frac{wx^2}{2y}\right)^2} = wx\sqrt{1 + \left(\frac{x}{2y}\right)^2} \qquad (5.5.3)$$

The above equation gives the tension for a cable subject to a uniform loading per unit horizontal distance; where a cable is subject to a uniform loading per unit length, e.g. its own weight, then the above equation gives a reasonable approximation when the cable sags only a small amount and the loading per unit length is effectively the same as the loading per unit horizontal distance.

Example 5.5.2

A light cable supports a load of 100 N per metre of horizontal length and is suspended between two points on the same level 240 m apart. If the sag is 12 m, determine the tension at the fixture points.

$$T = wx\sqrt{1 + \left(\frac{x}{2y}\right)^2}$$

$$= 100 \times 120\sqrt{1 + \left(\frac{120}{2 \times 12}\right)^2} = 61 \text{ kN}$$

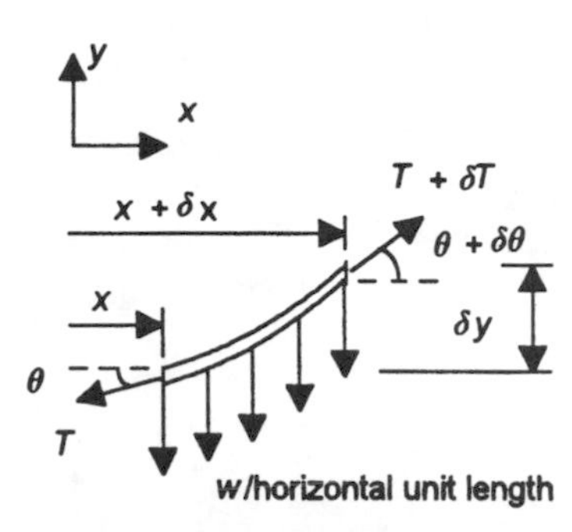

Figure 5.5.5 *Element of cable*

Mathematics in action

Cables with distributed loads

The following is a more general derivation of the relationships for cables when subject to distributed loads. It involves deriving a differential equation and then solving it for different loading conditions.

Consider the free-body diagram for a small element of a cable with a distributed load (Figure 5.5.5). The element has a length δx and is a distance x from the point of maximum sag. The load on the element is $w\delta x$. Equilibrium of the vertical components of the forces gives:

$$(T + \delta T)\sin(\theta + \delta\theta) - T\sin\theta - w\,\delta x = 0$$

Using the relationship $\sin(A + B) = \sin A\cos B + \cos A\sin B$ and making the approximation $\sin\delta\theta = \delta\theta$ and $\cos\delta\theta = 1$:

$$(T + \delta T)(\sin\theta + \cos\theta\,\delta\theta) - T\sin\theta - w\,\delta x = 0$$

Neglecting $\delta\theta\,\delta T$ term then gives:

$$T\cos\theta\,\delta\theta + \delta T\sin\theta - w\,\delta x = 0$$

Since $d(T\sin\theta)/d\theta = T\cos\theta + \sin\theta\,dT/d\theta$ then the above expression can be written as:

$$\delta(T\sin\theta) = w\,\delta x$$

Equilibrium of the horizontal components gives:

$$(T + \delta T)\cos(\theta + \delta\theta) - T\cos\theta = 0$$

Using the relationship $\cos(A + B) = \cos A\cos B - \sin A\sin B$ and making the approximation $\sin\delta\theta = \delta\theta$ and $\cos\delta\theta = 1$:

$$(T + \delta T)(\cos\theta - \sin\theta\,\delta\theta) - T\cos\theta = 0$$

Neglecting $\delta\theta\,\delta T$ term then gives:

$$-T\sin\theta\,\delta\theta + \delta T\cos\theta = 0$$

If we let T_0 equal the horizontal component of the tension in the cable then $T_0 = T\cos\theta$ and $dT_0/d\theta = -T\sin\theta + \cos\theta\,dT/d\theta$. Thus the above expression is just stating that the horizontal component of the tension remains unchanged.

Substituting $T = T_0/\cos\theta$ in $\delta(T\sin\theta) = w\,\delta x$ gives:

$$\delta(T_0\tan\theta) = w\,\delta x$$

As T_0 is constant:

$$T_0\,\delta(\tan\theta) = w\,\delta x$$

But $\tan\theta = \delta y/\delta x$ and so, in the limit, we can write:

$$\frac{d^2y}{dx^2} = \frac{w}{T_0} \qquad (5.5.4)$$

This is the basic differential equation which can be solved for the different types of loading conditions.

Consider the solution of this differential equation for a cable that has a uniformly distributed load per unit horizontal length. Integration with respect to x gives:

$$\frac{dy}{dx} = \frac{wx}{T_0} + A$$

If we choose the co-ordinate axes such that the point of maximum sag is the origin then $dy/dx = 0$ at $x = 0$ and so $A = 0$. Integrating again with respect to x gives:

$$y = \frac{wx^2}{2T_0} + B$$

$y = 0$ at $x = 0$ and so $B = 0$. Thus:

$$y = \frac{wx^2}{2T_0} \qquad (5.5.5)$$

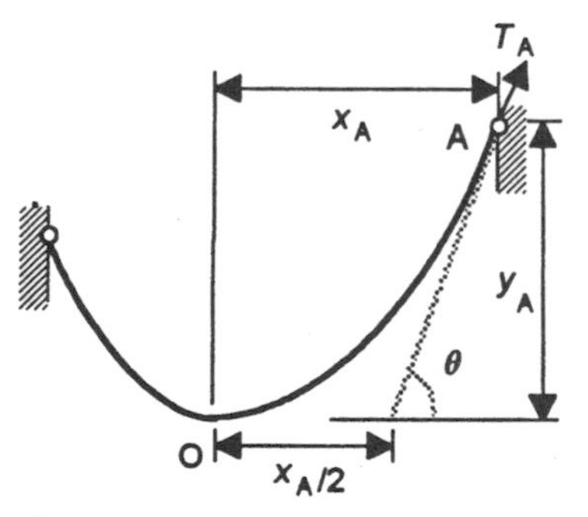

Figure 5.5.6 *Cable supporting a distributed load*

This is the equation of a parabola. The maximum tension in the cable will occur at the end A of the cable on the side the greater distance from the point of maximum sag since it will be supporting a greater proportion of the load (Figure 5.5.6). The tension T_A in the cable at A will have a vertical component $T_A \sin\theta$ which must equal the total load on the cable in the distance x_A. Since we have a uniform load per horizontal distance we have $T_A \sin\theta = wx_A$. But, for a parabola:

$$\sin\theta = \frac{y_A}{\sqrt{y_A^2 + (x_A/2)^2}}$$

Hence:

$$T_A \frac{y_A}{\sqrt{y_A^2 + (x_A/2)^2}} = wx_A$$

$$T_A = wx_A\sqrt{1 + \left(\frac{x_A}{2y_A}\right)^2} \qquad (5.5.6)$$

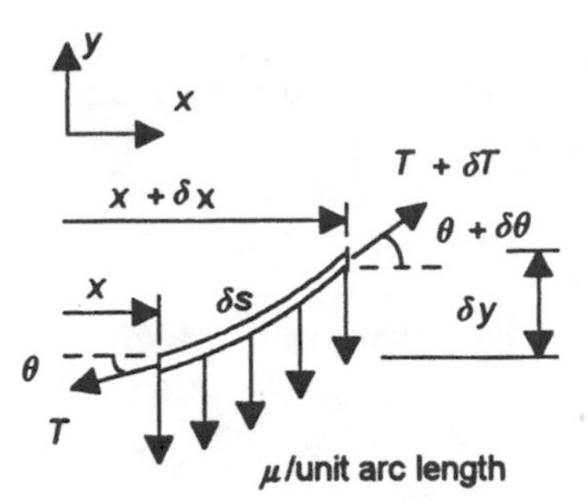

Figure 5.5.7 *Element of cable*

Now consider a cable with a uniform weight μ per unit length. For an element of this cable of length δs (Figure 5.5.7) the vertical load is $\mu\ \delta s$. If this arc gives a horizontal displacement of δx the vertical load per unit horizontal distance is $\mu\ \delta s/\delta x$. Thus the differential equation becomes:

$$\frac{d^2y}{dx^2} = \frac{\mu}{T_0}\frac{ds}{dx} \quad (5.5.7)$$

To a reasonable approximation $(\delta s)^2 = (\delta x)^2 + (\delta y)^2$ and so:

$$\left(\frac{ds}{dx}\right)^2 = 1 + \left(\frac{dy}{dx}\right)^2$$

and thus we can write:

$$\frac{d^2y}{dx^2} = \frac{\mu}{T_0}\sqrt{1 + \left(\frac{dy}{dx}\right)^2}$$

If we let $p = dy/dx$ then we can write the above equation as:

$$\frac{1}{\sqrt{1 + p^2}}\frac{dp}{dx} = \frac{\mu}{T_0}$$

To integrate with respect to x we can use the substitution $p = \sinh u$ and since $1 + \sinh^2 u = \cosh^2 u$ and $dp/du = \cosh u$ we obtain:

$$\frac{1}{\cosh u}\cosh u\frac{du}{dx} = \frac{\mu}{T_0}$$

$$u = \frac{\mu}{T_0}x + A$$

$$p = \sinh\left(\frac{\mu}{T_0}x\right) + A$$

Since $dy/dx = p = 0$ when $x = 0$ then $A = 0$. Thus we have:

$$\frac{dy}{dx} = \sinh\left(\frac{\mu}{T_0}x\right)$$

Integrating with respect to x gives:

$$y = \frac{T_0}{\mu}\cosh\left(\frac{\mu}{T_0}x\right) + B$$

When $x = 0$ we have $y = 0$ and thus $B = -T_0/\mu$ to give the equation of the curve of the cable as:

$$y = \frac{T_0}{\mu}\left(\cosh\frac{\mu}{T_0}x - 1\right) \quad (5.5.8)$$

This is the equation of a catenary.

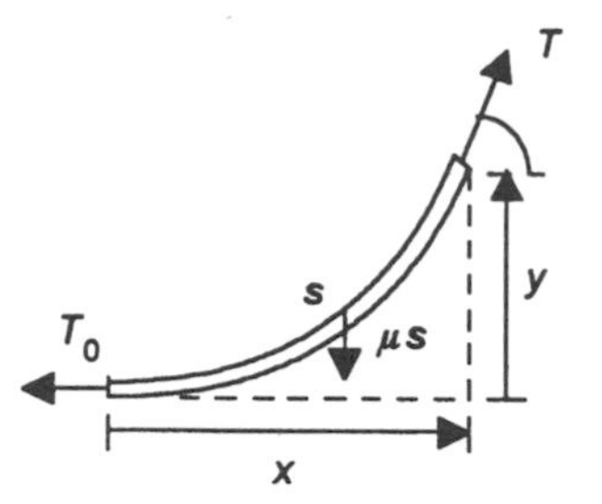

Figure 5.5.8 *Segment of cable*

For a segment of the cable measured from the lowest sag point where the tension is horizontal and T_0 (Figure 5.5.8), then for equilibrium the horizontal component of the tension T must equal T_0, i.e. $T\cos\theta = T_0$. The vertical component must equal the weight of the element of cable and so $T\sin\theta = \mu s$. Hence $\tan\theta = \mu s/T_0$. To a reasonable approximation $dy/dx = \tan\theta$ and thus we can write $dy/dx = \mu s/T_0$. But $dy/dx = \sinh(\mu x/T_0)$ and so:

$$s = \frac{T_0}{\mu}\sinh\frac{\mu}{T_0}x$$

Since $T\cos\theta = T_0$ and $T\sin\theta = \mu s$, then:

$$(T\cos\theta)^2 + (T\sin\theta)^2 = T^2 = T_0^2 + (\mu s)^2$$

$$T^2 = T_0^2 + \left(T_0\sinh\frac{\mu x}{T_0}\right)^2 = T_0^2\left(1 + \sinh^2\frac{\mu x}{T_0}\right)$$

$$= T_0^2\cosh^2\frac{\mu x}{T_0}$$

$$T = T_0\cosh\frac{\mu x}{T_0}$$

Using the equation derived above for y (5.5.8), we can write this equation as:

$$T = T_0 + \mu y \qquad (5.5.9)$$

Example 5.5.3

A cable is used to support a uniform loading per unit horizontal distance between the support points of 120 N/m. If the two support points are at the same level, a distance of 300 m apart, and the cable sags by 60 m, what will be the tension in the cable at its lowest sag point and at the supports?

The uniform loading per unit horizontal distance means that the cable will assume a parabolic curve. Using $y = wx^2/2T_0$ for $y = 60$ m we have the tension at the lowest sag point $T_0 = 120 \times 150^2/(2 \times 60) = 22\,500$ N $= 22.5$ kN. The tension at the supports is given by:

$$T = wx\sqrt{1 + \left(\frac{x}{2y}\right)^2} = 120 \times 150\sqrt{1 + \left(\frac{150}{2 \times 60}\right)^2}$$

and so is 28.8 kN.

Example 5.5.4

A cable has a weight per unit length of 25 N/m and is suspended between two points at the same level, a distance of 20 m apart, and the sag at its midpoint is 4 m. Determine the tension in the cable at its midpoint and at its fixed ends.

$$y = \frac{T_0}{\mu}\left(\cosh \frac{\mu}{T_0} x - 1\right)$$

$$4 = \frac{T_0}{25}\left(\cosh \frac{25 \times 10}{T_0} - 1\right)$$

We might guess a value and try it. Suppose we try $T_0 = 200$ N, then, using the cosh function on a calculator, we obtain $8(1.9 - 1) = 7.2$. With $T_0 = 250$ N we obtain $10(1.5 - 1) = 5.4$. With $T_0 = 300$ N we obtain $12(1.37 - 1) = 4.4$. With 350 N we obtain $14(1.27 - 1) = 3.7$. With 325 N we obtain $13(1.31 - 1) = 4.0$. The tension is thus 325 N. For the tension at a fixed end we can use $T = T_0 + \mu y = 325 + 25 \times 4 = 425$ N.

Problems 5.5.1

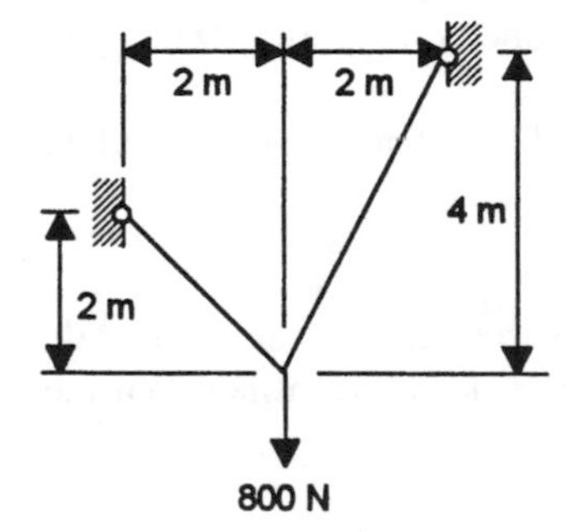

Figure 5.5.9 *Problem 1*

(1) A point load of 800 N is suspended from a cable arranged as shown in Figure 5.5.9. Determine the tensions in the cable either side of the load.

(2) A light cable spans 100 m and carries a uniformly distributed load of 10 kN per horizontal metre. Determine the tension at the fixture points if the sag is 15 m.

(3) A suspension bridge has a span of 120 m between two equal height towers to which the suspension cable is attached. The cable supports a uniform load per horizontal unit distance of 12 N/m and the maximum sag is 8 m. Determine the minimum and maximum tensions in the cable.

(4) Determine the minimum tension in a cable which has a weight per unit length of 120 N/m and spans a gap of 300 m, the points at which the cable is suspended being at the same level and 60 m above the lowest point of the cable.

(5) A light cable spans 100 m and carries a uniformly distributed load of 10 kN per horizontal metre. If the maximum tension allowed for the cable is 1000 kN, what will be the maximum sag that can be permitted?

(6) A light cable spans 90 m and carries a uniformly distributed load of 5 kN per horizontal metre. What will be the maximum tension in the cable if it has a sag of 4 m?

(7) A suspension bridge has a span of 130 m between two towers to which the suspension cable is attached, one point of suspension being 8 m and the other 4 m above the lowest point of the cable. The cable supports a uniform load per horizontal unit distance of 12 kN/m. Determine the minimum and maximum tensions in the cable.

(8) A cable is suspended from two equal height points 100 m apart and has a uniform weight per unit length of 150 N/m. If the sag in the cable is 0.3 m, what is the minimum cable tension?

5.6 Friction

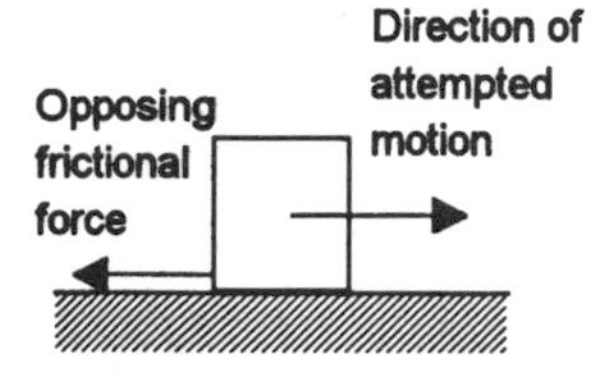

Figure 5.6.1 *Frictional force opposes attempted motion*

If you try to push a box across a horizontal floor and only apply a small force then nothing is likely to happen. So, since there is no resultant force giving acceleration, there must be another force which cancels out your push. The term *frictional force* is used to describe this tangential force that arises when two bodies are in contact with one another and that is in a direction that opposes the motion of one body relative to the other (Figure 5.6.1). If you apply a big enough force to move the box and eventually it moves, the frictional forces that are occurring and being overcome by your push result in a loss of energy which is dissipated as heat.

In some machines we want to minimize frictional forces and the consequential loss of energy, e.g. in bearings and gears. In other machines we make use of frictional forces, e.g. in brakes where the frictional forces are used to slow down motion and belt drives where without frictional forces between the drive wheel and the belt no movement would be transferred from wheel to belt. In walking we depend on friction between the soles of our shoes and the ground in order to move – think of the problems of walking on ice when the frictional forces are very low. When you use a ladder with the base resting on horizontal ground and the other end up a vertical wall, you rely on friction to stop the bottom of the ladder sliding away when you attempt to climb up it.

There are two basic types of frictional resistance encountered in engineering:

(1) *Dry friction*
This is encountered when one solid surface is attempting to slide over another solid surface and there is no lubricant between the two surfaces.

(2) *Fluid friction*
This is encountered when adjacent layers in a fluid are moving at different velocities.

In this chapter we will consider only dry friction, fluid friction being discussed in Chapter 3, Section 3.2 'Hydrodynamics – fluids in motion'.

Dry friction

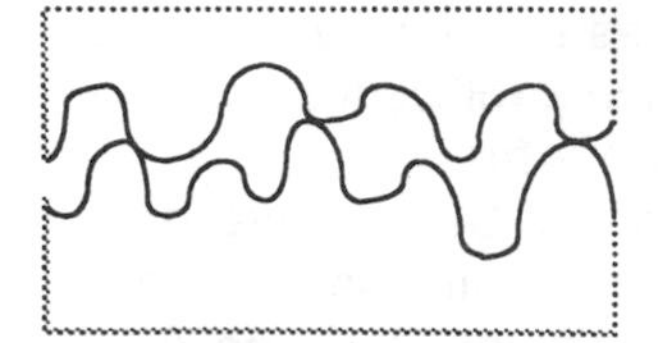

Figure 5.6.2 *Surfaces in contact, even when apparently smooth are likely to be only in contact at a few points*

When two solid surfaces are in contact, because surfaces are inevitably undulating, contact takes place at only a few points (Figure 5.6.2) and so the real area of contact between surfaces is only a small fraction of the apparent area of contact. It is these small areas of real contact that have to bear the load between the surfaces. With metals the pressure is so high that appreciable plastic deformation of the metal at the points of contact occurs and cold welding results. Thus when surfaces are slid over one another, these junctions have to be sheared. The frictional force thus arises from the force to shear the junctions and plough the 'hills' of one surface through those of the other.

The term *static friction* is used when the bodies are at rest and the frictional force is opposing attempted motion; the term *kinetic friction* is used when the bodies are moving with respect to each other and the frictional force is opposing motion with constant velocity.

The laws of friction

If two objects are in contact and at rest and a force applied to one object does not cause motion, then this force must be balanced by an opposing and equal size frictional force so that the resultant force is zero. If the applied force is increased a point is reached when motion just starts. When this occurs there must be a resultant force acting on the object and thus the applied force must have become greater than the frictional force. The value of the frictional force that has to be overcome before motion starts is called the *limiting frictional force*. The following are the basic laws of friction:

- Law 1. The frictional force is always in such a direction as to oppose relative motion and is always tangential to the surfaces in contact.
- Law 2. The frictional force is independent of the apparent areas of the surfaces in contact.
- Law 3. The frictional force depends on the surfaces in contact and its limiting value is directly proportional to the normal reaction between the surfaces.

The *coefficient of static friction* μ_s is the ratio of the limiting frictional force F to the normal reaction N:

$$\mu_s = \frac{F}{N} \tag{5.6.1}$$

The *coefficient of kinetic friction* μ_k is the ratio of the kinetic frictional force F to the normal reaction N:

$$\mu_k = \frac{F}{N} \tag{5.6.2}$$

Typical values for steel sliding on steel are 0.7 for the static coefficient and 0.6 for the kinetic coefficient.

Example 5.6.1

If a block of steel of mass 2 kg rests on a horizontal surface, the coefficient of static friction being 0.6, what horizontal force is needed to start the block in horizontal motion? Take g to be 9.8 m/s^2.

Figure 5.6.3 shows the forces acting on the block. The normal reaction force $N = mg$. The maximum, i.e. limiting, frictional force is $F = \mu N = 0.6 \times 2 \times 9.8 = 11.8$ N. Any larger force will give a resultant force on the block and cause it to accelerate while a smaller force will be cancelled out by the frictional force and give no resultant force and hence no acceleration. Thus the force which will just be on the point of starting the block in motion is 11.8 N.

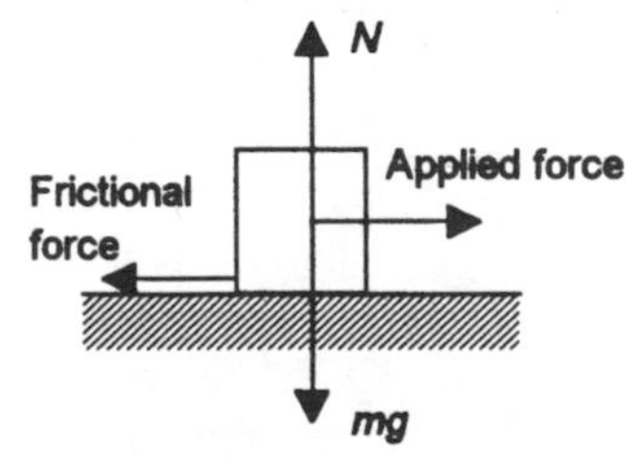

Figure 5.6.3 *Example 5.6.1*

Example 5.6.2

What acceleration is produced when a horizontal force of 30 N acts on a block of mass 2.5 kg resting on a rough horizontal surface if the coefficient of kinetic friction is 0.5? Take g to be 9.8 m/s^2.

The normal reaction $N = mg$ and so the frictional force $F = \mu N = 0.5 \times 2.5 \times 9.8 = 12$ N. The resultant horizontal force acting on the object is thus $30 - 12 = 18$ N and hence the resulting acceleration a is (resultant force)/m = 18/2.5 = 7.2 m/s^2.

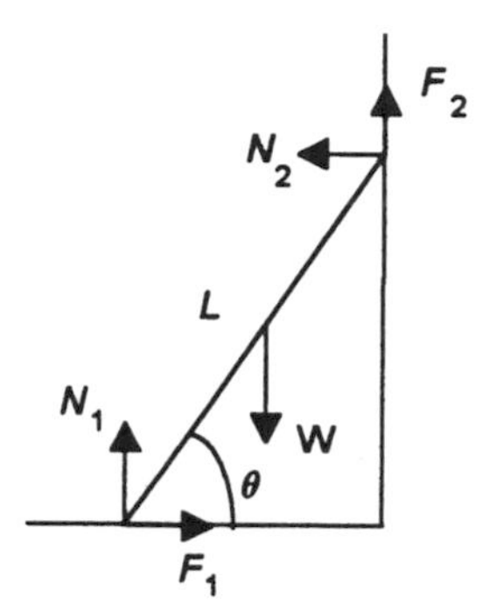

Figure 5.6.4 *Example 5.6.3*

Example 5.6.3

One end of a uniform ladder, of length L and weight W, rests against a rough vertical wall and the other end rests on rough horizontal ground (Figure 5.6.4). If the coefficient of static friction at each end is the same, determine the inclination of the ladder to the horizontal when it is on the point of slipping.

There will be frictional forces where the ladder is in contact with the vertical wall (F_2) and where it is in contact with the horizontal ground (F_1). Taking moments about the foot of the ladder:

$$\tfrac{1}{2}WL\cos\theta = N_2 L\sin\theta + F_2 L\cos\theta$$

and since $F_2 = \mu_s N_2$:

$$\tfrac{1}{2}W\cos\theta = N_2(\sin\theta + \mu_s\cos\theta)$$

$$\tfrac{1}{2}W = N_2(\tan\theta + \mu_s)$$

For equilibrium in the vertical direction $N_1 + F_2 = W$ and so:

$$N_1 + \mu_s N_2 = W$$

For equilibrium in the horizontal direction $F_1 = N_2$ and so:

$$\mu_s N_1 = N_2$$

Thus the above two equations give:

$$N_2(\mu_s + 1/\mu_s) = W$$

Eliminating N_2 then gives:

$$\frac{1}{2} = \frac{\tan\theta + \mu_s}{\mu_s + 1/\mu_s}$$

$$\tan\theta = \frac{1 - \mu_s^2}{2\mu_s}$$

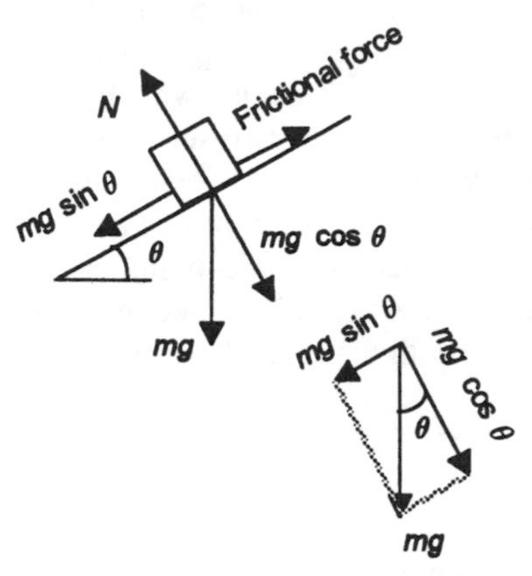

Figure 5.6.5 *Body resting on an inclined plane*

Body on rough inclined plane

Consider a block of material resting on an inclined plane (Figure 5.6.5) and the force necessary to just prevent sliding down the plane. The normal reaction force N is equal to the component of the weight of the block which is at right angles to the surfaces of contact and is thus $mg \cos \theta$. The limiting frictional force F must be equal to the component of the weight of the block down the plane and so is $F = mg \sin \theta$. Hence, the maximum angle such a plane can have before sliding occurs is given by:

$$\mu_s = \frac{F}{N} = \frac{mg \sin \theta}{mg \cos \theta} = \tan \theta \qquad (5.6.3)$$

This angle is sometimes called the *angle of static friction*.

If we have an angle less than the angle of static friction then no sliding occurs. If the angle is greater than the angle of static friction then sliding occurs. Then the resultant force acting down the plane is $mg \sin \theta - \mu_k mg \cos \theta$ and there will be an acceleration a of:

$$a = \frac{mg \sin \theta - \mu_k mg \cos\theta}{m} = g \sin \theta - \mu_k g \cos \theta \qquad (5.6.4)$$

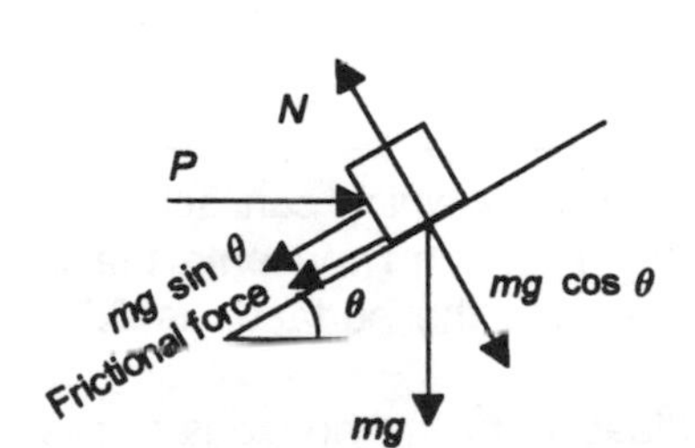

Figure 5.6.6 *Example 5.6.4*

Example 5.6.4

Determine the minimum horizontal force needed to make a body of mass 2 kg begin to slide up a plane inclined to the horizontal at 30° if the coefficient of static friction is 0.6. Take g to be 9.8 m/s^2.

Figure 5.6.6 shows the forces acting on the object. As the body is just on the point of sliding, there will be no resultant force acting on it. Thus for the components of the forces at right angles to the plane:

$$P \sin \theta + mg \cos \theta = N$$

and for the components parallel to the plane:

$$mg \sin \theta + F = P \cos \theta$$

Since $F = \mu_s N$, then:

$$mg \sin \theta + \mu_s(P \sin \theta + mg \cos \theta) = P \cos \theta$$

$$P = \frac{mg(\sin \theta + \cos \theta)}{\cos \theta - \mu_s \sin \theta}$$

$$= \frac{2 \times 9.8(\sin 30° + \cos 30°)}{\cos 30° - 0.6 \sin 30°} = 47.4 \text{ N}$$

Rolling resistance

A cylinder rolling over the ground will experience some resistance to rolling, though such resistance is generally much smaller than the resistance to sliding produced by friction. This rolling resistance arises

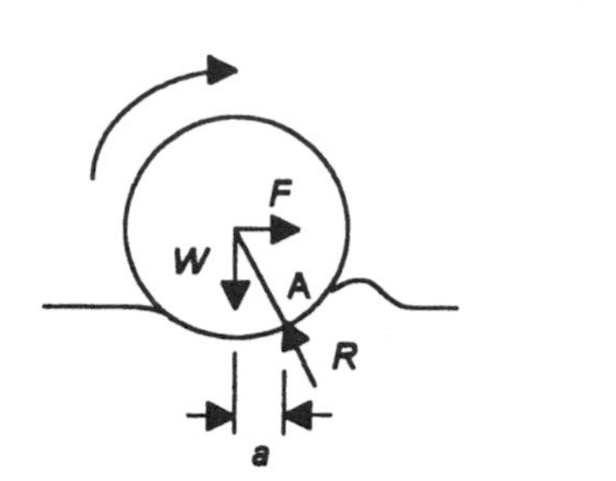

Figure 5.6.7 *Rolling resistance*

mainly as a result of deformation occurring at the point of contact between it and the surface over which it is rolling (Figure 5.6.7). As the cylinder rolls to the right, ahead of it a ridge of material is pushed along. Think of the problem of pushing a garden roller across a lawn when the ground is 'soggy'. Over the contact area between the two surfaces there is an upward directed pressure (think of the ground being like a stretched rubber membrane with the cylinder pressing into it) and this gives rise to a resultant force R which acts at some point on the circumference of the cylinder in a radial direction. If we take moments about point A, for a cylinder of radius r with the force R at an angle h to the vertical:

$$Fr \cos \theta = Wa$$

For small angles we can make the approximation $Fr \approx Wa$ and thus, for a cylinder rolling along a horizontal surface, the force F necessary to overcome rolling resistance is found to be proportional to the normal force W and can be written as:

$$F = \mu_r W \qquad (5.6.5)$$

where μ_r is the *coefficient of rolling resistance.*

Problems 5.6.1

(1) What is the maximum frictional force which can act on a block of mass 2 kg when resting on a rough horizontal plane if the coefficient of friction between the surfaces is 0.7? Take g to be 9.8 m/s^2.

(2) If the angle of a plane is raised from the horizontal, it is found that an object just begins to slide down the plane when the angle is 30°. What is the coefficient of static friction between the object and surface of the plane?

(3) A body of mass 3 kg rests on a rough inclined plane at 60° to the horizontal. If the coefficient of static friction between the body and the plane is 0.35, determine the force which must be applied parallel to the plane to just prevent motion down the plane. Take g to be 9.8 m/s^2.

(4) A ladder, of weight 250 N and length 4 m, rests with one end against a smooth vertical wall and the other end on the rough horizontal ground a distance of 1.6 m from the base of the wall. Determine the frictional force acting on the base of the ladder.

(5) An object of weight 120 N rests on a rough plane which is inclined so that it rises by 4 m in a horizontal distance of 5 m. What is the coefficient of static friction if a force of 60 N parallel to the plane is just sufficient to prevent the object sliding down the plane?

(6) An object of mass 40 kg rests on a rough inclined plane at 30° to the horizontal. If the coefficient of static friction is 0.25, determine the force parallel to the plane that will be necessary to prevent the object sliding down the plane. Take g to be 9.8 m/s^2.

(7) A body of mass m rests on a rough plane which is at 30° to the horizontal. The body is attached to one end of a light inextensible cord which passes over a pulley at the top of the inclined plane and hangs vertically with a mass M attached to the end of the cord. If M is greater than m, show that the bodies will move with an acceleration of $(M - m)g/(M + m)$ if the coefficient of dynamic friction between the body on the plane and its surface is $1/\sqrt{3}$.

(8) A body with a weight W rests on a rough plane inclined at angle θ to the horizontal. A force P is applied to the body at an angle ϕ to the slope. Show that value of this force necessary to prevent sliding down the plane is $W \sin(\theta - \alpha)/\cos(\phi - \alpha)$ where a is the angle of static friction and hence that the minimum value of P is when $\phi = \alpha$.

(9) A uniform pole rests with one end against a rough vertical surface and the other on the rough horizontal ground. Determine the maximum angle at which the pole may rest in this position without sliding if the coefficient of static friction between the ends of the pole and vertical and horizontal surfaces is 0.25.

(10) A body of mass 60 kg rests on a rough plane inclined at 45° to the horizontal. If the coefficient of static friction is 0.25, determine the force necessary to keep the body from sliding if the force is applied (a) parallel to the inclined plane, (b) horizontally. Take g to be 9.8 m/s^2.

(11) A body of mass 40 kg rests on a rough plane inclined at 45° to the horizontal. If the coefficient of static friction is 0.4 and the coefficient of kinetic friction is 0.3, determine (a) the force parallel to the plane required to prevent slipping, (b) the force parallel to the plane to pull the body up the plane with a constant velocity. Take g to be 9.8 m/s^2.

(12) A ladder of mass 20 kg and length 8 m rests with one end against a smooth vertical wall and the other end on rough horizontal ground and at an angle of 60° to the horizontal ground. If the coefficient of static friction between the ladder and the ground is 0.4, how far up the ladder can a person of mass 90 kg climb before the ladder begins to slide? Take g to be 9.8 m/s^2.

5.7 Virtual work

So far in this chapter the equilibrium of bodies has been determined by drawing a free-body diagram and considering the conditions necessary for there to be no resultant force and no resulting moment. There is, however, another method we can use called the *principle of virtual work*. This is particularly useful when we have interconnected members that allow relative motion between the connected parts and we want to find the position of the parts which will give equilibrium.

Before considering the principle of virtual work we will start off with a review of the work done by forces and couples.

Work

The *work* done by a force is the product of the force and the displacement in the direction of that force. Thus, for the force F in

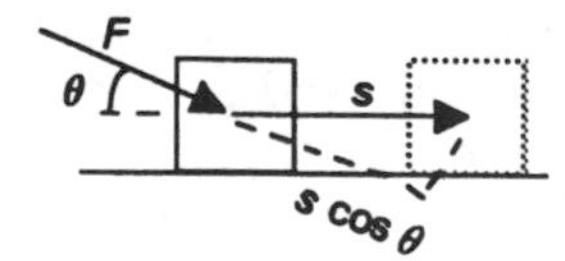

Figure 5.7.1 *Work is product of the force and the displacement in the direction of the force*

Figure 5.7.1, the displacement in the direction of the force is $s \cos \theta$ and so the work done in moving the object through a horizontal distance s is $Fs \cos \theta$. The same result is obtained if we multiply the component of the force in the direction of the displacement with the displacement.

The two forces of a couple do work when the couple rotates about an axis perpendicular to the plane of the couple. Thus, for the couple shown in Figure 5.7.2 where $M = Fr$, for a rotation by $\delta\theta$ then the point of application of each of the forces moves through $\frac{1}{2}r\,\delta\theta$ and so the total work done by the couple is $\frac{1}{2}Fr\,\delta\theta + \frac{1}{2}Fr\,\delta\theta = Fr\,\delta\theta$. Thus the work done by the couple is $M\,\delta\theta$.

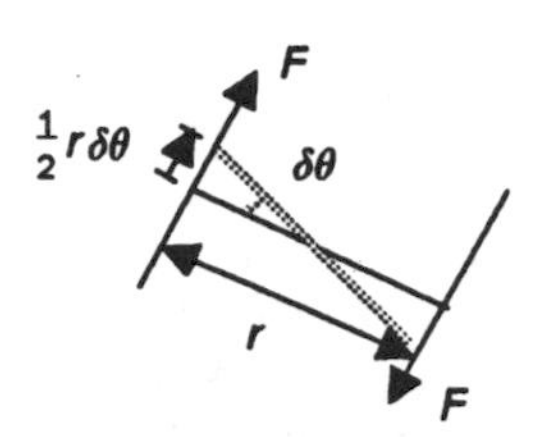

Figure 5.7.2 *Work done by a couple is the product of the moment and the angle rotated*

The principle of virtual work

The *principle of virtual work* may be stated as: if a system of forces acts on a particle which is in static equilibrium and the particle is given any virtual displacement consistent with the constraints imposed by the system then the net work done by the forces is zero. A virtual displacement is any arbitrary displacement which does not actually take place in reality but is purely conceived as taking place mathematically and is feasible given the constraints imposed by the system.

Thus if we have a body in equilibrium and subject to a number of forces F_1, F_2 and F_3, as in Figure 5.7.3, then, if it is given an arbitrary displacement which results in displacements δs_1, δs_2 and δs_3 in the direction of these forces, the virtual work principle gives:

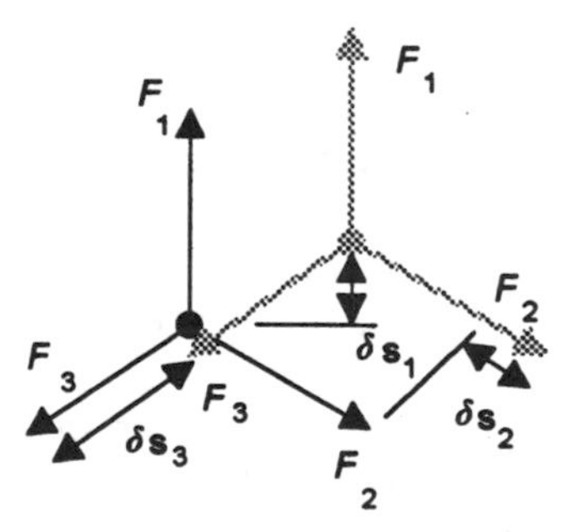

Figure 5.7.3 *Virtual work when there is a displacement*

$$F_1\,\delta s_1 + F_2\,\delta s_2 + F_3\,\delta s_3 = 0$$

If the resultant of the above forces is F then with a resultant displacement δs in the direction of the resultant force, the virtual work principle would mean that:

$$F\,\delta s = 0$$

This can only be the case if either F or δs is 0. Since δs need not be zero then we must have zero resultant force F and so the condition for static equilibrium.

In applying the principle of virtual work to a system of interconnected rigid bodies:

- Externally applied loads are forces capable of doing virtual work when subject to a virtual displacement.
- Reactive forces at fixed supports will do no virtual work since the constraints of the system mean that they are not capable of having virtual displacements.
- Internal forces in members always act in equal and oppositely directed pairs and so a virtual displacement will result in the work of one force cancelling out the work done by the other force. Thus the net work done by internal forces is zero.

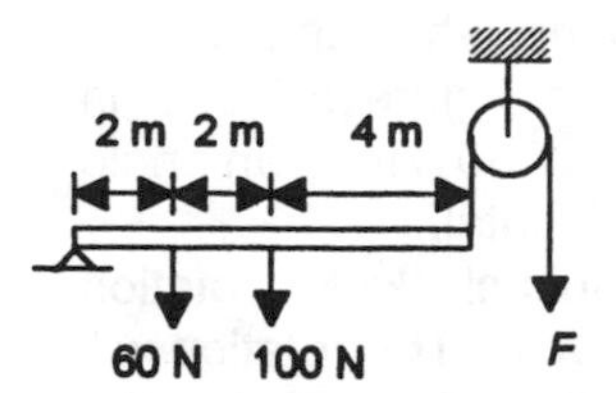

Figure 5.7.4 *Example 5.7.1*

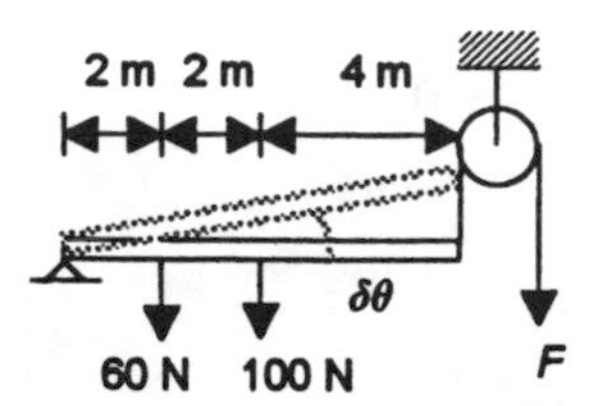

Figure 5.7.5 *Example 5.7.1*

Example 5.7.1

Determine the force F that has to be applied for the system shown in Figure 5.7.4 to be in equilibrium.

Let us give F a small virtual displacement of δs. As a consequence, the beam rotates through an angle $\delta\theta$ (Figure 5.7.5) and the points of applications of the 60 N and 100 N forces move. The vertical height through which the point of application of the 100 N force moves is $\delta h_{100} = 4 \sin \delta\theta$, which approximates to $\delta h_{100} = 4\ \delta\theta$. Likewise, for the 60 N force we have $\delta h_{60} = 2\ \delta\theta$. Hence, applying the principle of virtual work, we have:

$$F\,\delta s - 100 \times 4\ \delta\theta - 60 \times 2\ \delta\theta = 0$$

Since $\delta s/8 = \sin \delta\theta$, and this can be approximated to $\delta s/8 = \delta\theta$:

$$F \times 8\ \delta\theta - 100 \times 4\ \delta\theta - 60 \times 2\ \delta\theta = 0$$

Hence $F = 55$ N.

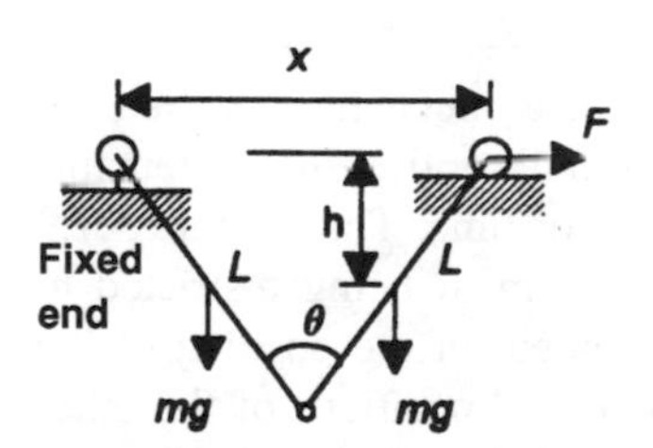

Figure 5.7.6 *Example 5.7.2*

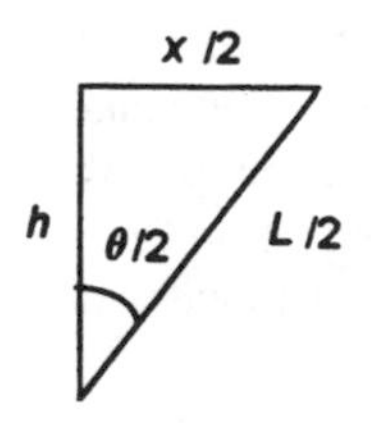

Figure 5.7.7 *Example 5.7.2*

Example 5.7.2

Determine the angle θ between the two plane pin-jointed rods shown in Figure 5.7.6 when the system is in equilibrium. Each rod has a length L and a mass m.

Suppose we give the force F a virtual displacement of δx in the direction of x. As a consequence, the weights mg will both be lifted vertically by δh. Applying the principle of virtual work:

$$F\,\delta x + 2mg\,\delta h = 0$$

Since the vertical height h of a weight equals $\frac{1}{2}L \cos \theta/2$ (Figure 5.7.7) then, as a result of differentiating, $\delta h = -\frac{1}{4}L \sin \theta/2\ \delta\theta$; the displacement δh is in the opposite direction to the force mg. For a positive displacement δx, i.e. one resulting in an increase in x, then there is a decrease in the height h. Thus the work resulting from the displacement of F is positive and that resulting from the displacement of mg is negative and this is taken account of by the sign given to δh. Since $x = 2L \sin \theta/2$ then $\delta x = L \cos \theta/2\ \delta\theta$ and so we can write:

$$FL \cos \theta/2\ \delta\theta - 2mg\,\tfrac{1}{4}L \sin \theta/2\ \delta\theta = 0$$

Thus:

$$\tan \frac{\theta}{2} = \frac{2F}{mg}$$

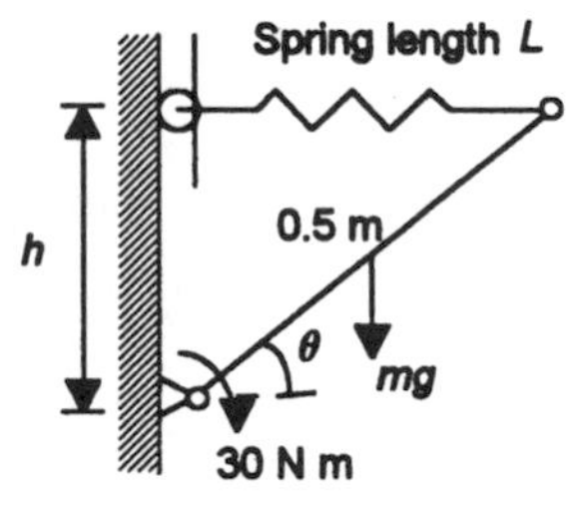

Figure 5.7.8 *Example 5.7.8*

Example 5.7.3

Determine the force acting in the spring when the rod, of mass 5 kg and length 0.5 m, shown in Figure 5.7.8 is in equilibrium under the action of a couple of 30 N m at an angle of 30°. The spring always remains horizontal.

Suppose we give the rod a small virtual clockwise rotation of $\delta\theta$. The work done by the couple is $M\,\delta\theta$. The rotation will result in the weight of the rod having a displacement δh in the direction of the weight. Since $h = 0.5 \sin\theta$ then $\delta h = 0.5 \cos\theta\,\delta\theta$ and the work done by the weight is $mg \times 0.5 \cos\theta\,\delta\theta$. Since $L = 0.5\cos\theta$ then $\delta L = -0.5 \sin\theta\,\delta\theta$. Hence the work done by the spring is $F\,\delta L = -0.5F \sin\theta\,\delta\theta$. Applying the principle of virtual work thus gives:

$$M\,\delta\theta + mg \times 0.5 \cos\theta\,\delta\theta - 0.5F \sin\theta\,\delta\theta = 0$$

Hence:

$$F = \frac{30 + 5 \times 9.8 \times 0.5 \cos 30°}{0.5 \sin 30°} = 205\text{ N}$$

Degrees of freedom

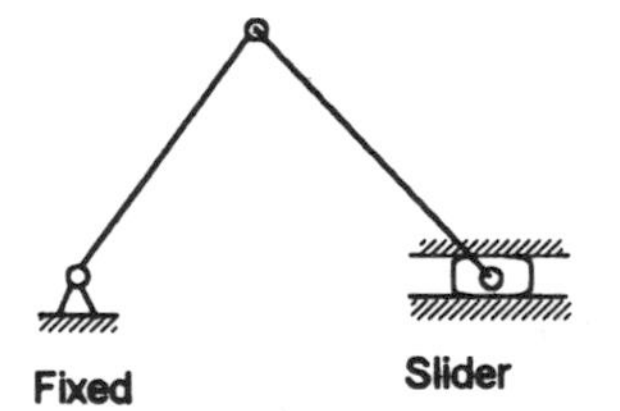

Figure 5.7.9 *One degree of freedom*

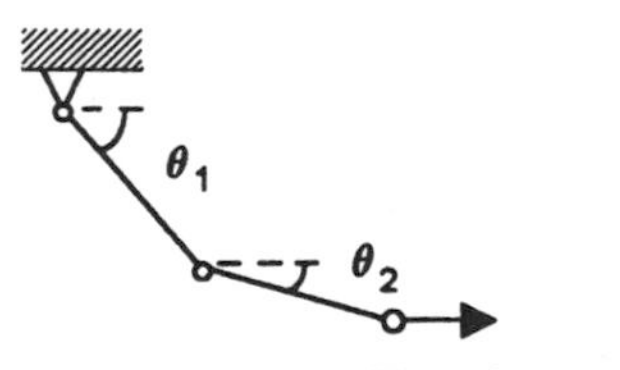

Figure 5.7.10 *Two degrees of freedom*

The term *degrees of freedom* is used with a mechanical system of interconnected bodies for the number of independent measurements required to locate its position at any instant of time. Thus Figure 5.7.9 shows a system having one degree of freedom, it being restricted to motion in just one direction and a measurement of the position of the slider is enough to enable all the interconnected positions of the other parts of the system to be determined. Figure 5.7.10 shows a system having two degrees of freedom; to specify the position of each of the two links then the two angles θ_1 and θ_2 have to be measured.

Because it takes n independent measurements to completely specify the positions of elements in a system with n degrees of freedom, it is possible to write n independent virtual work equations. We make a displacement for just one of the independent movements possible, keeping the others fixed, and then write a virtual work equation for that movement, repeating it for each of the other independent movements.

Problems 5.7.1

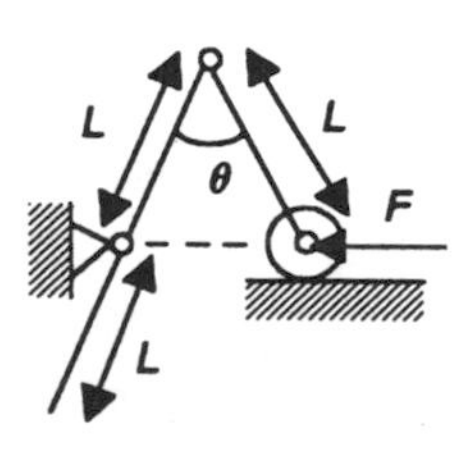

Figure 5.7.11 *Problem 1*

(1) For the two plane pin-jointed rods shown in Figure 5.7.11, use the principle of virtual work to determine the force required to give an angle between the rods of θ. The two rods are the same material and have the same uniform cross-sectional area but the left-hand rod is twice the length of the right-hand rod and so has twice the weight. The left-hand rod is pivoted at its midpoint.

(2) For the system of plane pin-jointed rods shown in Figure 5.7.12, use the principle of virtual work to determine the force F that has to be applied to give equilibrium.

(3) Figure 5.7.13 shows a rod of mass m and length L resting against a smooth vertical wall at one end and a smooth horizontal floor at the other. Use the principle of virtual work to determine the horizontal force F needed at its lower end to keep the rod in equilibrium at an angle θ to the horizontal.

(4) For the toggle press shown in Figure 5.7.14, by using the principle of virtual work, show that:

$$F_2 = \frac{2F_1 \, l \sin \theta}{L}$$

Hint: consider the effect of changing the angle θ to the horizontal of the arm by a small angle.

(5) How many degrees of freedom has the system shown in Figure 5.7.15?

(6) How many degrees of freedom has the system shown in Figure 5.7.16?

(7) Determine the angles θ_1 and θ_2 at which the two-bar linkage, each bar being the same length L, shown in Figure 5.7.17, will be in equilibrium when subject to the two indicated equal size couples. Neglect the weights of the links.

Hint: this linkage has two degrees of freedom and so two virtual work equations have to be written, one with angle θ_1 held fixed and θ_2 given a virtual increment and the other with θ_2 held fixed and θ_1 given a virtual increment.

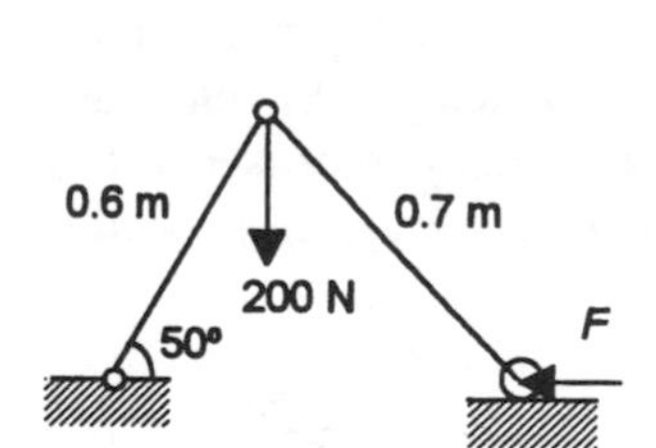

Figure 5.7.12 *Problem 2*

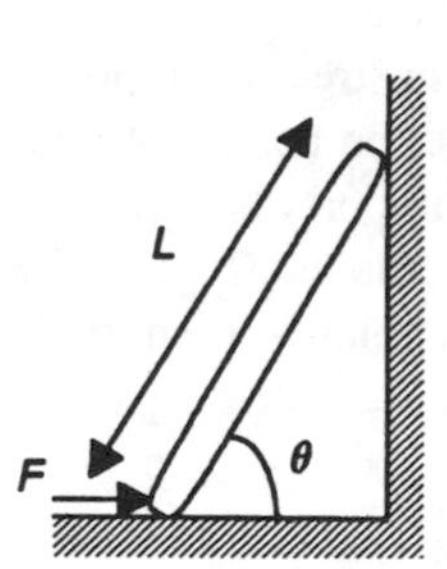

Figure 5.7.13 *Problem 3*

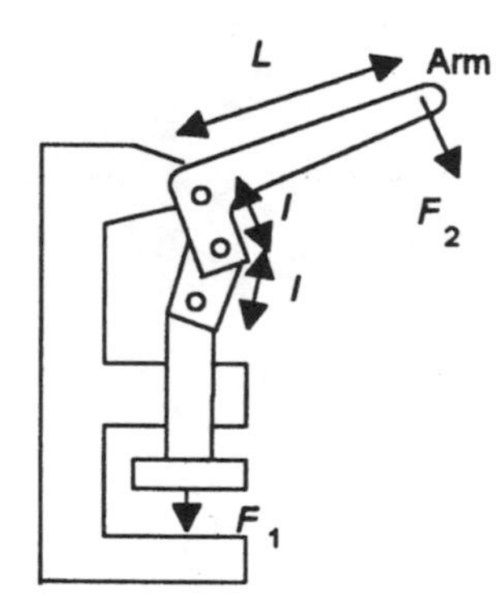

Figure 5.7.14 *Problem 4*

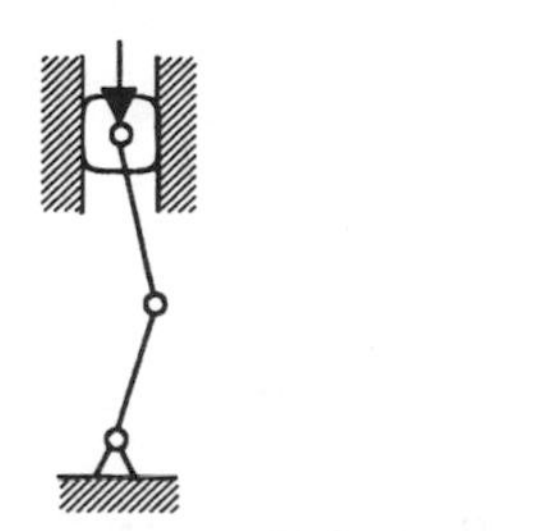

Figure 5.7.15 *Problem 5*

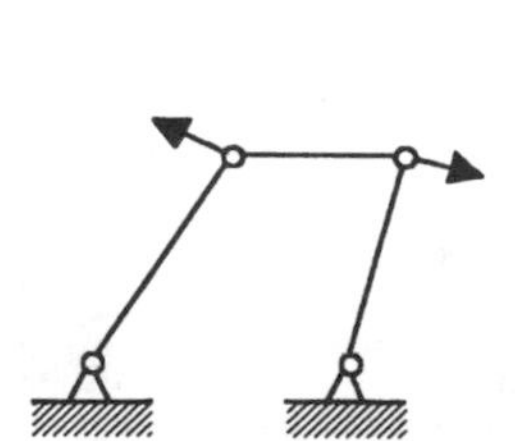

Figure 5.7.16 *Problem 6*

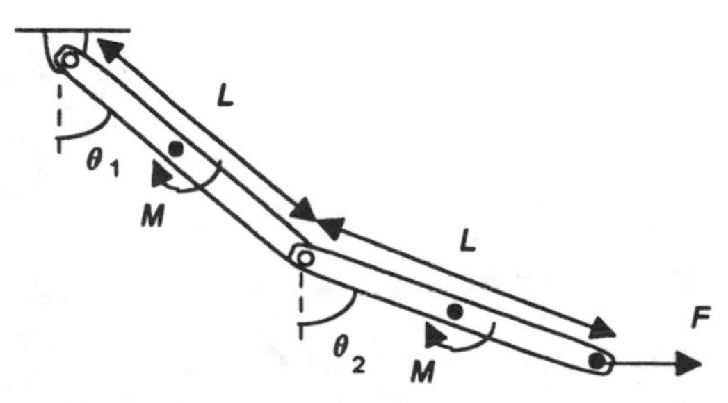

Figure 5.7.17 *Problem 7*

5.8 Case study: bridging gaps

Consider the problems involved in bridging gaps. It could be a bridge across a river or perhaps beams to carry a roof to bridge the gap between two walls.

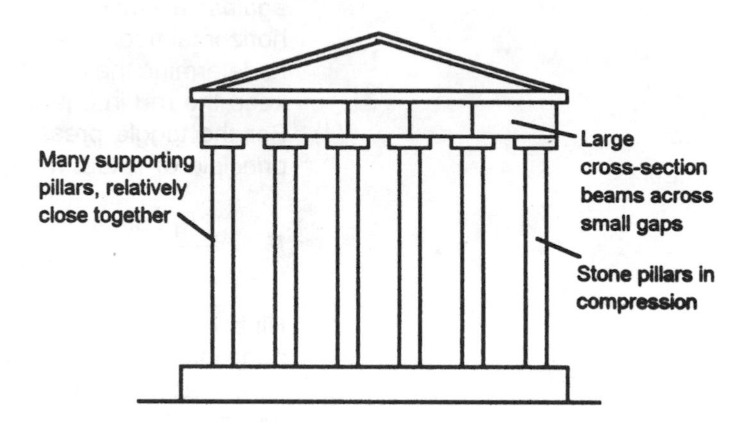

Figure 5.8.1 *The basic structure when stone beams are used: they need to have large cross-sections and only bridge small gaps*

The simplest solution is to just put a beam of material across the gap. The application of loads to the beam will result in bending, with the upper surface of the beam being in compression and the lower surface in tension. The pillars supporting the ends of the beam will be subject to compressive forces. Thus materials are required for the beam that will be strong under both tensile and compressive forces, and for the supporting pillars ones which will withstand compressive forces. Stone is strong in compression and weak in tension. While this presents no problems for use for the supporting pillars, a stone beam can present problems in that stone can be used only if the tensile forces on the beam are kept low. The maximum stress = My_{max}/I (see the general bending equation), where, for a rectangular section beam, y_{max} is half the beam depth d and $I = bd^3/3$, b being the breadth of the beam. Thus the maximum stress is proportional to $1/bd^2$ and so this means having large cross-section beams. We also need to have a low bending moment and so the supports have to be close together. Thus ancient Egyptian and Greek temples (Figure 5.8.1) tend to have many roof supporting columns relatively short distances apart and very large cross-section beams across their tops.

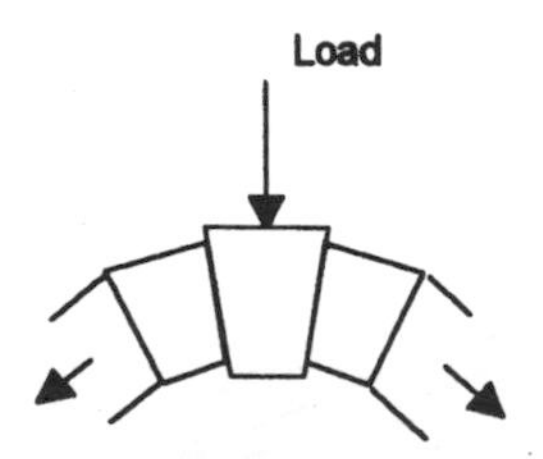

Figure 5.8.2 *The arch as a means of bridging gaps by putting the stone in compression*

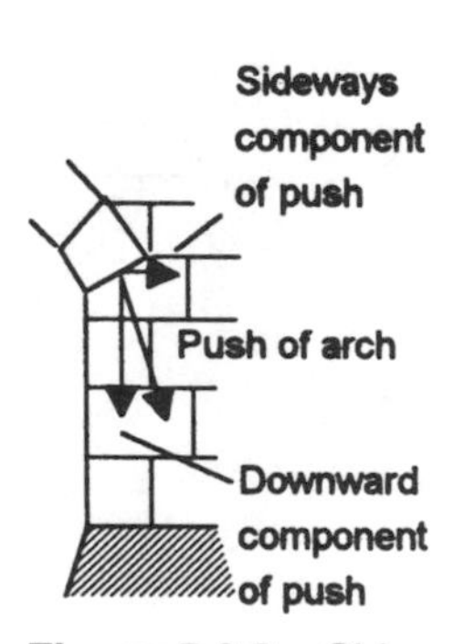

Figure 5.8.3 *Sideways push of arches*

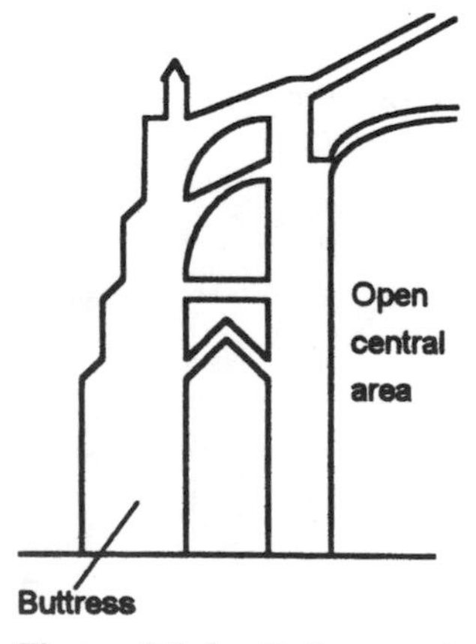

Figure 5.8.4 *Buttresses to deal with the sideways thrust of an arch*

One way of overcoming the weakness of stone in tension is to build *arches* (Figure 5.8.2), which enable large clear open spans without the need for materials with high tensile properties. Each stone in an arch is so shaped that when the load acts downwards on a stone it results in it being put into compression. The net effect of all the downward forces on an arch is to endeavour to straighten it out and so the supporting columns must be strong enough to withstand the resulting sideways push of the arch (Figure 5.8.3) and the foundations of the columns secure enough to withstand the base of the column being displaced. The most frequent way such arches collapse is the movement of the foundations of the columns.

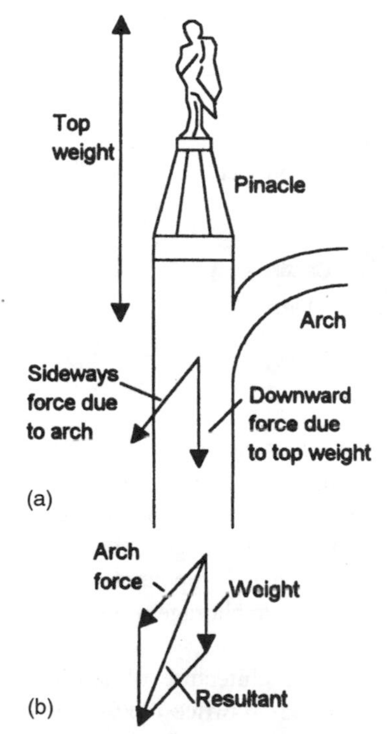

Figure 5.8.5 *Utilizing top weight with a buttress to give a resultant force in a more vertical direction*

Cathedrals use arches to span the open central area and thus methods have to be adopted to accommodate the sideways push of these arches. One method that is often used is to use buttresses (Figure 5.8.4). The sideways thrust of the arch has a force, the top weight of the buttress, added to it (Figure 5.8.5(a)) to give a resultant force which is nearer the vertical (Figure 5.8.5(b)). The heavier the top weight, the more vertical the resultant force, hence the addition of pinnacles and statues. As we progress down the wall, the weight of the wall above each point increases. Thus the line of action of the force steadily changes until ideally it becomes vertical at the base of the wall.

Both stone and brick are strong in compression but weak in tension. Thus arches are widely used in structures made with such materials and the term *architecture of compression* is often used for such types of structures since they have always to be designed to put the materials into compression.

The end of the eighteenth century saw the introduction into bridge building of a new material, cast iron. Like stone and brick, cast iron is strong in compression and weak in tension. Thus the iron bridge followed virtually the same form of design as a stone bridge and was in the form of an arch. The world's first iron bridge was built in 1779 over the River Severn; it is about 8 m wide and 100 m long and is still standing. Many modern bridges use reinforced and prestressed concrete. This material used the reinforcement to enable the concrete, which is weak in tension but strong in compression, to withstand tensile forces. Such bridges also use the material in the form of an arch in order to keep the material predominantly in compression.

The introduction of steel, which was strong in tension, enabled the basic design to be changed for bridges and other structures involving the bridging of gaps and enabled the *architecture of tension*. It was no longer necessary to have arches and it was possible to have small cross-section, long, beams. The result was the emergence of truss structures, this being essentially a hollow beam. Figure 5.8.6 shows one form of truss bridge. As with a simple beam, loading results in the upper part of

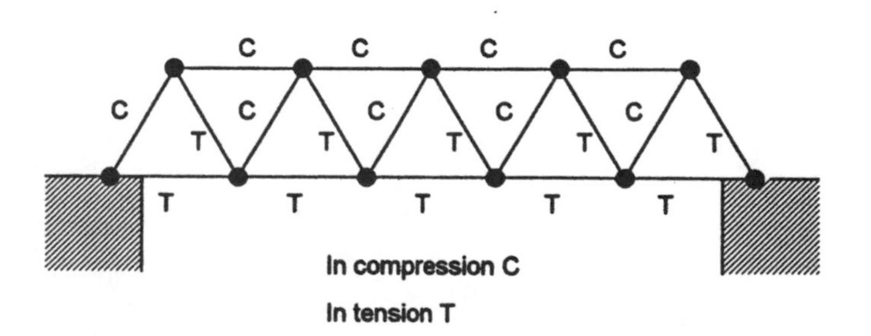

Figure 5.8.6 *The basic form of a truss bridge*

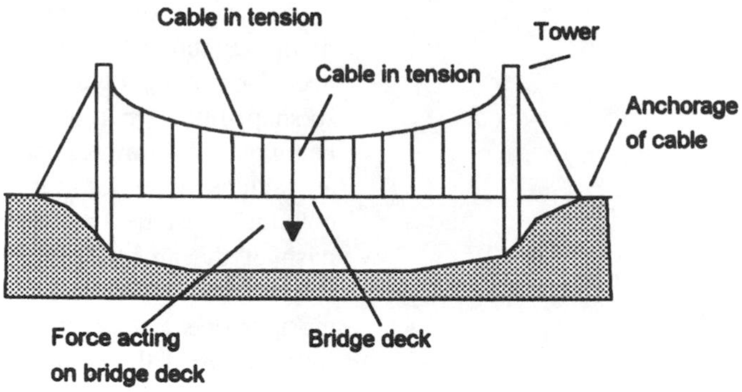

Figure 5.8.7 *The basic form of a suspension bridge; the cables are in tension*

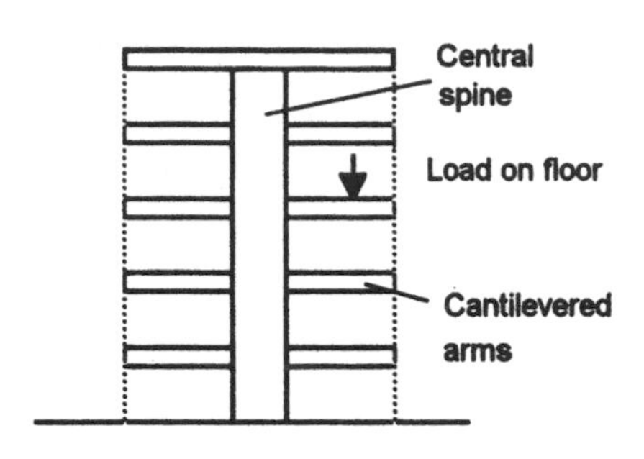

Figure 5.8.8 *Basic structure of a tower block as a series of cantilevered floors*

this structure being in compression and the lower part in tension; some of the diagonal struts are in compression and some in tension.

Suspension bridges depend on the use of materials that are strong in tension (Figure 5.8.7). The cable supporting the bridge deck is in tension. Since the forces acting on the cable have components which pull inwards on the supporting towers, firm anchorage points are required for the cables.

Modern buildings can also often use the architecture of tension. Figure 5.8.8 shows the basic structure of a modern office block. It has a central spine from which cantilevered arms of steel or steel-reinforced concrete stick out. The walls, often just glass in metal frames, are hung between the arms. The cantilevered arms are subject to the loads on a floor of the building and bend, the upper surface being in compression and the lower in tension.

Solutions to problems

Chapter 2

2.1.1

1. 678.6 kJ
2. 500 J/kgK

2.1.2

1. 195 kJ
2. 51.3°C
3. 3.19 kJ, 4.5 kJ
4. 712 kJ
5. 380.3 kJ

2.2.1

1. 24.2 kg
2. 1.05 m^3
3. 45.37 kg, 23.9 bar

2.2.2

1. 3.36 bar
2. 563.4 K
3. 0.31 m^3, 572 K
4. 0.276 m^3, 131.3°C
5. 47.6 bar, 304°C
6. 0.31 m^3, 108.9°C
7. 1.32
8. 1.24, 356.8 K
9. 600 cm^3, 1.068 bar, 252.6 K
10. 46.9 cm^3, 721.5 K

2.3.1

1. 3.2 kJ, 4.52 kJ
2. –52 kJ
3. 621.3 kJ
4. 236 K, 62.5 kJ
5. 0.0547, 0.0115 m^3, –33.7 kJ, –8.5 kJ
6. 1312 K, 318 kJ
7. 0.75 bar, 9.4 kJ
8. 23.13 bar, –5.76 kJ, –1.27 kJ

2.4.1

1. 734.5 K, 819.3 K, 58.5%
2. 0.602, 8.476 bar
3. 65%
4. 46.87 kJ/kg rejected, 180.1 kJ/kg supplied, 616.9 kJ/kg rejected
5. 61%, 5.67 bar

2.4.2

1. 1841 kW
2. 28.63 kW, 24.3 kW
3. 9.65 kW
4. 10.9 kW, 8.675 kW, 22.4%
5. 6.64 bar, 32%
6. 6.455 bar, 30.08%, 83.4%
7. 65%
8. 35%

2.5.1

1. 1330 kW
2. 125 kJ/kg
3. 5 kJ
4. 248.3 kW
5. 31 kW
6. 279 m/s
7. 2349 kJ/kg, 0.00317 m^2
8. 762 kW

2.5.2

1. 688 kW, 25.7%
2. 151 kJ/kg, 18.5%
3. 316 kW, 18%
4. 21%

2.6.1

1. 209.3, 2630.1, 2178.5, 3478, 2904, 2769 kJ/kg
2. 199.7°C
3. 1.8 m^3
4. 5 kg
5. 14218.2 kJ
6. 0.934

2.6.2

1. 4790 kW
2. 2995 kW
3. 839.8 kJ/kg, 210 kW
4. 2965.2 kJ/kg, 81.48°C
5. 604 kJ/kg, 0.0458 m^3/kg

2.6.3

1. 36.7%
2. 35%

2.6.4

1. 1410 kW, 4820 kW
2. 28.26%
3. 0.89, 32%, 0.85

2.6.5

1. 0.84
2. 0.8 dry
3. 0.65, 261 kJ/kg, –229.5 kJ/kg
4. 0.153 m^3, 0.787, 50 kJ
5. 0.976, 263 kJ
6. 15 bar/400°C, 152.1 kJ, 760 kJ

2.7.1

1. 0.148, 0.827
2. 323.2 kJ
3. 8.4 kW, 28.34 kW, 3.7
4. 0.97, 116.5 kJ/kg, 5.4, 6.45
5. 3.84
6. 6.7, 1.2 kg/min, 0.448 kW
7. 0.79, 114 kJ/kg, 3.1
8. 0.1486, 7.57 kW, 3.88

2.8.1

1. 1.2 kW
2. 2185 kJ
3. –11.2°C
4. 3.32 W/m^2, 0.2683°C
5. 72.8%, 4.17°C

2.8.2

1. 133.6 W
2. 101.3 W, 136.6°C, 178.8°C, 19.6°C
3. 2.58 kW
4. 97.9 mm, 149.1°C
5. 31 MJ/h, 0.97, 65°C

Chapter 3

3.1.1

3. 10.35 m
4. 0.76 m
5. (a) 6.07 m
   (b) 47.6 kPa
6. 2.25 m
7. 24.7 kPa
8. 16.96 kPa
9. 304 mm
10. 315 mm
11. 55.2 N
12. 7.07 MPa, 278 N
13. (a) 9.93 kN
    (b) 24.8 kN
    (c) 33.1 kN
14. 8.10 m
15. 9.92 kN
    2.81 kN
14. (a) 38.1 kN
    (b) 29.9 kN
15. 0.67 m
16. 466.2 kN, 42.57° below horizontal
17. (a) 61.6 kN
    (b) 35.3 kN m
18. 268.9 kN, 42.7° below horizontal
19. 1.23 MN, 38.2 kN
20. 6.373 kg, 2.427 kg

3.2.1

3. 1963
4. 0.063 75
5. 300 mm

6. very turbulent
7. $0.025\,m^3/s$, 25 kg/s
8. 0.11 m/s
9. 0.157 m/s, 1.22 m/s
10. 0.637 m/s, 7.07 m/s, 31.83 m/s
11. 134.7 kPa
12. $7.62 \times 10^{-6}\,m^3/s$
13. 208 mm
14. $0.015 \times 10^{-6}\,m^3/s$
15. 2.64 m, 25.85 kPa
16. 21.3 m
17. 12.34 m
18. 57.7 kPa
19. $0.13\,m^3/s$
20. 0.138 kg/s
21. $0.0762\,m^3/s$
22. 3.125 l/s
23. 3 m/s
24. 5.94 m
25. 194 kPa
26. 1000 km/hour
27. $233\,kN/m^2$
28. 0.75 m
29. (a) 2.64 m
    (b) 31.68 m
    (c) 310.8 kPa
30. 3.775 m, 91 kPa
31. 4.42 m
32. 0.000 136
33. 151 tonnes/hour
34. 1 in 1060
35. (a) 1 × 107
    (b) 0.0058
    (c) 1.95 kPa
36. 16.8 m
37. 200 mm
38. 2.35 N
39. 7.37 m/s
40. 3313 N
41. 9 kN m
42. 11.6 m/s
43. 0.36 N
44. 198 N, 26.1 N
45. 178.7 N, 20.9 N

Chapter 4

4.1.1

1. 20.4 m/s, 79.2 m
2. 192.7 m, 5.5 s, 35 m/s
3. 76.76 s, 2624 m
4. 1.7 m/s

5. 0.815 g
6. (a) 14 m/s
   (b) 71.43 s
   (c) 0.168 m/s
7. 15 m/s, 20 s
8. 16 m/s, 8 s, 12 s
9. 18 m/s, 9 s, 45 s, 6 s
10. 23 m/s, 264.5 m, 1.565 m/s^2
11. –4.27 m/s^2, 7.5 s
12. 430 m
13. 59.05 m/s
14. 397 m, 392 m, 88.3 m/s
15. 57.8 s, 80.45 s

4.2.1

1. 3.36 kN
2. 2.93 m/s^2
3. 345 kg
4. 4.19 m/s^2
5. 2.5 m/s^2, 0.24 m/s^2
6. 1454 m
7. 0.1625 m/s^2
8. –0.582 m/s^2, 1.83 m/s^2
9. –2.46 m/s^2, 1.66 m/s^2
10. 12.2 m/s^2, 1.37 km
11. 803 kg
12. 15.2 m/s^2
13. 14.7 m/s^2
14. 19.96 kN
15. 2.105 rad/s^2, 179 s
16. 7.854 rad/s^2, 298 N m
17. 28 500 N m
18. 22.3 kg m^2, 1050 N m
19. 796.8 kg m^2, 309 s
20. 7.29 m
21. 1309 N m
22. 772 N m
23. 9.91 s, 370 kN
24. 2.775 kJ
25. 4010 J
26. 259 kJ
27. 12.3 m/s
28. 51 m/s
29. 82%
30. 973.5 m
31. 17.83 m/s
32. 15.35 m/s
33. 70%
34. (a) 608 kN
    (b) 8670 m
    (c) 371 m/s

35. (a) 150 W, 4.05 kW, 32.4 kW, 30 kJ, 270 kJ, 1080 kJ
    (b) 4.32 kW, 16.56 kW, 57.42 kW, 863.8 kJ, 1103.8 kJ, 1913.8 kJ
36. 417.8 m/s
37. 589 kW
38. 7.85 kW, 20.4 m
39. 14.0 m/s
40. 12.5 m/s
41. (a) 10.5 m
    (b) 9.6 m/s
42. 5.87 m/s
43. 12 892.78 m/s
44. 6.37 m/s
45. 0.69 m/s right to left
46. 4.42 m/s
47. –5.125 m/s, 2.375 m/s
48. 5.84 m/s, 7.44 m/s
49. –3.33 m/s

Chapter 5

5.1.1

1. 372 N at 28° to 250 N force
2. (a) 350 N at 98° upwards to the 250 N force, (b) 191 N at 99.6° from 100 force to right
3. (a) 200 N, 173 N, (b) 73 N, 90 N, (c) 200 N, 173 N
4. 9.4 kN, 3.4 kN
5. 100 N
6. 14.1 N; vertical component 50 N, horizontal component 20 N
7. (a) 100 N m clockwise, (b) 150 N m clockwise, (c) 1.41 kN m anticlockwise
8. 25 N downwards, 222.5 N m anticlockwise
9. 26 N vertically, 104 N vertically
10. 300 N m
11. 2.732 kN m clockwise
12. $P = 103.3$ N, $Q = 115.1$ N, $R = 70.1$ N
13. 36.9°, 15 N m
14. 216.3 N, 250 N, 125 N
15. (a) 55.9 mm, (b) 21.4 mm, (c) 70 mm
16. 4.7 m
17. $4\sqrt{2}r/3\pi$ radially on central radius
18. $r/2$ on central radius
19. From left corner (40 mm, 35 mm)
20. $2r\sqrt{2}/\pi$ from centre along central axis
21. 73 mm centrally above base
22. As given in the problem
23. 32.5°
24. 34.3 kN, 25.7 kN
25. (a) 225 kN, 135 kN, (b) 15.5 kN, 11.5 kN
26. 7.77 kN, 9.73 kN

5.2.1

1. (a) Unstable, (b) stable, (c) stable
2. (a) Unstable, (b) unstable, (c) stable, (d) redundancy
3. F_{ED} +70 kN, F_{AG} −80 kN, F_{AE} −99 kN, F_{BH} +80 kN, F_{CF} +140 kN, F_{DE} +140 kN, F_{EF} +60 kN, F_{FG} −85 kN, F_{GH} +150 kN, F_{AH} = −113 kN, reactions 70 kN and 80 kN vertically
4. 8 kN, 7 kN at 8.2° to horizontal, F_{BH} +3.5 kN, F_{CH} −1.7 kN, F_{GH} −3.5 kN, F_{BG} +3.5 kN, F_{FG} +5.8 kN, F_{FD} −6.4 kN, F_{EF}−1.2 kN
5. (a) F_{BG} −54.6 kN, F_{CG} +27.3 kN, F_{FG} +54.6 kN, F_{AF} −14.6 kN, F_{EF} +14.6 kN, F_{AE} −14.6 kN, F_{ED} +47.3 kN,
   (b) F_{AE} −21.7 kN, F_{CG} −30.3 kN, F_{BF} −13.0 kN, F_{FG} −4.3 kN, F_{DE} +10.8 kN, F_{DG} +15.2 kN, F_{EF} + 4.3 kN,
   (c) F_{BE} +22.6 kN, F_{CG} +5.7 kN, F_{GD} −4.0 kN, F_{DF} −5.7 kN, F_{EF} −16.0 kN,
   (d) F_{AE} −3.2 kN, F_{BF} −1.8 kN, F_{BG} −1.8 kN, F_{CH} −3.9 kN, F_{EF} −1.4 kN, F_{GH} −2.1 kN, F_{ED} + 2.25 kN, F_{DH} + 2.75 kN, F_{FG} + 2.5 kN
6. −12.7 kN
7. 4.8 kN
8. +28.8 kN, +5.3 kN
9. −14.4 kN, +10 kN
10. −35 kN
11. +3.5 kN, −9 kN
12. 20 kN, 10 kN, F_{AD} −23.1 kN, F_{AE} −23.1 kN, F_{AG} −46.2 kN, F_{AI} −34.6 kN, F_{AK} −23.1 kN, F_{AM} −11.5 kN, F_{DE} +23.1 kN, F_{EF} −23.1 kN, F_{FG} +23.1 kN, F_{GH} +11.5 kN, F_{HI} −11.5 kN, F_{IJ} +11.5 kN, F_{JK} −11.5 kN, F_{KL} +11.5 kN, F_{LM} −11.5 kN, F_{MN} + 11.5 kN, F_{AN} −11.5 kN, F_{CD} +11.5 kN, F_{CF} +34.6 kN, F_{BH} +40.4 kN, F_{BJ} +28.9 kN, F_{BL} +17.3 kN, F_{BN} +5.8 kN
13. −7.5 kN, +4.7 kN

5.3.1

1. 20 MPa
2. −0.0003 or −0.03%
3. 1.0 mm
4. 160 kN
5. 0.015 mm, 0.0042 mm
6. 2.12 mm
7. 1768 mm^2
8. 0.64 mm
9. F_{FG} = 480 kN, F_{DC} = 180 kN, A_{FG} = 2400 mm^2, A_{DC} = 900 mm^2
10. 67 mm
11. (a) 460 MPa, (b) 380 MPa, (c) 190 GPa
12. 3.6 MPa, 51.5 MPa
13. 102 MPa, 144 MPa
14. 25 mm
15. 80 mm
16. 34 MPa, 57 MPa
17. 56.3
18. 96 MPa compressive
19. 1.6 MPa

20. 144 MPa compression
21. 47.7 MPa, 38.1 MPa
22. 14.2 MPa, 27.5 MPa
23. 22.9 MPa, 58.6 MPa
24. 29 mm
25. 235 kN
26. 0.00176
27. 16 mm
28. 31.4 kN
29. 251 kN
30. 111 MPa
31. 40 J
32. 21.6 J
33. $F^2h/(4EA \cos^3 \theta)$

5.4.1

1. (a) –50 N, + 50 N m, (b) +50 N, +75 N m
2. (a) +1 kN, –0.5 kN m, (b) + 1 kN, –1.0 kN m
3. (a) +2 kN, –2 kN m, (b) +1 kN, –0.5 kN m
4. See Figure S.1

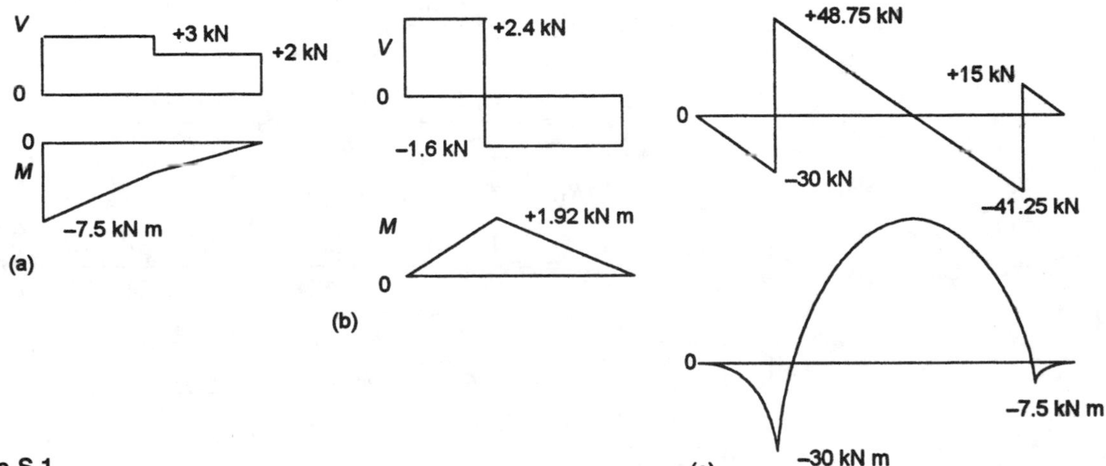

Figure S.1

5. See Figure S.2
6. As given in the problem
7. +48 kN m at 4 m from A
8. +9.8 kN m at 2.3 m from A
9. –130 kN m, 6 m from A
10. 31.2 kN m, 5.2 m from left
11. ±128.8 MPa
12. 600 N m
13. 478.8 kNm
14. 2.95×10^{-4} m^3
15. 141 MPa
16. 79 mm
17. 8.7×10^4 mm^4
18. 101.1×10^6 mm^4

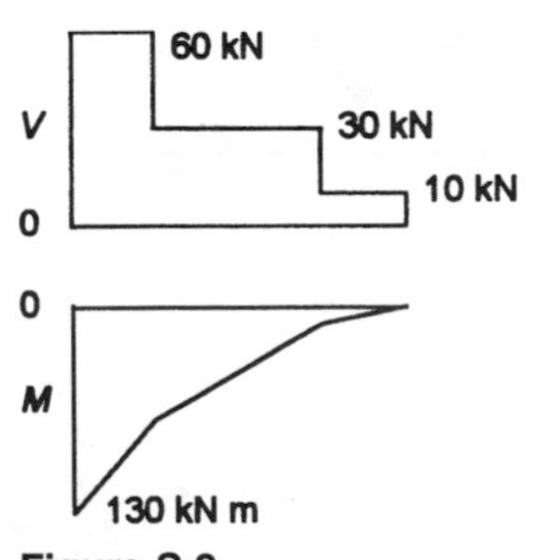

Figure S.2

19. 137.5 mm
20. 165.5 mm
21. $11.7 \times 10^6\,mm^4$, $2.2 \times 10^6\,mm^4$
22. 158 mm
23. $d/4$
24. (a) 71 mm, $74.1 \times 10^6\,mm^4$, (b) 47.5 mm, $55.3 \times 10^4\,mm^4$
25. $247.3 \times 10^6\,mm^4$, $126.3 \times 10^6\,mm^4$
26. 31.4 MPa
27. 7.0 MPa, 14 MPa
28. As given in the question
29. 2.3 kN
30. 15 mm
31. 3.75 mm
32. As given in the problem
33. As given in the problem
34. $FL^3/48EI + 5wL^4/384EI$
35. $0 \le x \le a$: $y = \dfrac{1}{EI}\left[\dfrac{Fx^3}{6} + \left(\dfrac{Fa^2}{2} - \dfrac{FaL}{2}\right)x\right]$,

$$a \le x \le a + b:\ y = \frac{1}{EI}\left(\frac{Fax^2}{2} - \frac{FaLx}{2} + \frac{Fa^3}{6}\right)$$

36. (a) $y = -\dfrac{1}{EI}\left(\dfrac{R_1\,x^3}{6} - \dfrac{F_1\,\{x-a\}^3}{6} - \dfrac{F_2\,\{x-b\}^3}{6} + Ax\right)$

$$A = -\frac{R_1\,L^2}{6} + \frac{F_1\,(L-a)^3}{6L} + \frac{F_2(L-b)^3}{6L}$$

$$R_1L = F_1(L_1 - a) + F_2(L - b)$$

(b) $y = -\dfrac{1}{EI}\left(\dfrac{R_1x^3}{6} - \dfrac{w\{x-a\}^4}{24} + \dfrac{w\{x-a-b\}^4}{24} + Ax\right)$

$$A = -\frac{R_1\,L^2}{6} + \frac{w(L-a)^4}{24L} + \frac{w(L-a-b)^4}{24L}$$

$$2R_1\,L = w(b-a)(2L-a-b)$$

37. $7\,wL^4/384EI$
38. 28.96 mm
39. $19wL^4/2048EI$

5.5.1

1. 375 N, 596 N
2. 651 kN
3. 2.70 kN, 2.79 kN
4. 23 kN
5. 14.4 m
6. 1285 kN
7. 2655 kN, 2701 kN
8. 630 kN

5.6.1

1. 13.7 N
2. 0.58
3. 20.3 N
4. 54.6 N
5. 0.16
6. 111 N
7. As given in the problem
8. As given in the problem
9. 28° to vertical
10. (a) 312 N, (b) 353 N
11. (a) 166 N, (b) 194 N
12. 5.9 m

5.7.1

1. $F = \frac{1}{4}mg \tan \frac{1}{2}\theta$
2. 105.6 N
3. $mg/(2 \tan \theta)$
4. As given in the problem
5. One
6. One
7. $\theta_1 = \cos^{-1}(2M/FL)$, $\theta_2 = \cos^{-1}(M/FL)$

Index